# Lecture Notes in Physics

Volume 934

# The Lecture Notes in Physics

The series Lecture Notes in Physics (LNP), founded in 1969, reports new developments in physics research and teaching-quickly and informally, but with a high quality and the explicit aim to summarize and communicate current knowledge in an accessible way. Books published in this series are conceived as bridging material between advanced graduate textbooks and the forefront of research and to serve three purposes:

- to be a compact and modern up-to-date source of reference on a well-defined topic
- to serve as an accessible introduction to the field to postgraduate students and nonspecialist researchers from related areas
- to be a source of advanced teaching material for specialized seminars, courses and schools

Both monographs and multi-author volumes will be considered for publication. Edited volumes should, however, consist of a very limited number of contributions only. Proceedings will not be considered for LNP.

Volumes published in LNP are disseminated both in print and in electronic formats, the electronic archive being available at springerlink.com. The series content is indexed, abstracted and referenced by many abstracting and information services, bibliographic networks, subscription agencies, library networks, and consortia.

Proposals should be sent to a member of the Editorial Board, or directly to the managing editor at Springer:

Lisa Scalone
Springer Nature
Physics Editorial Department
Tiergartenstrasse 17
69121 Heidelberg, Germany
Lisa.Scalone@springernature.com

More information about this series at http://www.springer.com/series/5304

Yoshio Kuramoto

# Quantum Many-Body Physics

## A Perspective on Strong Correlations

Springer

Yoshio Kuramoto
Department of Physics
Tohoku University
Sendai, Japan

ISSN 0075-8450 ISSN 1616-6361 (electronic)
Lecture Notes in Physics
ISBN 978-4-431-55392-2 ISBN 978-4-431-55393-9 (eBook)
https://doi.org/10.1007/978-4-431-55393-9



This Springer imprint is published by the registered company Springer Japan KK, part of Springer Nature.
The registered company address is: Shiroyama Trust Tower, 4-3-1 Toranomon, Minato-ku, Tokyo 105-6005, Japan

# Preface

The condensed matter consists of enormous number of interacting electrons together with neutralizing positive ions. As a consequence, there emerges an abundant variety of phenomena in condensed matter. Some of these can be understood in terms of single-electron picture that replaces the inter-electron Coulomb interaction by its average, or by a classical mean field. Realistic description of energy bands in simple metals and semiconductors is one of the most successful examples of the mean field theory. Other phenomena, however, require treatment that goes beyond the classical mean field scheme. Examples of the latter include high-temperature superconductivity, heavy electrons, and fractional quantum Hall effect.

The main purpose of this book is to provide a compact tutorial for basic concepts and tools of quantum many-particle physics, which focuses on correlation effects caused by mutual interactions. The book tries to explain important concepts in depth for serious reading. Since the quantum many-particle physics has expanded to a large complex, it may be most important for a textbook to select the material for learning in a reasonable span of time, a year or so.

The book has grown out of my lectures on condensed matter theory mainly delivered in the graduate school of Tohoku University, Sendai, Japan. The original version of the book was published in Japanese. In writing this English version, I have made substantial revision, adding new topics and improving descriptions. The readership is assumed to be of late undergraduate level and graduate level. The book intends to be useful also for those who have already learned condensed matter physics, but try to acquire coherent image of the quantum many-particle systems. Within a compact size, the book intends to let the reader acquire not only the fundamental concepts but also useful theoretical tools of many-particle physics. The description makes minimum use of technically sophisticated setup. Although the concept of Green functions appear in the early stage (Chap. 3), it is not linked with perturbation expansion with the use of Feynman diagrams. The reader can proceed until the second last chapter without encountering the full-fledged field-theoretical formalism combined with Feynman diagrams. Instead the book uses mostly a more intuitive (younger) cousin, called Goldstone diagram, which is faithful to each perturbation process. I thus hope that the book is more accessible to the readership with the background of experimental physics. The book touches on some recent topics such as room-temperature superconductivity at high pressure so that the reader can feel the atmosphere of the research frontier. Most materials, however, are selected for their importance in longer time span.

In each topic, concise but self-contained account is provided so that the reader need not refer to other textbooks and original papers. In particular, I explain in detail those items which may well arouse confusion during learning. Such items are selected according to the experience of my own confusion, and secondly of the graduate students around me. The following two concepts thread through most topics discussed in the book. The first is the renormalization which guides our attention to relevant energy range by successive elimination of high-energy states. The second is the quasi-particle, by which one concentrates on excitations from the ground state as a collection of smaller number of virtual particles. Both concepts are essential for extracting manageable degrees of freedom out of a huge number inherent in many-particle systems.

The content of the book is roughly classified into three parts. The first part provides a review of basic theoretical concepts and tools. In order to learn the quantum many-particle physics, it is indispensable to have reasonable knowledge of both quantum mechanics and statistical physics. The first part is designed to review the basics of these subjects in the course of reading. As a start, the first chapter deals with perturbation theory. The treatment also aims at concise review of elementary quantum mechanics. We reformulate the perturbation theory in such a manner that evolves into the renormalization theory. As specific examples, we take hydrogen and oxygen molecules, and discuss their spin and orbital states. In Chap. 2, we discuss basic properties of electrons in solids. Under the periodic potential, a single electron itinerate on many atomic sites by forming the energy band. However, with Coulomb interactions among electrons, each electron may localize spatially by losing its itinerant character. In order to deal with many particles, we need statistical physics as an indispensable tool. Among various subjects in statistical physics, Chap. 3 discusses the linear response theory, which is heavily used in the rest of the book.

In the second part, we deal basically with mean field treatment of itinerant electrons. As the simplest framework that can deal with their interaction effects, the Fermi liquid theory is outlined in Chap. 4. It is known that the Fermi liquid theory becomes exact in the limit of low excitation energies, in spite of its approximate nature as a mean field theory. We proceed in Chap. 5 to superconductivity. Since a lot of textbooks and monographs are already available about superconductivity, we try to make least duplication with these books. Hence, we focus on the effect of Coulomb repulsion among electrons and ensuing structure of Cooper pairs. The emphasis on this aspect reflects our hope to promote better understanding of high-temperature superconductivity in cuprates and iron pnictides, both of which are subject of recent active research. On the other hand, the hydride superconductors with transition temperatures near 300 K are briefly mentioned in the context of the isotope effect.

In the third part, we discuss quantum fluctuation effects beyond the mean field picture. The conflict between itinerant and localized characters of interacting electrons leads to interesting physical phenomena. In most cases, these properties can be fully understood only by going beyond the mean field treatment. Chapter 6 discusses Kondo effect due to magnetic impurities in a metal. The renormalization effect appears most dramatically in this problem. We proceed then to low-dimensional systems, where fluctuation effects are essential. Chapter 7 explains a powerful method called bosonization which provides exact solution for certain models such as the Tomonaga–Luttinger model and the Kondo model with special value of interaction parameters. In Chap. 8, we first explain one-particle physics such as Aharonov–Bohm and Berry phases, and proceed to Laughlin states in two dimensions. The interaction effects appear there as exotic statistics which are neither Fermi nor Bose. We then discuss a one-dimensional model which we can solve exactly and obtain free quasi-particles with exotic statistics. Chapter 9 deals concisely with many-body perturbation theory with the use of path integral formalism. The diagram method invented by Feynman is the most convenient in systematic perturbation analysis to infinite order. Then in the final part, Chap. 10 deals with the dynamical effective field theory which becomes exact in the limit of large number of neighboring sites, as in large spatial dimensions. Interestingly, the high-dimensional limit connects to the zero-dimensional one which is nothing but an impurity system. Hence, Kondo effect emerges again as an essential ingredient in systems in large dimensions.

To use the book as a lecture material, only the first and second parts may be selected depending on the situation, since the content after Chap. 6 is more advanced. We assume the second-quantization method as known by the reader. We use the natural units with $\hbar = c = k_B = 1$ except for cases stated otherwise, as in estimating numerical values in conventional unit systems. Each chapter has problems and their solutions to help in the understanding of the reader. Some problems with asterisk (*) are more advanced. It is advised that the reader tries to solve the problem before reading the solution.

Since the description of the book is mostly self-contained, the number of references is limited to a minimum. The list refers to original papers on the subject, however old they may be, together with a few of comprehensive reviews and monographs. I find it more useful for beginners to combine the original work and good reviews in a concise manner, rather than to give the exhaustive list. In addition, the book does not try to be encyclopedic. Let us make explicit some important topics that we do *not* discuss. First to mention is the topological insulator and related topics. Since topological aspect of the electronic structure is primarily of single-electron property, we just refer to a good textbook for the topic.[1] We do discuss, however, elementary topological concepts such as the winding number and the Berry phase in the context of many-particle physics. Other omissions include

[1] A recent textbook on the subject: D. Vanderbilt, *Berry Phases in Electronic Structure Theory*, (Cambridge University Press, 2018).

quantum spin systems such as Kitaev spin liquid,[2] quantum phase transitions,[3] disordered systems,[4] and nonequilibrium systems such as optical lattices.[5] The selection of advanced topics in the book obviously reflects my own taste and limited competence.

Finally, I would thank many people in various generations who have kindly helped me to acquire my viewpoint on physics. The first occasion for me to study seriously the physics around Kondo effect was in Germany many years ago. I owe very much to E. Müller-Hartmann, J. Zittartz, D. Wohlleben (deceased), and F. Steglich for introducing me to the field. My further education on the subject owes mainly to working for original results with my colleagues and previous students, including J. Akimitsu, M. Arikawa, N. Fukushima, H. Harima, C. Horie, S. Hoshino, K.-I. Imura, T. Inoshita, K. Itakura, K. Iwasa, R. Kadono, T. Kasuya, Y. Kato, N. Kawakami, C.-I. Kim. A. Kiss, H. Kojima, H. Kono, K. Kubo, T. Kuromaru, H. Kusunose, S.W. Lovesey, S. Mao, K. Miyake, Y. Murakami, O. Narayan, K. Nomura, J. Otsuki, S. Ozaki, R. Peters, Y. Saiga, O. Sakai, A. Sakuma, G. Sakurai, H.J. Schmidt, T. Seki, N. Shibata, Y. Shimizu, S. Suzuki, G. Uimin, S. Watanabe, T. Watanabe, A. Yamakage, T. Yamamoto, S. Yamazaki, H. Yokoyama, and many others to whom I would express my sincere gratitude. Lastly, I thank the editors of the book for their constructive criticism which has motivated me to upgrade the content substantially from translation of the Japanese edition.

Yoshio Kuramoto

[2] An example of a recent monograph on related subject: T.D. Stanescu, *Introduction to Topological Quantum Matter & Quantum Computation*, (CRC Press, 2017).

[3] A representative monograph on the subject: S. Sachdev, *Quantum Phase Transitions*, Second Edition, (Cambridge University Press, 2011).

[4] A collection of review papers: *50 Years of Anderson Localization*, edited by E. Abrahams (World Scientific, 2010).

[5] An example of the monograph: M. Lewenstein, A. Sanpera, and V. Ahufinger, *Ultracold Atoms in Optical Lattices: Simulating quantum many-body systems*, (Oxford University Press, 2012).

# Contents

# Chapter 1
# Perturbation Theory and Effective Hamiltonian

**Abstract** Perturbation theory is one of the standard subjects in elementary quantum mechanics. The connection to renormalization emerges naturally if one adopts a generalized view on perturbation theory. While the ordinary perturbation theory derives approximate eigenenergy and the corresponding eigenfunction, the generalized view derives an operator called the effective Hamiltonian. In this chapter we reformulate the perturbation theory by introducing the concept of model space. As a simple application of the effective Hamiltonian, we discuss the spin structure of hydrogen and oxygen molecules.

## 1.1 Projection onto Model Space

When considering physical phenomena, it is seldom that all energy ranges are relevant. In most cases, information on limited range of energy is sufficient. For example, the specific heat at low temperature is determined only by excitation spectra extending up to the energy range corresponding to the temperature. In such cases, instead of dealing with all states in the Hamiltonian, we may focus only at low-energy states. The effective Hamiltonian serves precisely for such purpose.

Let us start with the Schrödinger equation $H\psi = E\psi$, with eigenfunction $\psi$ for the whole system. The model space is defined so that eigenfunctions in the model space have the same eigenenergies as those of original states within a specified energy range. The fundamental requirement for the effective Hamiltonian $H_{\text{eff}}$ in the model space is that each eigenvalue $E$ is reproduced within the restricted energy range. Namely, introducing the projection operator $P$ onto the model space, we require

$$H_{\text{eff}} P\psi = E P\psi, \tag{1.1}$$

which defines $H_{\text{eff}}$. It is possible to derive $H_{\text{eff}}$ explicitly (at the formal level). For this purpose we split the Hamiltonian as $H = H_0 + V$ where the unperturbed part $H_0$ can be diagonalized within the model space, which means $[H_0, P] = 0$. We introduce another projection operator $Q$ so that the relation $P + Q = 1$ holds.


Y. Kuramoto, *Quantum Many-Body Physics*, Lecture Notes in Physics 934,
https://doi.org/10.1007/978-4-431-55393-9_1

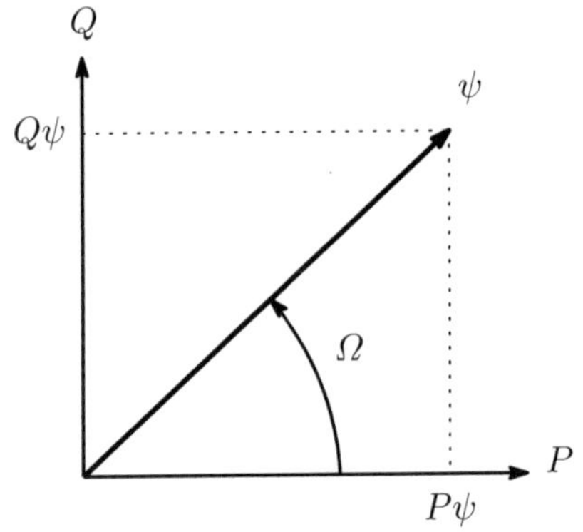


**Fig. 1.1** Projection of the eigenfunction $\psi$ to the model space $P$ and its complementary space $Q$. The wave operator $\Omega$ restores $\psi$ from $P\psi$

Namely, the states projected by $Q$ are orthogonal to those in the model space. We may then rewrite the Schrödinger equation as

$$(E - H_0)Q\psi = QV\psi, \tag{1.2}$$

where we have used the relation $[H_0, Q] = 0$.

Next we introduce an operator $\Omega$ which is called the wave operator [1]. As illustrated in Fig. 1.1, $\Omega$ recovers the original eigenstate $\psi$ by operating on a state $P\psi$ in the model space. As is clear from Fig. 1.1, there are apparently infinite number of different states that give the same $P\psi$, provided $Q\psi$ is arbitrary. However, the condition that $\psi$ is an eigenfunction of $H$ makes the strong constraint on $Q\psi$, which leads to the recovery of $\psi$ by $\Omega$.

From the definition of the wave operator $\Omega$, we obtain the relation

$$\Omega P\psi = P\psi + Q\psi = P\psi + (E - H_0)^{-1}QV\psi, \tag{1.3}$$

where the last term in the rightmost side results from eliminating $Q\psi$ with use of Eq. (1.2). We substitute $\psi = P\psi + Q\psi$ in the last term, and eliminate $Q\psi$ using Eq. (1.2) again. Repeating this process, we arrive at the closed form of $\Omega$ as

$$\Omega(E) = \sum_{n=0}^{\infty} \left[(E - H_0)^{-1}QV\right]^n. \tag{1.4}$$

Thus the effective Hamiltonian is obtained in the power series as

$$H_{\text{eff}}(E) = PH\Omega(E)P, \tag{1.5}$$

with use of $P^2 = P$.

Note that $H_{\text{eff}}(E)$ contains the exact energy $E$, which is to be obtained as the eigenvalue. This characteristic becomes most obvious in the simplest case where the model space consists of a single state $|0\rangle$. Namely, we obtain the condition to determine $H_{\rm eff}(E) = E$ as

$$E = \langle 0|H_0 + V|0\rangle + \langle 0|V(E - H_0)^{-1}QV|0\rangle + \ldots, \tag{1.6}$$

which is different from the more familiar framework, called the Rayleigh–Schrödinger perturbation theory, where the right-hand side (RHS) would have the unperturbed energy $E_0$ instead of $E$.

An advantage of the present framework, often called the Brillouin–Wigner perturbation theory, is demonstrated by a simple example where the second-order perturbation theory gives the exact result. Let us take a single-particle system with energy levels $\epsilon_0$ and $\epsilon_1$, which are connected by the matrix element $V$. Using the creation ($c_i^\dagger$) and annihilation ($c_i$) operators for each level $i = 0, 1$, the Hamiltonian is written as

$$H = \epsilon_0 c_0^\dagger c_0 + \epsilon_1 c_1^\dagger c_1 + V\big(c_0^\dagger c_1 + c_1^\dagger c_0\big). \tag{1.7}$$

The diagonalization of the Hamiltonian gives the energy levels as

$$E_\pm = \frac{1}{2}(\epsilon_0 + \epsilon_1) \pm \sqrt{\frac{1}{4}(\epsilon_1 - \epsilon_0)^2 + V^2}. \tag{1.8}$$

Amazingly, these eigenenergies are reproduced by the Brillouin–Wigner perturbation theory in the lowest order. The confirmation is the subject of Problem 1.1.

## 1.2 Rearrangement of Perturbation Series

In many cases other than those demonstrated above, it is more convenient to deal with the effective Hamiltonian which does not involve the unknown energy to be derived. The goal is achieved by adopting the Rayleigh–Schrödinger perturbation theory to the formalism of effective Hamiltonians [1]. We start from the Schrödinger equation in the form

$$(E - H_0)\psi = E\psi - H_0 \Omega P\psi = V\Omega P\psi, \tag{1.9}$$

where we have used the property $\Omega P\psi = \psi$ for an eigenstate $\psi$ of the operator $H_0 + V$. We apply the operator product $\Omega P$ from the left of Eq. (1.9) to obtain

$$\Omega(E - H_0)P\psi = E\psi - \Omega H_0 P\psi = \Omega P V \Omega P\psi, \tag{1.10}$$

using the relation $PH_0 = H_0 P$. Subtraction of Eq. (1.10) from Eq. (1.9) gives

$$[\Omega, H_0]P\psi = (1 - \Omega P)V\Omega P\psi. \tag{1.11}$$

We now expand the wave operator as $\Omega = \Omega_0 + \Omega_1 + \Omega_2 + \ldots$ where $\Omega_n$ is $O(V^n)$ and $\Omega_0 = 1$. By comparing the terms with the same order in $V$ in both sides of Eq. (1.11), we obtain

$$[\Omega_n, H_0] = QV\Omega_{n-1} - \sum_{j=1}^{n-1} \Omega_j PV\Omega_{n-j-1}. \tag{1.12}$$

Using this result iteratively, the $n$-th order term $\Omega_n$ can be derived successively from lower order ones. Furthermore, using the relation $H_n = PV\Omega_{n-1}P$, we obtain the expansion $H_{\text{eff}} = P(H_0 + V)P + H_2 + H_3 + \ldots$ explicitly. Examples of lower order terms are given by

$$\langle a|H_2|b\rangle = \langle a|V(\epsilon_b - H_0)^{-1}QV|b\rangle, \tag{1.13}$$

$$\langle a|H_3|b\rangle = \langle a|V\frac{1}{\epsilon_b - H_0}QV\frac{1}{\epsilon_b - H_0}QV|b\rangle - \sum_c \langle a|V\frac{1}{\epsilon_b - H_0}\frac{1}{\epsilon_c - H_0}QV|c\rangle\langle c|V|b\rangle, \tag{1.14}$$

where states such as $|a\rangle, |b\rangle, |c\rangle$ all belong to the model space. Equation (1.13) is familiar in the elementary quantum mechanics as the perturbation theory for degenerate levels. In $H_3$, the first term in the RHS of Eq. (1.14) takes the same form as in the Brillouin–Wigner framework except for the energy denominator, while the second term is specific to the Rayleigh–Schrödinger framework.

## 1.3 Hydrogen Molecule

Before discussing complicated behavior of electrons in solids, it is useful to consider the electronic state in a molecule. Let us first take the simplest model for the hydrogen molecule $H_2$. This example demonstrates the usefulness of the effective Hamiltonian approach.

The energy levels of excited states such as $2s, 2p$ are lying much higher than the low-lying states of our interest. Hence we focus only on wave functions of $1s$ sates. When two hydrogen atoms come close, it becomes possible for $1s$ electrons to go from one proton to the other. Taking account of this hopping effect by the parameter $t$, we consider the following model:

$$H = -t\sum_\sigma \left(c_{1\sigma}^\dagger c_{2\sigma} + c_{2\sigma}^\dagger c_{1\sigma}\right) + U\left(n_{1\uparrow}n_{1\downarrow} + n_{2\uparrow}n_{2\downarrow}\right), \tag{1.15}$$

where $c_{i\sigma}^\dagger$ is the creation operator of the $1s$ electron with spin $\sigma$, which is bound to a proton specified by $i = 1, 2$. Each proton is assumed to be fixed at its respective position, which is called the site. The configuration $(1s)^2$ around each proton has the extra energy $U$ associated with the Coulomb repulsion between the electrons. We take $t > 0$ for the transfer energy, and take the $1s$ level as the origin of energy.

This model of $H_2$ corresponds to the two-site version of the Hubbard model [2–4], which will be explained in Chap. 2 in more detail. The two-site model can be solved exactly for arbitrary magnitude of $U$. However, in order to obtain clear physical image, we consider two limits $t \gg U$ and $0 < t \ll U$ separately. If we have $U = 0$, the eigenstate consists of the antisymmetrized product of one-electron states, each of which has the eigenenergy $\pm t$. The wave function with energy $-t$ is called the bonding orbital since the charge density tends to accumulate in the center of protons, while the other wave function with energy $t$ is called the anti-bonding orbital since the charge density is zero in the center. Both are examples of molecular orbitals, which have been used most popularly in chemistry [5]. If two electrons have different spin states, the Pauli exclusion principle allows both electrons to enter bonding orbitals. The resultant ground state has energy $-2t$ and spin 0 (singlet). The effect of Coulomb repulsion is negligible as long as $U \ll t$.

In the opposite limit of $t = 0$, the electronic state is reduced to that of two hydrogen atoms. Then the ground state has the four-fold spin degeneracy, which is broken by even tiny $t \neq 0$. The resultant energy gain with the singlet state causes bonding of the two atoms to hydrogen molecule. It is amazing that essentially the same idea was already put forward by Heitler and London [6] in the primary stage of quantum mechanics. We take the effective Hamiltonian approach which is convenient for discussing the spin configuration. With $t$ regarded as the perturbation parameter and the model space taken as $(1s)^1$ at each site, we construct the effective Hamiltonian in the second order in $t$. Among the four combinations up and down spins, only the singlet pair of electrons has intermediate states where either of hydrogen atoms has two electrons. Figure 1.2 represents pictorially the perturbation process corresponding to Eq. (1.13). This is the simplest example of Goldstone diagrams [1, 7].

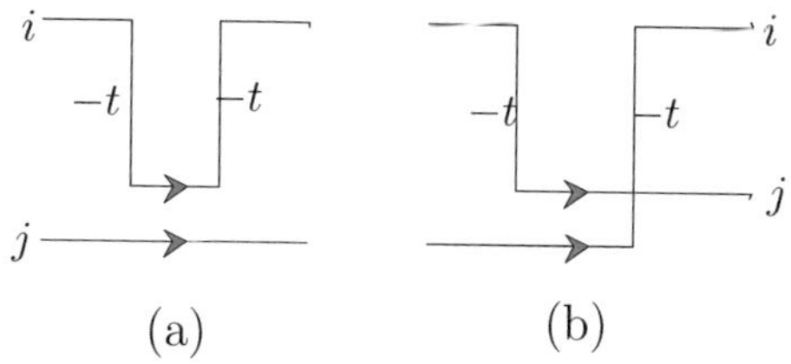


**Fig. 1.2** Examples of second-order perturbation processes by $t$. In **(a)**, the $1s$ electron bound to the proton site $i$ hops virtually to the other site $j$ and comes back to $i$, while **(b)** shows the exchange of $1s$ electrons at $i$ and $j$

On the other hand, the triplet pair does not have a configuration with two electrons per hydrogen atom, hence no shift in the second-order energy. Thus the effective Hamiltonian is given by

$$H_{\text{eff}} = J\left(\boldsymbol{S}_1 \cdot \boldsymbol{S}_2 - \frac{1}{4}\right), \tag{1.16}$$

with $J = 4t^2/U$. Since we have $J > 0$, the ground state is the spin singlet with energy $-J$. This is because only the singlet can take the intermediate state $(1s)^2$ without violating the Pauli principle. Hence it gains the second-order perturbation energy. Problem 1.3 deals with a more complicated case of oxygen molecule, which may also help to derive Eq. (1.16).

As we have seen, the contrasting limiting pictures called molecular orbital and Heitler–London both give the spin-singlet ground state. Since it is possible to solve this simple model for arbitrary value of $U/t$, we can see how these limiting states are connected to each other. The ground state energy $E_0$ is exactly derived as

$$E_0 = \frac{1}{2}U - \sqrt{\frac{1}{4}U^2 + 4t^2}. \tag{1.17}$$

The derivation is the subject of Problem 1.2. The limiting values at $U/t \to 0, \infty$ are given, respectively, by $-2t, -4t^2/U$. Namely, the molecular orbital and Heitler–London pictures are connected continuously. In the case of a crystalline solid, the former goes over to the itinerant picture, while the latter (Heitler–London) goes over to the localized picture. In contrast with the molecule system, however, the smooth connection between the two limiting states is not guaranteed because of collective interaction effects.

## 1.4 Oxygen Molecule

The oxygen molecule $O_2$ has spin 1, and is paramagnetic. This section clarifies the origin in terms of a simplified model. A neutral oxygen atom has the electronic configuration $(2p)^4$. This is interpreted as having two holes in the $2p$ orbitals. We take the axial direction of the diatomic molecule along the $z$ axis. Then the anti-bonding orbital $2p_z$ has a hole per oxygen, namely two holes per $O_2$. The remaining hole at each oxygen atom enters into either $2p_x$ or $2p_y$ orbital. These orbitals extend perpendicularly to the bonding axis, and are energetically degenerate, which are illustrated in Fig. 1.3.

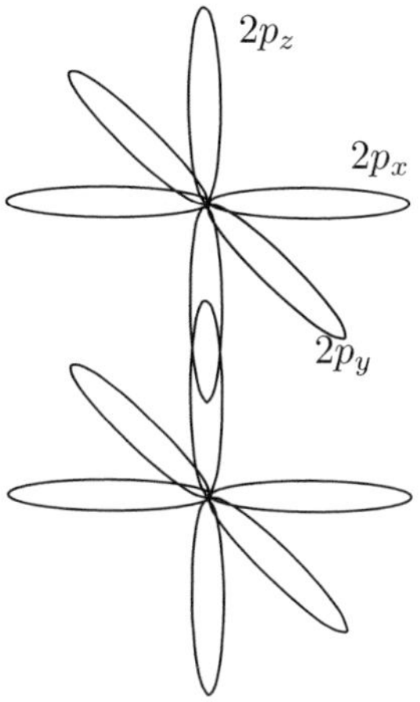


**Fig. 1.3** Illustration of $2p$ hole wave functions in $O_2$. Either of $2p_x, 2p_y$ orbitals has a hole per oxygen atom

In order to discuss the paramagnetism of $O_2$, we focus on holes in $2p_x, 2p_y$ orbitals with the spin degrees of freedom remaining. We take the following model:

$$H = -t \sum_{\alpha\sigma} \left( p^\dagger_{1\alpha\sigma} p_{2\alpha\sigma} + p^\dagger_{2\alpha\sigma} p_{1\alpha\sigma} \right) + H^{(1)}_{int} + H^{(2)}_{int}, \tag{1.18}$$

$$H^{(i)}_{int} = U \sum_{\alpha} n_{i\alpha\uparrow} n_{i\alpha\downarrow} + U' n_{ix} n_{iy} - J_{\rm H} \left( \boldsymbol{S}_{ix} \cdot \boldsymbol{S}_{iy} + \frac{3}{4} n_{ix} n_{iy} \right), \tag{1.19}$$

where $\alpha$ $(= x, y)$ distinguishes two kinds of $2p$ orbitals, and $p^\dagger_{i\alpha\sigma}$ $(p_{i\alpha\sigma})$ represents the hole creation (annihilation) operator at site $i$ $(= 1, 2)$, orbital $\alpha$, and spin $\sigma$. The corresponding spin and number operators are given by $\boldsymbol{S}_{i\alpha}$ and $n_{i\alpha}$ (or $n_{i\alpha\sigma}$ for a spin component), respectively. The Coulomb repulsion is given by $U$ for two holes in the same orbital, and $U'$ $(< U)$ for those in different orbitals. The positive parameter $J_{\rm H}$ represents the Hund's rule which favors parallel spins for two holes in different orbitals $2p_x, 2p_y$. Note that the axial symmetry of the molecule allows the hopping $-t$ only between the same kind of orbitals.

Assuming $U > U' \gg t > 0$, we derive energies of all (=16) eigenstates. Use of the pseudo-spin $\boldsymbol{\tau}$ for orbital degrees of freedom simplifies the calculation. Namely, using the vector $\boldsymbol{\sigma}$ composed of the Pauli matrices, we define the pseudo-spin for a site $i$

$$\boldsymbol{\tau}_i = \frac{1}{2} \sum_{\alpha\beta} \sum_{\rho} p^\dagger_{i\alpha\rho} \boldsymbol{\sigma}_{\alpha\beta} p_{i\beta\rho},$$

and the total pseudo-spin $\boldsymbol{L}_\tau \equiv \boldsymbol{\tau}_1 + \boldsymbol{\tau}_2$ of the molecule. Then by analogy with the case of two spins, we can represent the $2 \times 2 = 4$ orbital states in terms of the orbital singlet $L_\tau = 0$ and the orbital triplet $L_\tau = 1$ with three states.

Let us introduce the orbital projection operators as follows:

$$P(L_\tau = 0) = -\boldsymbol{\tau}_1 \cdot \boldsymbol{\tau}_2 + \frac{1}{4}, \quad P\left(L_\tau^z = \pm 1\right) = 2\tau_1^z \tau_2^z + \frac{1}{2},$$

$$P\left(L_\tau = 1, L_\tau^z = 0\right) = \boldsymbol{\tau}_1 \cdot \boldsymbol{\tau}_2 - 2\tau_1^z \tau_2^z + \frac{1}{4}.$$

Note that the sum of these three operators gives the identity operator. Using analogous projection operators for spin degrees of freedom, we obtain the effective Hamiltonian:

$$\begin{aligned} H_{\text{eff}} = &-\frac{4t^2}{U' - J_{\text{H}}} P\left(S = 1, L_\tau = 0\right) - \frac{4t^2}{U} P(S = 0, L_\tau^z = \pm 1) \\ &- \frac{4t^2}{U'} P\left(S = 0, L_\tau = 1, L_\tau^z = 0\right). \end{aligned} \tag{1.20}$$

The derivation of energy of each spin and orbital configuration is the subject of Problem 1.3 with the solution given by Eq. (1.28). Each projection operator onto a given configuration $(S, L_\tau)$ is the product of spin and orbital projection operators. An example is given by

$$P(S = 1, L_\tau = 0) = \left(\boldsymbol{S}_1 \cdot \boldsymbol{S}_2 + \frac{3}{4}\right)\left(-\boldsymbol{\tau}_1 \cdot \boldsymbol{\tau}_2 + \frac{1}{4}\right). \tag{1.21}$$

Then $H_{\text{eff}}$ in Eq. (1.20) is written in terms of the product of spin and orbital operators involving two sites. According to Eq. (1.20), the two orbital states with $L_\tau = 1$ favor the singlet state of spins. On the other hand, the orbitals singlet ($L_\tau = 0$) favor paramagnetism ($S = 1$) by the effect of $J_{\text{H}}$.

Replacing the oxygen pair by a pair of transition metal ions in the crystal, $H_{\text{eff}}$ in Eq. (1.20) can be applied to systems such as a perovskite $LaMnO_3$ which has orbitally degenerate electrons in the 3*d* shell [8]. The 3*d* orbital has the five-fold orbital degeneracy under the rotational symmetry, but is split to two levels with three- and two-fold degeneracies under the cubic symmetry. With further lowering of the symmetry, it may happen that the orbital degeneracy disappears completely. If the energy gain by this lowering is more than the loss of the elastic energy, the crystal deforms spontaneously. This is called the cooperative Jahn–Teller effect. The name comes from the work of Jahn and Teller who discussed the distortion of molecules with orbital degeneracy [9]. Thus, consideration of orbital degrees of freedom is necessary for understanding rich phenomena of magnetic ordering in perovskites and other transition metal systems.

## 1.5 Itinerant vs. Localized Limits in a Molecule

The picture of molecular orbitals is connected to the energy band picture for electrons in crystals, while the Heitler–London picture has more similarity to localized electron picture in solids. We have seen in $H_2$ molecule that the spin-singlet state changes continuously from the one limit to the other. We now consider the fate of the paramagnetic ($S \neq 0$) ground state as the transfer $t$ grows. In this connection we mention the sulfur molecule, $S_2$, where each S atom has $(3p)^4$ configuration. Because the $3p$ orbital is more extended than the $2p$ orbital of oxygen, $t/U$ in $S_2$ is larger than that in $O_2$. It is known that the $S_2$ molecule is also paramagnetic with spin 1.

Let us start with the extreme case $t \gg U, U', J_H > 0$ in the model given by Eq. (1.18). It is legitimate to keep only the bonding orbitals of a hole with energy $-t$. With two holes in these orbitals, the Pauli principle allows either $S = 1, L_\tau = 0$ or $S = 0, L_\tau = 1$. Hence in the non-interacting limit the ground state is six-fold degenerate. With finite interactions $J_H$, the degenerate energy level splits into different spin states with $S = 1$ and $S = 0$. The latter with $L_\tau = 1$ splits in general further into levels with $L_\tau^z = \pm 1$ and with $L_\tau^z = 0$ because of the different Coulomb repulsions $U$ and $U'$. Absence of the SU(2) symmetry for pseudo-spins allows the difference. The shift of each energy level by interactions is summarized as

$$\Delta E\big(S=0,\ L_\tau^z=\pm 1\big) = \frac{U}{2},\quad \Delta E\big(S=0,\ L_\tau=1,\ L_\tau^z=0\big) = \frac{U'}{2},$$
$$\Delta E(S=1,\ L_\tau=0) = \frac{U'}{2} - J_H. \tag{1.22}$$

Namely, the ground state in the weak-coupling limit has spin triplet and orbital singlet ($S = 1, L_\tau = 0$), which has the same quantum numbers as those of the ground state in the localized limit.

If one hypothetically increases the number of atoms in a molecule, the system connects to a crystalline solid. Then the state with finite spin connects to ferromagnetism, which can change continuously from the itinerant to localized limits. These examples suggest important role of the orbital degeneracy to realize ferromagnetism.

## Problems

**1.1** Derive the effective Hamiltonian for the ground state of Eq. (1.7), and its eigenvalue.

**1.2** Diagonalize the pair Hamiltonian $H_2$ given by Eq. (1.15), and derive the exact ground state energy given by Eq. (1.17).

**1.3** Derive the energy of a hole pair in Eq. (1.18) by performing second-order perturbation expansion with respect to $t$.

## Solutions to Problems

**Problem 1.1**
The model space $P$ is chosen as $|0\rangle\langle 0|$. With only the single state involved, the effective Hamiltonian is a scalar with energy $E$. According to Eq. (1.4), the perturbation series terminates at $O(V^2)$, resulting in

$$\langle 0|H_{\text{eff}}|0\rangle = \epsilon_0 + \frac{V^2}{E-\epsilon_1} = E. \tag{1.23}$$

This quadratic equation for $E$ has two solutions:

$$E = \frac{1}{2}(\epsilon_0+\epsilon_1) \pm \sqrt{\frac{1}{4}(\epsilon_1-\epsilon_0)^2 + V^2}. \tag{1.24}$$

By continuity with vanishing $V$, we should choose the solution with minus sign. The result agrees with the exact ground state energy derived by diagonalization of the $2 \times 2$ Hamiltonian matrix. Note that the other solution with the plus sign in Eq. (1.24) tends to $\epsilon_1$ with vanishing $V$. The latter solution gives the exact energy of the excited state, which belongs to the orthogonal space $|1\rangle\langle 1|$ projected by $Q$.

**Problem 1.2**
We construct the two-electron states $|1\rangle, |2\rangle, |3\rangle$ with spin 0 by the product of spin states $|\uparrow\rangle$ and $|\downarrow\rangle$ at each site as

$$|1\rangle = |\uparrow\downarrow\rangle \otimes |0\rangle, \quad |2\rangle = |0\rangle \otimes |\uparrow\downarrow\rangle, \tag{1.25}$$

$$|3\rangle = \frac{1}{\sqrt{2}}\left(|\uparrow\rangle \otimes |\downarrow\rangle - |\downarrow\rangle \otimes |\uparrow\rangle\right). \tag{1.26}$$

Since the charge distribution of $|3\rangle$ is symmetric about the center of two sites, the hopping $t$ mixes $|3\rangle$ only with another symmetric state defined by $|4\rangle = (|1\rangle + |2\rangle)/\sqrt{2}$. Therefore with the basis set $|3\rangle$ and $|4\rangle$, the two-site Hamiltonian for the spin singlet is represented by a $2 \times 2$ matrix as

$$\begin{pmatrix} 0 & -2t \\ -2t & U \end{pmatrix}. \tag{1.27}$$

The lower eigenvalue the matrix corresponds to the result given by Eq. (1.17).

**Problem 1.3**

The Pauli exclusion principle decides whether two holes can enter the same site. Hence according to the spin and orbital quantum numbers, the two-hole intermediate state has the energy specified by $U, U', J_H$. The Pauli principle also demands different parities between the orbital and spin exchanges, which guarantees the antisymmetry of the wave function. As a result, states with $S = 1, L_\tau = 1$ or with $S = 0, L_\tau = 0$ are forbidden. For example, the Coulomb repulsion is $U$ in the case of $L_\tau^z = \pm 1$, while it is $U'$ with $L_\tau = 1,\ L_\tau^z = 0$ since two holes are in different orbitals. Note that there is no rotational symmetry in the pseudo-spin space.

Figure 1.2 in p.5 illustrates the second-order perturbation processes by $t$, also in the presence of orbital degeneracy. With the help of these figures, the energy $E_{\text{pair}}$ of a hole pair is derived as

$$E_{\text{pair}} = \begin{cases} -4t^2/(U' - J_H), & (S = 1,\ L_\tau = 0) \\ -4t^2/U, & (S = 0,\ L_\tau = 1,\ L_\tau^z = 1) \\ -4t^2/U', & (S = 0,\ L_\tau = 1,\ L_\tau^z = 0) \end{cases}, \tag{1.28}$$

where the factor 4 in front of $t^2$ accounts for equivalent hopping processes from $j$ to $i$. From Eq. (1.28), we conclude that the most favorable state for the interaction energy is the spin triplet configuration with $(S = 1, L_\tau = 0)$.

## References


1. Lindgren, I., Morrison, J.: Atomic Many-Body Theory. Springer, Berlin (1986)
2. Hubbard, J.: Proc. Royal. Soc. Lond. **276**, 238 (1963)
3. Kanamori, J.: Prog. Theor. Phys. **30**, 275 (1963)
4. Gutzwiller, M.C.: Phys. Rev. Lett. **10**, 159 (1963)
5. Mulliken, R.S.: Phys. Rev. **41**, 49 (1932)
6. Heitler, W., London, F.: Z. Phys. **44**, 455 (1927)
7. Goldstone, J.: Proc. Roy. Soc. Lond. **A239**, 267 (1956)
8. Imada, M., Fujimori, A., Tokura, Y.: Rev. Mod. Phys. **70**, 1039 (1998)
9. Jahn, H.A., Teller, E.: Proc. R. Soc. Lond. A. **161**, 220 (1937)

# Chapter 2
# Itinerant and Localized Characters of Electrons

**Abstract** Crystalline solids are composed of a huge number ($\sim 10^{23}$) of electrons and nuclei, which make up quantum many-body systems with rich properties. For intuitive understanding of the electronic property of actual solids, it is necessary to extract only essential ingredients out of a large number of degrees of freedom. In this chapter we take a simplified model where positively charged ions form a simple lattice. We discuss basic effects of kinetic energy of electrons, potential energy from ions, and mutual Coulomb interaction energy among electrons. Special attention is paid how the attractive and repulsive Coulomb interactions from ions and other electrons combine to work on each electron. Depending on relative importance of kinetic and mutual interaction energies, electrons take either itinerant states forming energy bands, or localized states around each ion. The discussion in this chapter is restricted to absolute zero of temperature.

## 2.1 Model of Electrons in Solids

Crystalline solids consist of many electrons and nuclei with opposite charges. In this Chapter we focus only on electrons in solids, and neglect all dynamics of positive ions for simplicity. Namely, we assume unit point charge for each of positive ions that are fixed at lattice sites $\boldsymbol{R}_i$ $(i = 1, 2, \cdots, N)$ with $N \gg 1$. The number of electrons is also taken to be $N$ so that the system is neutral as a whole. The operators $\Psi_\sigma^\dagger(\boldsymbol{r})$ and $\Psi_\sigma(\boldsymbol{r})$, respectively, represent creation and annihilation of electrons at position $\boldsymbol{r}$ with spin $\sigma$. Then with $v(\boldsymbol{r}) = e^2/|\boldsymbol{r}|$, the Hamiltonian is given by

$$H = \sum_\sigma \int \mathrm{d}\boldsymbol{r} \Psi_\sigma^\dagger(\boldsymbol{r}) \left[ -\frac{\Delta}{2m} - \sum_i v(\boldsymbol{r} - \boldsymbol{R}_i) \right] \Psi_\sigma(\boldsymbol{r}) + V_\mathrm{C} + V_\mathrm{I}, \tag{2.1}$$


Y. Kuramoto, *Quantum Many-Body Physics*, Lecture Notes in Physics 934,
https://doi.org/10.1007/978-4-431-55393-9_2

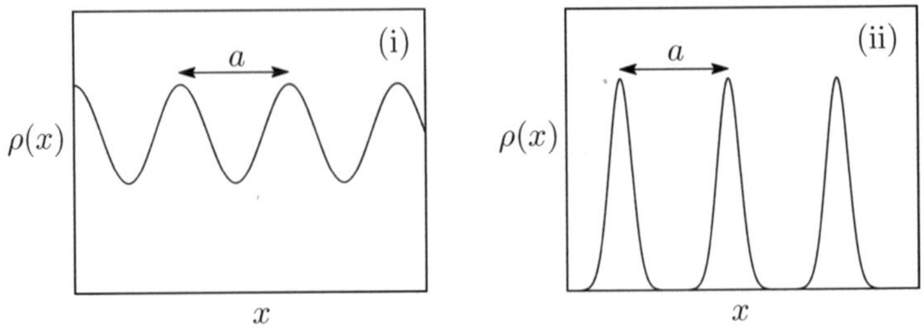


**Fig. 2.1** Illustration of the density $\rho(x)$ of electrons in a one-dimensional lattice for (i) $a < a_B$ and (ii) $a > a_B$

where $\Delta$ is the Laplacian and $V_C$ represents the Coulomb repulsion among electrons, while $V_I$ among ionic charges. They are explicitly written as

$$V_C = \frac{1}{2}\int d\boldsymbol{r} \int d\boldsymbol{r}' \sum_{\sigma,\sigma'} v(\boldsymbol{r}-\boldsymbol{r}')\Psi_\sigma^\dagger(\boldsymbol{r})\Psi_{\sigma'}^\dagger(\boldsymbol{r}')\Psi_{\sigma'}(\boldsymbol{r}')\Psi_\sigma(\boldsymbol{r}), \tag{2.2}$$

$$V_I = \frac{1}{2}\sum_{i\neq j} v(\boldsymbol{R}_i - \boldsymbol{R}_j). \tag{2.3}$$

The wave function of an electron bound to an ion has the extension of the order of $a_B = 1/(me^2) \sim 0.5$ Å, which is called the Bohr radius. Figure 2.1 illustrates the density profile of electrons in one-dimensional lattice with the lattice parameter $a$ being either small or large as compared with $a_B$. Although most of actual solids have comparable values for $a$ and $a_B$, it is instructive to consider the limiting cases such as (i) $a \ll a_B$, and (ii) $a \gg a_B$.

## 2.2 Formation of Energy Bands

Let us first consider the case (i) $a \ll a_B$. The wave functions of nearby sites are overlapping strongly. As a result, the ground state wave function deviates much from collection of the atomic $1s$ orbitals, and mixing with excited orbitals such as $2s$ and $2p$ becomes substantial. The wave function of the crystal may even be closer to a plane wave rather than atomic orbitals if the kinetic energy dominates over the periodic potential.

In such a case, the interaction energy $V_C$ is mostly determined by the average $\langle n_\sigma(\boldsymbol{r})n_{\sigma'}(\boldsymbol{r}')\rangle$, but is less influenced by $\langle \delta n_\sigma(\boldsymbol{r})\delta n_{\sigma'}(\boldsymbol{r}')\rangle$ where

$$\delta n_\sigma(\boldsymbol{r}) = \Psi_\sigma^\dagger(\boldsymbol{r})\Psi_\sigma(\boldsymbol{r}) - \langle \Psi_\sigma^\dagger(\boldsymbol{r})\Psi_\sigma(\boldsymbol{r})\rangle$$

represents the density fluctuation operator. The neglect of fluctuation contribution corresponds to the approximation

$$\begin{aligned}&\Psi_\sigma^\dagger(\boldsymbol{r})\Psi_{\sigma'}^\dagger(\boldsymbol{r}')\Psi_{\sigma'}(\boldsymbol{r}')\Psi_\sigma(\boldsymbol{r})\\ \Rightarrow\quad & n_\sigma(\boldsymbol{r})\langle n_{\sigma'}(\boldsymbol{r}')\rangle + \langle n_\sigma(\boldsymbol{r})\rangle n_{\sigma'}(\boldsymbol{r}') - \langle n_\sigma(\boldsymbol{r})\rangle\langle n_{\sigma'}(\boldsymbol{r}')\rangle,\end{aligned} \tag{2.4}$$

which amounts to replacing the inter-electron interaction by an effective potential felt by each electron. This scheme is called the Hartree approximation. The effective potential is an example of the molecular field. We shall later explain an improved approximation that takes account of another contribution, called the exchange term, in addition to Eq. (2.4).

In the Hartree approximation, the original Hamiltonian $H$ is replaced by

$$H_{\mathrm{H}} = \int \mathrm{d}\boldsymbol{r} \sum_\sigma \Psi_\sigma^\dagger(\boldsymbol{r}) h_{\mathrm{H}}(\boldsymbol{r}) \Psi_\sigma(\boldsymbol{r}) + E_{\mathrm{D}}, \tag{2.5}$$

where $E_{\mathrm{D}} \equiv N\epsilon_{\mathrm{D}}$ is the constant term given by

$$\epsilon_{\mathrm{D}} = \frac{1}{2N}\sum_{i\neq j} v(\boldsymbol{R}_i - \boldsymbol{R}_j) - \frac{1}{2N}\int \mathrm{d}\boldsymbol{r}\int \mathrm{d}\boldsymbol{r}' v(\boldsymbol{r}-\boldsymbol{r}')\sum_{\sigma,\sigma'}\langle n_\sigma(\boldsymbol{r})\rangle\langle n_{\sigma'}(\boldsymbol{r}')\rangle. \tag{2.6}$$

In the thermodynamic limit $N \to \infty$, each term in Eq. (2.6) is divergent because of the long range of the $1/r$ interactions. However, the sum $\epsilon_{\mathrm{D}}$ converges to a finite value. This is because the ionic repulsion given by the first term in the RHS cancels with the second term owing to the charge neutrality. The origin of the second term with the negative sign becomes clear in the following discussion of $h_{\mathrm{H}}(\boldsymbol{r})$.

The motion of each electron is controlled by the one-body Hamiltonian

$$h_{\mathrm{H}}(\boldsymbol{r}) = -\frac{\Delta}{2m} + v_{\mathrm{H}}(\boldsymbol{r}), \tag{2.7}$$

where the Hartree potential $v_{\mathrm{H}}(\boldsymbol{r})$ is given by

$$v_{\mathrm{H}}(\boldsymbol{r}) = \int \mathrm{d}\boldsymbol{r}' v(\boldsymbol{r}-\boldsymbol{r}')\left[\sum_\sigma \langle n_\sigma(\boldsymbol{r}')\rangle - \sum_j \delta(\boldsymbol{r}'-\boldsymbol{R}_j)\right]. \tag{2.8}$$

The first term in the square bracket originates from the repulsive interaction in $V_{\mathrm{C}}$, while the second term represents the attractive potential provided by positive ions. The potential $v_{\mathrm{H}}(\boldsymbol{r})$ becomes finite only by combination of both terms, each of which is divergent in the thermodynamic limit as in the case of $\epsilon_{\mathrm{D}}$. Note that $V_{\mathrm{C}}$ is counted twice in the integral of Eq. (2.8), which is corrected by the final term in Eq. (2.6) with negative sign. Provided the charge density has the periodicity $\langle n_\sigma(\boldsymbol{r}'+\boldsymbol{R})\rangle = \langle n_\sigma(\boldsymbol{r}')\rangle$ for any lattice vector $\boldsymbol{R}$, the Hartree potential also has the same periodicity: $v_{\mathrm{H}}(\boldsymbol{r}+\boldsymbol{R}) = v_{\mathrm{H}}(\boldsymbol{r})$.

The one-body Schrödinger equation is given by

$$h_{\rm H}(\boldsymbol{r})\phi_n(\boldsymbol{r}) = E_n\phi_n(\boldsymbol{r}), \tag{2.9}$$

where the eigenvalue $E_n$ is indexed as $E_1, E_2, \ldots$ from lower to higher energies. For definiteness, we take the cubic shape of the system with each edge length $L$, and impose the periodic boundary condition along each direction. Strictly speaking, we have to modify the Coulomb interaction in the periodic form as well. However, we use the original form $e^2/r$ assuming $L$ is much larger than the characteristic length which determines the property of the system. Since $v_{\rm H}(\boldsymbol{r})$ has the crystalline periodicity, each eigenfunction $\phi_n$ is given as a Bloch state characterized by the wave number $\boldsymbol{k}$ in the Brillouin zone. As we shall show below, the $N$-body state in fact satisfies the assumed periodicity in $\langle n_\sigma(\boldsymbol{r}')\rangle$.

With use of the complete set $\{\phi_n\}$, the field operator $\Psi_\sigma(\boldsymbol{r})$ of electrons is expanded as

$$\Psi_\sigma(\boldsymbol{r}) = \sum_n c_{n\sigma}\phi_n(\boldsymbol{r}), \tag{2.10}$$

where $c_{n\sigma}$ is the annihilation operator of the corresponding state. In the Hartree approximation, the $N$-particle ground state $|g\rangle$ is constructed by starting from the vacuum $|0\rangle$ and filling the electronic states successively from the lowest level:

$$|g\rangle = \prod_\sigma \prod_{n=1}^{N/2} c_{n\sigma}^\dagger |0\rangle. \tag{2.11}$$

Here we have assumed $N$ even. In the first-quantized notation, the state $|g\rangle$ is described by a Slater determinant. With the coordinates $\boldsymbol{r}_i$ for spin up electrons and $\boldsymbol{r}'_i$ for spin down ones with $i = 1, 2, \ldots, M \equiv N/2$, we obtain

$$\begin{aligned}
&\langle \boldsymbol{r}_1 \ldots \boldsymbol{r}_M; \boldsymbol{r}'_1 \ldots \boldsymbol{r}'_M | g\rangle \equiv \Psi_g(\boldsymbol{r}_1 \ldots \boldsymbol{r}_M; \boldsymbol{r}'_1 \ldots \boldsymbol{r}'_M) \\
&= \frac{1}{M!} \begin{vmatrix} \phi_1(\boldsymbol{r}_1) & \cdots & \phi_1(\boldsymbol{r}_M) \\ \vdots & \ddots & \vdots \\ \phi_M(\boldsymbol{r}_1) & \cdots & \phi_M(\boldsymbol{r}_M) \end{vmatrix} \cdot \begin{vmatrix} \phi_1(\boldsymbol{r}'_1) & \cdots & \phi_1(\boldsymbol{r}'_M) \\ \vdots & \ddots & \vdots \\ \phi_M(\boldsymbol{r}'_1) & \cdots & \phi_M(\boldsymbol{r}'_M) \end{vmatrix}.
\end{aligned} \tag{2.12}$$

From the antisymmetry of the determinant against exchange of two columns, the wave function vanishes with $\boldsymbol{r}_i = \boldsymbol{r}_j$ or $\boldsymbol{r}'_i = \boldsymbol{r}'_j$, while it can remain with $\boldsymbol{r}_i = \boldsymbol{r}'_j$. This property corresponds to the Pauli exclusion principle.

The electronic density is the same for both spin directions $\sigma$, and is given by

$$\langle n_\sigma(\boldsymbol{r}')\rangle = \sum_{n=1}^{M} |\phi_n(\boldsymbol{r}')|^2, \tag{2.13}$$

where $|\phi_n(\boldsymbol{r}')|^2$ has the crystalline periodicity, and depends obviously on the Hartree potential $v_{\mathrm{H}}(\boldsymbol{r})$. On the other hand, Eq. (2.8) shows that $v_{\mathrm{H}}(\boldsymbol{r})$ in turn is specified by $\langle n_\sigma(\boldsymbol{r}')\rangle$. Specifically, if $\langle n_\sigma(\boldsymbol{r}')\rangle$ has the crystalline periodicity, the same applies to $v_{\mathrm{H}}(\boldsymbol{r})$. Namely, the density and $v_{\mathrm{H}}(\boldsymbol{r})$ are determined self-consistently.

In the ground state of $N$ electrons, they occupy the one-electron states with energies $E_1, E_2, \ldots, E_{N/2}$, each of which is doubly degenerate by spin degrees of freedom. If the lowest energy band has no overlap with other energy bands, this lowest band is half-filled by $N$ electrons, since each band can accommodates up to $2N$ electrons. The set of $\boldsymbol{k}$, which has the highest energy of the occupied states, forms the Fermi surface. The volume $V_F$ surrounded by the Fermi surface is related to the electron density $n = N/L^3$ as

$$V_F/(2\pi)^3 = n/2. \tag{2.14}$$

The Hartree approximation successfully gives a finite self-consistent potential in the simplest manner. However, it has a defect that becomes serious if one looks more carefully into quantum mechanical nature of electrons. Namely, $v_{\mathrm{H}}$ is finite even for $N = 1$, which means a self-interaction in contradiction with the basic principle of quantum mechanics. Although crystalline solids have $N \gg 1$, the self-interaction spoils the accuracy especially for low density of electrons. The defect of self-interaction can be remedied by including the other terms (called exchange) in approximating $V_{\mathrm{C}}$:

$$\begin{aligned}\Psi_\sigma^\dagger(\boldsymbol{r})\Psi_{\sigma'}^\dagger(\boldsymbol{r}')\Psi_{\sigma'}(\boldsymbol{r}')\Psi_\sigma(\boldsymbol{r}) \Rightarrow &\text{(direct terms)} - \langle\Psi_\sigma^\dagger(\boldsymbol{r})\Psi_{\sigma'}(\boldsymbol{r}')\rangle\Psi_{\sigma'}^\dagger(\boldsymbol{r}')\Psi_\sigma(\boldsymbol{r})\\ &- \Psi_\sigma^\dagger(\boldsymbol{r})\Psi_{\sigma'}(\boldsymbol{r}')\langle\Psi_{\sigma'}^\dagger(\boldsymbol{r}')\Psi_\sigma(\boldsymbol{r})\rangle + \langle\Psi_\sigma^\dagger(\boldsymbol{r})\Psi_{\sigma'}(\boldsymbol{r}')\rangle\langle\Psi_{\sigma'}^\dagger(\boldsymbol{r}')\Psi_\sigma(\boldsymbol{r})\rangle,\end{aligned} \tag{2.15}$$

where "direct terms" means those in the RHS of Eq. (2.4). Note that the averages in Eq. (2.15) are taken over non-local quantities with different coordinates. The framework to consider the exchange terms as well is called the Hartree–Fock approximation. The Hartree–Fock approximation can deal with magnetically ordered states. Namely, the effective potential may depend on spins through the spin-dependent exchange term $\langle\Psi_{\sigma'}^\dagger(\boldsymbol{r}')\Psi_\sigma(\boldsymbol{r})\rangle$. In an ordered state, the effective potential can break the crystalline symmetry spontaneously. The framework with a broken symmetry is often called the unrestricted Hartree–Fock approximation.

In the Hartree–Fock approximation, the effective single-electron Hamiltonian is represented by

$$H_{\mathrm{HF}} = \int \mathrm{d}\boldsymbol{r} \int \mathrm{d}\boldsymbol{r}' \sum_{\sigma,\sigma'} \Psi_\sigma^\dagger(\boldsymbol{r}) h_{\sigma\sigma'}(\boldsymbol{r},\boldsymbol{r}')\Psi_{\sigma'}(\boldsymbol{r}') + E_{\mathrm{DX}}, \tag{2.16}$$

where the constant term $E_{\mathrm{DX}}$ is given by

$$E_{\mathrm{DX}} = E_{\mathrm{D}} + \frac{1}{2}\int \mathrm{d}\boldsymbol{r} \int \mathrm{d}\boldsymbol{r}' v(\boldsymbol{r}-\boldsymbol{r}') \sum_{\sigma,\sigma'} \langle\Psi_{\sigma'}^\dagger(\boldsymbol{r}')\Psi_\sigma(\boldsymbol{r})\rangle\langle\Psi_\sigma^\dagger(\boldsymbol{r})\Psi_{\sigma'}(\boldsymbol{r}')\rangle. \tag{2.17}$$

The effective potential, with exchange terms included, acts not only on the density operator $n_\sigma(\boldsymbol{r})$ but also on the non-local part. Therefore, the one-body Hamiltonian $h_{\sigma\sigma'}(\boldsymbol{r}, \boldsymbol{r}')$ also becomes non-local. Explicitly, we obtain

$$h_{\sigma\sigma'}(\boldsymbol{r}, \boldsymbol{r}') = h_{\mathrm{H}}(\boldsymbol{r})\delta_{\sigma,\sigma'}\delta(\boldsymbol{r} - \boldsymbol{r}') - v(\boldsymbol{r} - \boldsymbol{r}')\langle \Psi^\dagger_{\sigma'}(\boldsymbol{r}')\Psi_\sigma(\boldsymbol{r})\rangle. \tag{2.18}$$

The resultant the Schrödinger equation is given by

$$\begin{aligned} E_{n\sigma}\phi_n(\boldsymbol{r}, \sigma) &= \int \mathrm{d}\boldsymbol{r}' \sum_{\sigma'} h_{\sigma\sigma'}(\boldsymbol{r}, \boldsymbol{r}')\phi_n(\boldsymbol{r}', \sigma') \\ &= h_{\mathrm{H}}(\boldsymbol{r})\phi_n(\boldsymbol{r}) - \int \mathrm{d}\boldsymbol{r}' \sum_{\sigma'} v(\boldsymbol{r} - \boldsymbol{r}')\langle \Psi^\dagger_{\sigma'}(\boldsymbol{r}')\Psi_\sigma(\boldsymbol{r})\rangle\phi_n(\boldsymbol{r}', \sigma'), \end{aligned} \tag{2.19}$$

which is a complicated integro-differential equation. In the case of small number of electrons, the exchange compensates partially the Hartree potential that has been overestimated. The inspection is the subject of Problem 2.1. The Hartree–Fock approximation can also describe a magnetically ordered state with spin-dependent average in Eq. (2.18). For real materials, however, it requires a lot of computational resource to derive the non-local average self-consistently. Hence it is usual to employ a local approximation for the exchange terms to make the calculation feasible. In deriving energy bands of actual solids, most calculations use the so-called density functional theory [1], combined with the local approximation. We refer to recent reviews [2, 3] and an extensive monograph [4] for highly sophisticated details of practical calculation.

Without symmetry breaking, the eigenfunction $\phi_n(\boldsymbol{r}, \sigma)$ of $H_{\mathrm{HF}}$ is a Bloch function. Then the $N$-electron ground state is described by the Slater determinant as in the case of the Hartree approximation. It often happens that self-consistent solutions are not unique. For example, a solution corresponding to the paramagnetic state exists even though the magnetic state is more stable. It is necessary in such a case to examine the change of energy against variation of the solution. Even if the stationary solution gives the local minimum against infinitesimal variation, it is possible that other solution gives the global minimum of the energy.

With $a \ll a_{\mathrm{B}}$, the magnitude of average interaction energy among electrons becomes much smaller than the band width. Then the picture of almost free (non-interacting) Bloch electrons emerges as the zero-th order approximation. Here the difference between the Hartree and Hartree–Fock approximation is small. However, it does not exclude the importance of fluctuation effects due to the mutual interaction. Since the excitation energy in the metallic state has no gap, perturbation theory with a vanishing energy denominator may encounter divergence. In one spatial dimension, the divergence commonly occurs even in the lowest-order perturbation theory. As a result, the starting ground state in the Hartree–Fock approximation becomes unstable. Another case with important fluctuation effects is the system where an impurity spin is screened out by the surrounding conduction electrons, which is known as Kondo effect. These fluctuation effects will be discussed in later Chapters.

With increasing interactions in any dimension, the ground state may have ordered phases such as magnetic or superconducting states. If there is no symmetry breaking, the size of the Fermi surface is independent of the strength of interaction, and remains the same as given by Eq. (2.14) [5]. This situation is described by the Fermi-liquid theory as explained in more detail later.

## 2.3 Localized Orbitals and Hopping of Electrons

In the case (ii) $a \gg a_{\rm B}$, we start from the atomic picture of the ground state in which each electron is in the $1s$ orbital, and the influence from other sites is a small perturbation. In this case the ground state should be insulating. According to Mott [6], such insulating state can be found in actual materials, although the electronic state is much more complicated than the hydrogen lattice. In the energy band picture, the insulating state needs either even number of electrons per unit cell, or a symmetry breaking order so that the new unit cell contains even number of electrons. In reality, certain materials such as CoO, with odd electron number per unit cell, remains insulating even above the Neél temperature. On the other hand, systems such as $V_2O_3$ change into a metal from the antiferromagnetic insulator if one applies pressure, or replaces constituent atoms to cause equivalent chemical pressure. The class of such systems is often called Mott insulators [6, 12].

We inspect how to deal with the case of $a \gg a_{\rm B}$ more systematically. If we start from the paramagnetic ground state, the many-electron wave function given by the Slater determinant of Bloch states has substantial probability of both up- and down-spin electrons present in each $1s$ orbital. However, with $a \gg a_{\rm B}$, the energy loss by the double occupation should be much larger than the gain by forming Bloch states. Hence there should be either up or down spin but not both at each site.

The Hartree approximation cannot deal with the situation properly because $v_H(\boldsymbol{r})$ cannot distinguish spins. The exchange potential, on the other hand, cancels most of the direct Coulomb interaction from nearby electrons with the same spin. As a result the Hartree–Fock Hamiltonian reproduces the limit of weakly coupled hydrogens as long as the spin is completely polarized. Therefore we consider artificially the situation where all electrons have spin up [7]. This observation enables us to define the basis functions which are necessary to establish the localized picture. Suppose we have solved the Hartree–Fock equation assuming a fully polarized ground state. Then the lowest energy band is full, which consists dominantly of the $1s$ state at each site. The next lowest bands separated by the energy gap consist of $2s$ and $2p$ states. In the case of $a \gg a_{\rm B}$, the gap should be about 0.75 Ryd, which is the difference between the $1s$ and $2s, 2p$ states. From the

Bloch functions $\psi_{\boldsymbol{k}}(\boldsymbol{r})$ of the lowest band we construct the Wannier function by [7]

$$w_i(\boldsymbol{r}) = \frac{1}{\sqrt{N}} \sum_{\boldsymbol{k}} \psi_{\boldsymbol{k}}(\boldsymbol{r}) \exp(-\mathrm{i}\boldsymbol{k} \cdot \boldsymbol{R}_i) \tag{2.20}$$

which is similar to the $1s$ wave function at site $\boldsymbol{R}_i$. In contrast to the atomic wave function, however, the Wannier functions constitute an orthogonal set. The annihilation operator $c_{i\sigma}$ corresponding to this Wannier state is defined with use of the field operator $\Psi_\sigma(\boldsymbol{r})$ by

$$c_{i\sigma} = \int \mathrm{d}\boldsymbol{r}\, w_i^*(\boldsymbol{r}) \Psi_\sigma(\boldsymbol{r}). \tag{2.21}$$

The eigenenergy $E_{\boldsymbol{k}}$ of the first band is Fourier transformed to give

$$t_{ij} = \frac{1}{\sqrt{N}} \sum_{\boldsymbol{k}} E_{\boldsymbol{k}} \exp[\mathrm{i}\boldsymbol{k} \cdot (\boldsymbol{R}_i - \boldsymbol{R}_j)]. \tag{2.22}$$

Here the diagonal element $t_{ii} \equiv \epsilon_a$ represents the $1s$ level with a possible shift, and $t_{ij}$ with $i \neq j$ is the hopping energy between the sites $i, j$. Explicitly, $t_{ij}$ is given by

$$t_{ij} = \int \mathrm{d}\boldsymbol{r} \int \mathrm{d}\boldsymbol{r}' w_i^*(\boldsymbol{r}) h_{\uparrow\uparrow}(\boldsymbol{r}, \boldsymbol{r}') w_j(\boldsymbol{r}'). \tag{2.23}$$

We are now ready to represent the original Hamiltonian given by Eq. (2.1) in terms of the Wannier basis, which has the form analogous to the tight-binding approximation. Since we are interested in the ground state and low-lying excitations, we consider the subspace consisting of $1s$ states given by Eq. (2.20), but with arbitrary spin configurations. We introduce a projection operator $P_{1s}$ to this subspace, and obtain

$$P_{1s} H P_{1s} = H_1 + H_2 + E_{\mathrm{DX}}, \tag{2.24}$$

$$H_1 = \sum_{i\sigma} \epsilon_a c_{i\sigma}^\dagger c_{i\sigma} + \sum_{i \neq j, \sigma} \tilde{t}_{ij}^{(\sigma)} c_{i\sigma}^\dagger c_{j\sigma}, \tag{2.25}$$

$$H_2 = \frac{1}{2} \sum_{ijlm} \sum_{\sigma\sigma'} \langle ij|v|ml\rangle c_{i\sigma}^\dagger c_{j\sigma'}^\dagger c_{m\sigma'} c_{l\sigma}, \tag{2.26}$$

where the matrix element of the Coulomb interaction is given by

$$\langle ij|v|ml\rangle = \int \mathrm{d}\boldsymbol{r} \int \mathrm{d}\boldsymbol{r}' v(\boldsymbol{r} - \boldsymbol{r}') w_i^*(\boldsymbol{r}) w_j^*(\boldsymbol{r}') w_m(\boldsymbol{r}') w_l(\boldsymbol{r}). \tag{2.27}$$

Note that we use the ordering convention of basis in the bra as defined by $\langle ij| = |ji\rangle^\dagger$ throughout the book. The transfer $\tilde{t}_{ij}^{(\sigma)}$ may be different from the one given by Eq. (2.23) since the latter is defined for the fully polarized state. In the Hartree–Fock approximation, we obtain

$$\tilde{t}_{ij}^{(\sigma)} = t_{ij} + \sum_{m\sigma'} \left( \langle im|v|mj\rangle \delta n_{m\sigma'} - \langle im|v|jm\rangle \delta n_{m\sigma} \delta_{\sigma\sigma'} \right), \tag{2.28}$$

where $\delta n_{m\sigma'}$ denotes the deviation of the occupation number from the fully polarized state. For qualitative argument, we identify $\tilde{t}_{ij}^{(\sigma)}$ with $t_{ij}$, neglecting the dependence on spin polarization.

Since $w_i(\boldsymbol{r})$ is localized around $\boldsymbol{R}_i$, most of $\langle ij|v|ml\rangle$ is short-ranged for different sites. However, terms of the type $\langle ij|v|ji\rangle$ remain significant even for a pair of distant sites $i$ and $j$. This type is called the Coulomb integral and is responsible for charge fluctuations like the plasma oscillation. Another important contribution comes from the type $\langle ij|v|ij\rangle$ for neighboring sites. The matrix element is called the exchange integral, which acts in favor of parallel arrangement of spins at neighboring sites.

In order to go beyond the mean field approximation, we take the drastic approximation. Namely, among $\langle ij|v|ml\rangle$ in Eq. (2.27), only the largest term is kept with $i = j = l = m$. Then we obtain

$$H_2 \to U \sum_i (n_{i\uparrow} n_{i\downarrow} - n_{i\uparrow} - n_{i\downarrow}) \equiv H_U - U N_e, \tag{2.29}$$

where $U = \langle ii|v|ii\rangle$, and $n_{i\sigma} = c_{i\sigma}^\dagger c_{i\sigma}$. The model $H_1 + H_U$ is often called the Hubbard model [8–10]. Namely, we define $H_{\rm Hub}$ as

$$H_{\rm Hub} = \sum_{i\sigma} \epsilon_a c_{i\sigma}^\dagger c_{i\sigma} + \sum_{i\neq j,\sigma} t_{ij} c_{i\sigma}^\dagger c_{j\sigma} + U \sum_i n_{i\uparrow} n_{i\downarrow}. \tag{2.30}$$

The electron number $N_e$ can be different from the number $N$ of the lattice sites. In real systems, the difference is caused by doping, or by reservoir bands which are neglected in the model. The fully polarized Hartree–Fock state becomes an eigenstate of the Hubbard model. However, this ferromagnetic state should have higher kinetic energy than that of paramagnetic and antiferromagnetic states. In one dimension, for example, the ground state does not have a spin polarization according to available exact solution [11]. Moreover, the ground state is insulating if the occupation number of electrons is unity per site. This paramagnetic insulating state is consistent with the picture proposed by Mott [6].

Let us return to more general forms of the Coulomb interaction than in Eq. (2.29). The simplest way to see the effect of each term in Eqs. (2.25) and (2.26) on the spin configuration is to take $N = 2$, which makes possible the analogy with the hydrogen molecule. In the two-body interaction $H_2$, we keep $U \equiv \langle 11|v|11\rangle$, $K \equiv \langle 12|v|21\rangle$,

$J_d \equiv \langle 12|v|12\rangle$ and equivalent ones, all of which are positive. In this case exact solution of $H_1 + H_2$ is easily obtained. Since the total spin is conserved, we classify the states into spin triplet ($S = 1$) and singlet ($S = 0$). The three triplet states have the energy $E_t = 2\epsilon_a + K - J_d$. Among the three singlet states, the lowest one has the energy

$$E_s = 2\epsilon_a + K + J_d + U_-/2 - \left(U_-^2/4 + 4t^2\right)^{1/2}, \tag{2.31}$$

with $t \equiv t_{12}$ and $U_- \equiv U - K - J_d$. The result can be derived by analogy with the simpler case of Eq. (1.17). The derivation is the subject of Problem 2.2. The double occupation of a site tends to zero in the limit $a \gg a_B$. In this limit $U$ is much larger than $|t|$, $J_d$, and $K$. Then we may expand the square root in Eq. (2.31) to first order in $t^2$. It is convenient to use the projection operators $P_s$ and $P_t$ to the singlet and triplet pairs as given by

$$P_s = \frac{1}{4} - \mathbf{S}_1 \cdot \mathbf{S}_2, \quad P_t = \frac{3}{4} + \mathbf{S}_1 \cdot \mathbf{S}_2. \tag{2.32}$$

The effective Hamiltonian describing these states is then given by

$$\begin{aligned} H_{\text{eff}} &= E_s P_s + E_t P_t = E_t + (E_s - E_t) P_s \\ &\sim 2\epsilon_a + K - J_d + \left(\frac{4t^2}{U_-} - 2J_d\right)\left(\mathbf{S}_1 \cdot \mathbf{S}_2 - \frac{1}{4}\right), \end{aligned} \tag{2.33}$$

which can also be derived by perturbation theory in $t$ as in Eq. (1.28). The coefficient of the spin-dependent term forms the effective exchange interaction. Two competing effects are present; the part with the transfer $t$ is called the superexchange interaction which favors the singlet pair. This is because the doubly occupied site can occur as the intermediate state by perturbation in terms of $t$. On the other hand, the Coulombic exchange $J_d$ favors the triplet pair. This gain of energy originates from the Pauli principle which discourages the overlap of wave functions with the same spin. Depending on the parameters in the system, either of them can dominate over the other.

In the Heitler–London theory for the bonding of hydrogen molecule, the stability of the singlet state is interpreted in terms of Coulomb attraction due to accumulated electrons in between the protons. The stabilization by the kinetic exchange represents the same accumulation effect in different terms. In the case of general $N$, the competing exchange mechanisms described above are also present. Since $J_d$ should decay faster than $t_{12}^2$ as $a/a_B$ increases, the kinetic exchange wins in the case (ii). As a result, antiferromagnetism has more chance to be realized for larger $a/a_B$ than ferromagnetism. If the average occupation number per site is less than unity, it is possible for electrons to propagate by using empty sites, thus avoiding the large Coulomb repulsion associated with the double occupation. Hence working with the model space without double occupation of any site, one often uses the following effective Hamiltonian:

$$H_{tJ} = H_t + J \sum_{\langle ij \rangle} \left( \boldsymbol{S}_i \cdot \boldsymbol{S}_j - \frac{1}{4} n_i n_j \right), \tag{2.34}$$

which is called the $t$-$J$ model [12]. Here $H_t$ takes the same form as in the Hubbard model, except that it works in the subspace without the double occupation. The exchange term gives $-J$ if the nearest-neighbor sites $\langle ij \rangle$ are occupied by the singlet pair, and becomes zero otherwise. The number operator $n_i$ for the site $i$ takes either 0 or 1. The $t$-$J$ model is regarded as the effective model for the Hubbard model if $J = 4t^2/U \ll t$. However, $t$ and $J$ are usually treated as independent parameters in the $t$-$J$ model.

## 2.4 Density of States of Electrons

The density of states $\rho(\epsilon)$ of electrons plays a fundamental role in determining the property of the system, as will be discussed in later chapters. Let us derive the explicit form of $\rho(\epsilon)$ for representative lattices, restricting to non-interacting electrons [13]. The Fourier transform of $t_{ij}$ gives the spectrum $\epsilon_{\boldsymbol{k}}$ as

$$\epsilon_{\boldsymbol{k}} = \frac{1}{\sqrt{N}} \sum_i t_{ij} \exp[-\mathrm{i}\boldsymbol{k} \cdot (\boldsymbol{R}_i - \boldsymbol{R}_j)], \tag{2.35}$$

where $N$ corresponds to the number of lattice sites $\boldsymbol{R}_i$. The density of states per spin and per unit cell is given by

$$\rho(\epsilon) = \frac{1}{N} \sum_{\boldsymbol{k}} \delta(\epsilon - \epsilon_{\boldsymbol{k}}). \tag{2.36}$$

For convenience we introduce the quantity $G(z)$ defined by

$$G(z) = \frac{1}{N} \sum_{\boldsymbol{k}} \frac{1}{\epsilon - \epsilon_{\boldsymbol{k}}}, \tag{2.37}$$

with $z$ being complex energy. We defer discussion of the detailed meaning of $G(z)$ to Sect. 3.7. Here it is sufficient to recognize that $G(z)$ is analytic function of $z$ with a cut along the real axis. The density of states $\rho(\epsilon)$ is obtained by

$$\rho(\epsilon) = -\frac{1}{\pi} \mathrm{Im} G(\epsilon + \mathrm{i}0_+), \tag{2.38}$$

with $0_+$ positive infinitesimal.

We begin with the one-dimensional system with only the nearest-neighbor hopping $t$, and the lattice constant $a$. The spectrum is given by

$$\epsilon_k = 2t\cos(ka). \tag{2.39}$$

We introduce the notation $D_1 \equiv 2|t|$ to represent the maximum of $\epsilon_k$. It is easy to derive $\rho(\epsilon) = \rho_1(\epsilon)$ analytically in this case. Introducing the complex variable $\zeta = \exp(\mathrm{i}\phi)$ with $\phi = ka$, we obtain the expression

$$G(z) = \int_0^{2\pi} \frac{\mathrm{d}\phi}{2\pi} \frac{1}{z - 2t\cos\phi} = -\oint_C \frac{\mathrm{d}\zeta}{2\pi \mathrm{i}} \frac{2}{\zeta^2 - \zeta z/t + 1}, \tag{2.40}$$

where the integration path $C$ is the unit circle in the complex plane of $\zeta$. The denominator vanishes at $\zeta = x \pm \sqrt{x^2 - 1}$ with $x = z/(2t)$. In the case of $x^2 > 1$, the path $C$ encloses a single pole at $\zeta_- = x - \sqrt{x^2 - 1}$. Evaluating the residue, we obtain

$$G(z) = \left(z^2 - D_1^2\right)^{-1/2}, \tag{2.41}$$

which tends to $1/z$ if $|z|/D_1 \gg 1$. The branch of the square root should be chosen to reproduce the correct asymptotic behavior and the analyticity of $G(z)$, which is regular except on the real axis. The resultant expression is valid for arbitrary value of $z$, including the cut $z \in [-D_1, D_1]$. Thus we obtain

$$\rho_1(\epsilon) = \frac{\theta\left(D_1^2 - \epsilon^2\right)}{\pi\sqrt{D_1^2 - \epsilon^2}}, \tag{2.42}$$

which is divergent at band edges. The inverse square-root divergence at the lower edge has the same feature as the free particle with the spectrum $\epsilon_k \propto k^2$. Figure 2.2a illustrates $\rho_1(\epsilon)$.

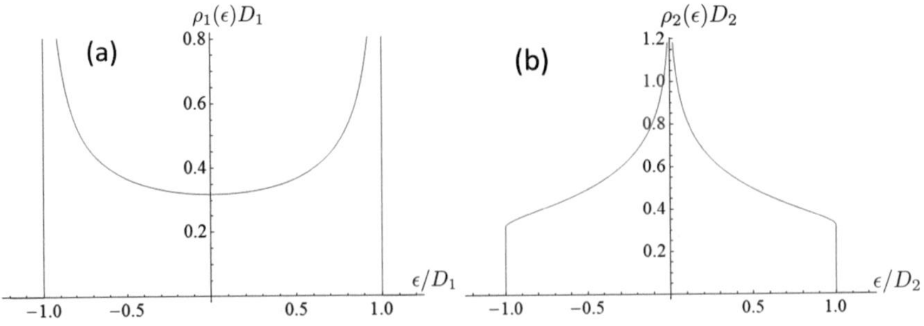


**Fig. 2.2** Density of states for (**a**) the one-dimensional lattice, and (**b**) the square lattice

We next consider the the square lattice where the spectrum is given by

$$\epsilon_{\boldsymbol{k}} = 2t\left[\cos(k_x a) + \cos(k_y a)\right], \tag{2.43}$$

with the energy range $|\epsilon_{\boldsymbol{k}}| \le D_2 \equiv 4|t|$. Although much more complicated than in the previous case, it is possible to obtain the density of states $\rho_2(\epsilon)$ analytically, which is the subject of Problem 3.7. The result is given by

$$\rho_2(\epsilon) = \frac{2}{\pi^2 D_2}\theta\left(D_2^2 - \epsilon^2\right)\mathcal{K}(\sqrt{1-(\epsilon/D_2)^2}), \tag{2.44}$$

where $\mathcal{K}(z)$ is the complete elliptic integral of the first kind, as defined by

$$\mathcal{K}(z) = \int_0^{\pi/2} \mathrm{d}\theta \frac{1}{\sqrt{1 - z^2 \sin^2\theta}}. \tag{2.45}$$

Figure 2.2b illustrates $\rho_2(\epsilon)$, which has the logarithmic divergence at $\epsilon = 0$, and discontinuity at $\epsilon = \pm D_2$. The origin of the divergence is the degeneracy $\epsilon_{\boldsymbol{k}} = 0$ along the lines $k_x \pm k_y = \pm\pi/a$. The discontinuity at $\epsilon = -D_2$ shares the same feature as that of the free particle with the spectrum $\epsilon_{\boldsymbol{k}} \propto k_x^2 + k_y^2$.

We turn to the simple cubic lattice where the spectrum is given by

$$\epsilon_{\boldsymbol{k}} = 2t\left[\cos(k_x a) + \cos(k_y a) + \cos(k_z a)\right], \tag{2.46}$$

with the energy range $|\epsilon_{\boldsymbol{k}}| \le 6|t| \equiv D_3$. The numerical result for the density of states $\rho_3(\epsilon)$ is shown in Fig. 2.3. The square-root rise from $\epsilon/D_3 = -1$ is common with that of the free particle with the spectrum $\epsilon_{\boldsymbol{k}} \propto k_x^2 + k_y^2 + k_z^2$. The cusps at $\epsilon/D_3 = \pm 1/3$ correspond to the van Hove singularity [14], which arises if the spectrum becomes locally flat around certain $\boldsymbol{k}$ in the Brillouin zone. In the present case, we obtain $\partial\epsilon_{\boldsymbol{k}}/\partial\boldsymbol{k} = 0$ at $k_i a = 0, \pi$ for $i = x, y, z$. The energy becomes $\epsilon_{\boldsymbol{k}} = \pm D_3/3$ at these points.

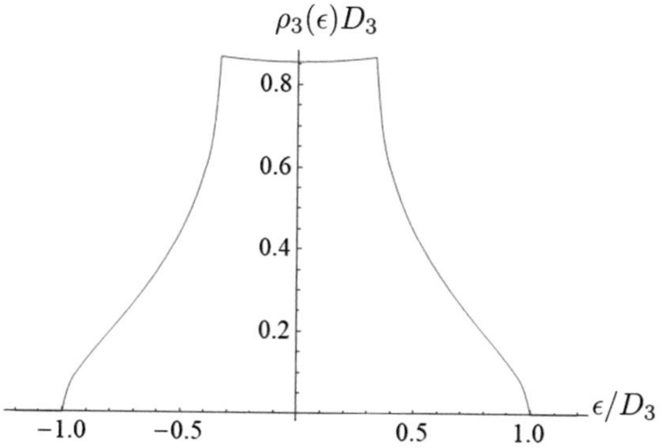


**Fig. 2.3** Density of states $\rho_3(\epsilon)$ for the simple cubic lattice with the nearest-neighbor hopping

## Problems

**2.1** A single electron does not interact with itself. Show that this feature is not respected in the Hartree approximation, but is indeed realized in the Hartree–Fock approximation.

**2.2** Derive the ground state energy given by Eq. (2.31) for the two-site system.

**2.3*** Derive the density of states for the square lattice as given by Eq. (2.44).

## Solutions to Problems

### Problem 2.1

The interaction term in the Hartree–Fock Hamiltonian (2.16) has the following operator part:

$$\Psi_1^\dagger \Psi_2 \left\langle \Psi_3^\dagger \Psi_4 \right\rangle - \Psi_1^\dagger \Psi_4 \left\langle \Psi_3^\dagger \Psi_2 \right\rangle, \tag{2.47}$$

where $\Psi_i$ is the abbreviated form for the faithful expression $\Psi_\sigma(\boldsymbol{r})$. The first term in Eq. (2.47) corresponds to the Hartree term, and the second one to the exchange (Fock) term. We shall show that these terms combine to give vanishing average for a single-electron system. The Hartree term alone leads to a finite value, which is unreasonable for a single electron because the electron should not interact with itself.

Let $\phi(\boldsymbol{r}, \sigma)$ be the wave function of a single-electron state $|1\rangle$. Then with the vacuum state $|0\rangle$, we obtain $\langle 0|\Psi_1|1\rangle = \phi(\boldsymbol{r}_1, \sigma_1)$ by definition. The average of the first term in Eq. (2.47) is written as

$$\left\langle \Psi_1^\dagger \Psi_2 \right\rangle \left\langle \Psi_3^\dagger \Psi_4 \right\rangle = \phi(\boldsymbol{r}_1, \sigma_1)^* \phi(\boldsymbol{r}_2, \sigma_2) \phi(\boldsymbol{r}_3, \sigma_3)^* \phi(\boldsymbol{r}_4, \sigma_4). \tag{2.48}$$

The average of the second term is given by the interchange of suffices 2 and 4, which results in the same expression as Eq. (2.48) with opposite sign. Hence we obtain $E_{\rm DX} = 0$ in Eq. (2.17).

### Problem 2.2

The two-site system has the Coulomb interaction $H_2$ as given by

$$H_2 = U \sum_{i=1}^{2} n_{i\uparrow} n_{i\downarrow} + K n_1 n_2 + J_d (P_s - P_t), \tag{2.49}$$

where $n_i = n_{i\uparrow} + n_{i\downarrow}$ and $P_s$, $P_t$ are projection operators onto singlet and triplet pairs, respectively. Following the argument in Problem 1.2, we take the basis set $|3\rangle$

and $|4\rangle$ for the singlet states. Including the kinetic energy part $H_1$, the Hamiltonian $H_1 + H_2$ for the singlet is given by the $2 \times 2$ matrix as

$$\begin{pmatrix} K + J_d & -2t \\ -2t & U \end{pmatrix} = K + J_d + \begin{pmatrix} 0 & -2t \\ -2t & U_- \end{pmatrix}, \tag{2.50}$$

where we put $U_- = U - K - J_d$ and the one-body energy $\epsilon_a$ is set to zero. Comparing the last matrix with that in the solution to Problem 1.2, we obtain Eq. (2.31) for the ground state energy.

**Problem 2.3***

We first derive the analytic expression of $G(z)$. Using the variables defined by

$$k_x a \equiv \phi_x = \phi_+ - \phi_-, \quad k_y a \equiv \phi_y = \phi_+ + \phi_-. \tag{2.51}$$

The integral is replaced by

$$\int_{-\pi}^{\pi} \frac{d\phi_x}{2\pi} \int_{-\pi}^{\pi} \frac{d\phi_y}{2\pi} \rightarrow \int_{-\pi}^{\pi} \frac{d\phi_+}{2\pi} \int_{-\pi}^{\pi} \frac{d\phi_-}{2\pi}. \tag{2.52}$$

The validity of the replacement is understood by Fig. 2.4. These integrals correspond to the angular average, and are abbreviated as $\langle \cdots \rangle_\phi$ for the average over $\phi$. By writing $\epsilon_k$ in the product form

$$\epsilon_{\boldsymbol{k}} = 4t \cos\phi_+ \cos\phi_-, \tag{2.53}$$

we obtain

$$G(z) = \left\langle \frac{1}{z - 4t \cos\phi_+ \cos\phi_-} \right\rangle_{\phi_\pm} = \left\langle \frac{1}{\sqrt{z^2 - D_2^2 \cos^2\phi_+}} \right\rangle_{\phi_+}, \tag{2.54}$$

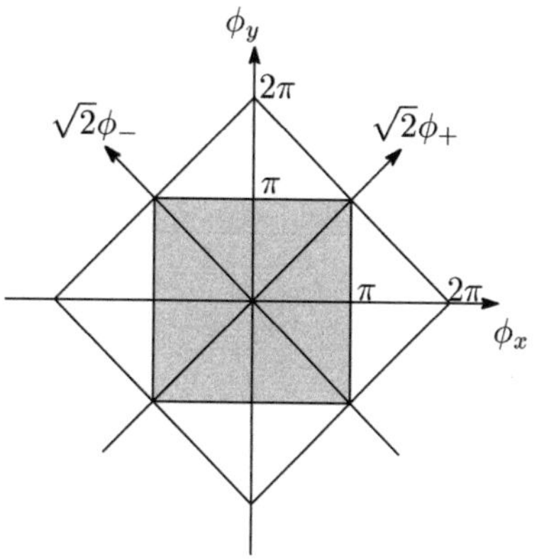


**Fig. 2.4** Integration regions of angular variables. The original one is indicated by the shaded square, while the larger square minus the original one gives the equivalent result by the periodicity of the integrand. The larger square corresponds to the integration range $[-\pi, \pi]$ for the variables $\phi_\pm$

with $D_2 = 4|t|$. The last integrand has the same from as the local Green function in the one-dimensional case, which is given by Eq. (2.40). To obtain the second equality we have taken the first average over $\phi_-$ by the contour integral as in the one-dimensional system, regarding $4t \cos\phi_+$ as the effective transfer. The second average over $\phi_+$ results in the complete elliptic integral with the definition Eq. (2.45). We thus obtain

$$G(z) = \frac{2}{\pi z} \mathcal{K}\left(\frac{D_2}{z}\right), \tag{2.55}$$

which tends to the correct asymptotic form $1/z$ since $\mathcal{K}(0) = \pi/2$.

Let us now derive the density of states $\rho_2(\epsilon)$ from Eq. (2.55) by setting $z \to \epsilon + i0_+$. For this purpose we introduce the auxiliary function

$$L(\zeta) = \int_0^{\pi/2} d\phi \left(1 - \zeta \sin^2\phi\right)^{-1/2}, \tag{2.56}$$

which is related to $\mathcal{K}$ by $L(\zeta^2) = \mathcal{K}(\zeta)$. If $\zeta$ is taken as real variable: $\zeta \to x$, it is obvious that $\mathrm{Im}\, L(x) = 0$ for $x < 1$. For $x > 1$, on the other hand, we decompose the integral into two ranges $[0, \phi_0]$ and $[\phi_0, \pi/2]$ with $x \sin^2\phi_0 = 1$. Then only the second range contribute to $\mathrm{Im}\, L(x)$ as

$$\mathrm{Im} L(x) = \pm \int_{\phi_0}^{\pi/2} d\phi \left(x \sin^2\phi - 1\right)^{-1/2}, \tag{2.57}$$

where the sign depends on the infinitesimal imaginary part in $x$. We change to the new integration variable $\theta$ so that the integral range becomes $[0, \pi]$. This is realized if we take

$$(x-1)\cos^2\theta = x \sin^2\phi - 1, \tag{2.58}$$

with $\phi_0 \leftrightarrow \theta = \pi/2$ and $\phi = \pi/2 \leftrightarrow \theta = 0$. Using the ensuing relations

$$\frac{d\phi}{d\theta} = -\frac{x-1}{x} \cdot \frac{\cos\theta \sin\theta}{\cos\phi \sin\phi}, \quad \frac{\sin^2\theta}{\cos^2\phi} = \frac{x}{x-1}, \tag{2.59}$$

we arrive at the remarkable identity:

$$\mathrm{Im} L(x) = \pm \frac{1}{\sqrt{x}} L\left(1 - \frac{1}{x}\right). \tag{2.60}$$

This is equivalent to

$$\frac{D_2}{\epsilon}\mathrm{Im}\mathcal{K}\left(\frac{D_2}{\epsilon + \mathrm{i}0_+}\right) = -\mathcal{K}\left(\sqrt{1 - \frac{\epsilon^2}{D_2^2}}\right)\theta(D_2^2 - \epsilon^2), \tag{2.61}$$

which gives Eq. (2.44).

## References


1. Kohn, W., Sham, L.J.: Phys. Rev. **140**, A1133 (1965)
2. Jones, R.O.: Rev. Mod. Phys. **87**, 897 (2015)
3. Marzari, N., Mostofi, A.A., Yates, J.R., Souza, I., Vanderbilt, D.: Rev. Mod. Phys. **84**, 1419 (2012)
4. Martin, R.M.: Electronic Structure: Basic Theory and Practical Methods. Cambridge University Press, Cambridge (2004)
5. Luttinger, J.M.: J. Math. Phys. **4**, 1154 (1963)
6. Mott, N.: Metal-Insulator Transitions, 2nd edn. Taylor & Francis, London (1990)
7. Anderson, P.W.: In Solid State Physics, vol. 14, p. 14. Academic Press, New York (1963)
8. J. Hubbard, Hubbard, J.: Proc. Royal. Soc. London **276**, 238 (1963)
9. Kanamori, J.: Prog. Theor. Phys. **30**, 275 (1963)
10. Gutzwiller, M.C.: Phys. Rev. Lett. **10**, 159 (1963)
11. E.H. Lieb, F.Y. Wu, Lieb, E.H., Wu, F.Y.: Phys. Rev. Lett. **20**, 1445 (1968)
12. M. Imada, A. Fujimori and Y. Tokura, Imada, M., Fujimori, A., Tokura, Y.: Rev. Mod. Phys. **70**, 1039 (1998)
13. Economou, E.N.: Green's Functions in Quantum Physics, 3rd edn. Springer, Berlin (2006)
14. Van Hove, L.: Phys. Rev. **89**, 1189 (1953)

# Chapter 3
# Linear Response and Green Functions

**Abstract** The linear response means a change in the system proportional to the magnitude of the external force. If the external force is infinitesimally small, the higher order effects can safely be neglected. Actually the dominance of linear response often extends to the practical range of the external force. Thus physical properties such as electrical conduction and magnetic susceptibility are conveniently described by the linear response theory. Furthermore, the linear response theory leads to a simple relation between dissipation of energy, such as the Joule heat, and fluctuation without external force. In this Chapter, we explain the linear response theory starting with basics of statistical physics, proceeding to practical applications such as deriving resistivity and magnetic relaxation. Various kinds of Green functions appear in the linear response theory. In addition to those related to response functions, we also discuss single-particle Green functions which describe propagation and damping of bosons and fermions.

## 3.1 Static Response

Let us first consider the case of static external field. To be specific, we take an isotropic system with the weak magnetic field $B$ applied along the $z$-axis. The magnetic moment $\boldsymbol{M}$ is induced in certain localized region in the system. The interaction Hamiltonian is given by

$$H_{\text{ext}} = -\boldsymbol{M} \cdot \boldsymbol{B} = -MB, \tag{3.1}$$

where the index for the $z$-component has been omitted in the rightmost side, The statistical average without external field is given in terms of eigenstates $|n\rangle$ of the system Hamiltonian $H$ with the eigenvalue $E_n$ as

$$\langle M \rangle = \sum_n \exp(-\beta E_n) M_{nn} / \sum_n \exp(-\beta E_n) \equiv \text{Tr}\,(\rho M)\,. \tag{3.2}$$


Y. Kuramoto, *Quantum Many-Body Physics*, Lecture Notes in Physics 934,
https://doi.org/10.1007/978-4-431-55393-9_3

Here $\rho$ is called the statistical operator or the density matrix, which is given by

$$\rho = \exp(-\beta H)/\mathrm{Tr}\exp(-\beta H) \equiv \exp(-\beta H)/Z, \tag{3.3}$$

with $Z$ being the partition function. The magnetization $M$ changes its sign under the time reversal in common with spin operator or orbital angular momentum. If the system described by $H$ has the time-reversal symmetry, every eigenstate $|n\rangle$ has a degenerate partner $|m\rangle$ such that $M_{mm} = -M_{nn}$. Hence we obtain $\langle M\rangle = 0$ in Eq. (3.2).

A magnetic field $B$ breaks the time reversal. Then we derive the statistical average $\langle M\rangle_{\mathrm{ext}}$ up to the lowest order in $B$. Here the suffix of the average emphasizes the external field described by $H_{\mathrm{ext}}$. The statistical operator of the total Hamiltonian $H_{\mathrm{tot}} = H + H_{\mathrm{ext}}$ is formally factorized as

$$\exp(-\beta H_{\mathrm{tot}}) = \exp(-\beta H)\mathcal{U}(\beta), \tag{3.4}$$

with

$$\mathcal{U}(\beta) = \exp(\beta H)\exp(-\beta H_{\mathrm{tot}}). \tag{3.5}$$

By taking derivative of both sides of Eq. (3.5) with respect to $\beta$, we obtain

$$\frac{\partial}{\partial\beta}\mathcal{U}(\beta) = e^{\beta H}(H - H_{\mathrm{tot}})e^{-\beta H_{\mathrm{tot}}} = -e^{\beta H}H_{\mathrm{ext}}e^{-\beta H}\mathcal{U}(\beta). \tag{3.6}$$

We now introduce the Matsubara picture by

$$H_{\mathrm{ext}}^{\mathrm{M}}(\beta) \equiv e^{\beta H}H_{\mathrm{ext}}e^{-\beta H} = H_{\mathrm{ext}}^{\mathrm{H}}(-\mathrm{i}\beta), \tag{3.7}$$

where $H_{\mathrm{ext}}^{\mathrm{H}}(-\mathrm{i}\beta)$ is the Heisenberg picture with the imaginary time $-\mathrm{i}\beta$.

We solve the differential equation (3.6) by iteration using the Matsubara picture. With the boundary condition $\mathcal{U}(\beta) = 1$ for $H_{\mathrm{ext}} = 0$, we integrate both sides to obtain

$$\mathcal{U}(\beta) = 1 - \int_0^\beta \mathrm{d}\tau\, H_{\mathrm{ext}}^{\mathrm{M}}(\tau) + O(H_{\mathrm{ext}}^2). \tag{3.8}$$

Then the perturbed statistical average is evaluated as

$$\begin{aligned}\langle M\rangle_{\mathrm{ext}} &= \mathrm{Tr}\left[\exp(-\beta H)\mathcal{U}(\beta)M\right]/\mathrm{Tr}\left[\exp(-\beta H)\mathcal{U}(\beta)\right]\\ &= -\int_0^\beta \mathrm{d}\tau\,\langle H_{\mathrm{ext}}^{\mathrm{M}}(\tau)M\rangle + O(H_{\mathrm{ext}}^2)\\ &= \chi B + O(H_{\mathrm{ext}}^2),\end{aligned} \tag{3.9}$$

where the magnetic susceptibility $\chi$ is given by

$$\chi = \int_0^\beta \mathrm{d}\tau \langle M^{\mathrm{M}}(\tau)M\rangle = \int_0^\beta \mathrm{d}\tau \langle e^{\tau H} M e^{-\tau H} M\rangle, \tag{3.10}$$

with $H_{\mathrm{ext}}$ given by Eq. (3.1). The correction to the linear response in Eq. (3.9) begins actually from $O(H_{\mathrm{ext}}^3)$, which is the subject of Problem 3.1.

We note that the average in Eq. (3.10) is taken for the system without external field. Namely, the linear response measured by $\chi$ is basically given by the squared average of the observable $M$, modified by integration over $\tau$. The squared average is closely related to the fluctuation of $M$. In the special case where $M$ is conserved, Eq. (3.10) gives the Curie law. Namely, if $[M, H] = 0$, $M^{\mathrm{M}}(\tau)$ is independent of $\tau$. Then we obtain

$$\chi = \beta\langle M^2\rangle = C/T, \tag{3.11}$$

where $C = \langle M^2\rangle$ is called the Curie constant.

## 3.2 Dynamic Response

We go on to the case with time-dependent external field such as oscillating magnetic and electric fields [1]. The discussion now becomes more delicate than the case of static fields. Namely, the dynamic linear response theory is built upon the following two assumptions:

(a) The system is in thermal equilibrium before the external field, described by the Hamiltonian $H_{\mathrm{ext}}(t)$, is applied. Namely, the statistical weight of each eigenstate $|n\rangle$ is given by $\rho_n = \exp(-\beta E_n)/Z$ in the remote past.
(b) The system develops adiabatically in the course of application of $H_{\mathrm{ext}}(t)$. In other words, an eigenstate $|n\rangle$ develops to $|n(t)\rangle$ without changing the statistical weight $\rho_n$. Then the statistical operator $\rho_{\mathrm{tot}}(t)$ at each time $t$ is given by

$$\rho_{\mathrm{tot}}(t) = |n(t)\rangle\rho_n\langle n(t)|. \tag{3.12}$$

We introduce external fields first at zero temperature, and derive the linear response. For this purpose we introduce the unitary operator $U(t)$ by the relation:

$$|\Psi_{\mathrm{g}}(t)\rangle = \exp(-\mathrm{i}Ht)U(t)|\Psi_{\mathrm{g}}(-\infty)\rangle, \tag{3.13}$$

where $|\Psi_{\mathrm{g}}(-\infty)\rangle$ is the eigenstate without $H_{\mathrm{ext}}(t)$. Substitution to the Schrödinger equation

$$\mathrm{i}\frac{\partial}{\partial t}|\Psi_{\mathrm{g}}(t)\rangle = [H + H_{\mathrm{ext}}(t)]\big|\Psi_{\mathrm{g}}(t)\rangle \tag{3.14}$$

leads to

$$\mathrm{i}\frac{\partial}{\partial t}U(t) = \exp(\mathrm{i}Ht)H_{\mathrm{ext}}(t)\exp(-\mathrm{i}Ht)U(t) \equiv H^{\mathrm{H}}_{\mathrm{ext}}(t)U(t), \tag{3.15}$$

where we have used the Heisenberg picture $H^{\mathrm{H}}_{\mathrm{ext}}(t)$. The assumption (a) is used as the boundary condition $U(t) \to 1$ as $t \to -\infty$. Then $U(t)$ is derived from Eq. (3.15) by integration over $t$. Namely, we first replace $U(t)$ in the RHS by 1, and derive the solution iteratively. The solution takes the form analogous to the case of static external fields:

$$U(t) = 1 - \mathrm{i}\int_{-\infty}^{t} \mathrm{d}t' H^{\mathrm{H}}_{\mathrm{ext}}(t') + O(H^2_{\mathrm{ext}}). \tag{3.16}$$

Then the expectation value for arbitrary observable $A$ at time $t$ is given by

$$\begin{aligned}\langle\Psi_{\mathrm{g}}(t)|A|\Psi_{\mathrm{g}}(t)\rangle &= \langle\Psi_{\mathrm{g}}(-\infty)|U^{\dagger}(t)A^{\mathrm{H}}(t)U(t)|\Psi_{\mathrm{g}}(-\infty)\rangle \\ &= \langle\Psi_{\mathrm{g}}(-\infty)|A^{\mathrm{H}}(t)|\Psi_{\mathrm{g}}(-\infty)\rangle \\ &\quad - \mathrm{i}\int_{-\infty}^{t} \mathrm{d}t' \langle\Psi_{\mathrm{g}}(-\infty)|[A^{\mathrm{H}}(t), H^{\mathrm{H}}_{\mathrm{ext}}(t')]|\Psi_{\mathrm{g}}(-\infty)\rangle + O(H^2_{\mathrm{ext}}).\end{aligned} \tag{3.17}$$

It is straightforward to extend the analysis to finite temperature. By the assumption (b), the average at time $t$ is given by

$$\mathrm{Tr}\,[A\rho_{\mathrm{tot}}(t)] = \sum_n \rho_n \langle n(t)|A|n(t)\rangle. \tag{3.18}$$

Hence the deviation $\delta\langle A^{\mathrm{H}}(t)\rangle$ from the equilibrium average $\langle A^{\mathrm{H}}(t)\rangle$ can be derived as

$$\delta\langle A^{\mathrm{H}}(t)\rangle = -\mathrm{i}\int_{-\infty}^{t} \mathrm{d}t' \langle[A^{\mathrm{H}}(t), H^{\mathrm{H}}_{\mathrm{ext}}(t')]\rangle + O(H^2_{\mathrm{ext}}), \tag{3.19}$$

where the contribution of each $|n(-\infty)\rangle$ to the statistical average has been taken into account by generalization of Eq. (3.17). Equation(3.19) is commonly called the Kubo formula. The same result can be obtained also from the equation of motion for the density matrix, which is the subject of Problem 3.2.

As an example of dynamic external field, we take the magnetic field $B\exp(-\mathrm{i}\omega t')$ that oscillates in time $t'$. The corresponding Hamiltonian is given by $H^{\mathrm{H}}_{\mathrm{ext}}(t') = -M^{\mathrm{H}}(t')B\exp(-\mathrm{i}\omega t')$. To be precise, the Hermiticity of the Hamiltonian requires that physical external field accompanies another Fourier component $\exp(\mathrm{i}\omega t')$. The effect of this component is immediately derived once we know the response against the component $\exp(-\mathrm{i}\omega t')$. Hence we discard the component $\exp(\mathrm{i}\omega t')$ unless we focus on the Hermite nature of the observable. We take the magnetization $M$ as an observable. Then Eq. (3.19) with $A = M$ leads to

$$\delta\langle M^{\rm H}(t)\rangle = \chi(\omega)\exp(-{\rm i}\omega t)B, \tag{3.20}$$

where $\chi(\omega)$ is the dynamical susceptibility given by

$$\chi(\omega) = {\rm i}\int_{-\infty}^{t} {\rm d}t' \langle [M^{\rm H}(t), M^{\rm H}(t')]\rangle \exp[{\rm i}\omega(t-t')]. \tag{3.21}$$

This is often called the Kubo formula for the dynamical susceptibility. In simple cases, one can derive the dynamical susceptibility explicitly as demonstrated later in Problem 3.3. Note the relation $t' \le t$ for the integration range, which represents the causality that the external field at $t'$ influences the observable only at later time. In other words, a future external field cannot influence the observable at present. As we explain later, the causality is reflected on the analytic property of the dynamical response function if it is regarded as a function of complex frequency.

## 3.3 Green Function and Its Spectral Representation

Throughout the section, the Heisenberg picture is used for the time dependence of physical quantities. Hence we omit the upper index H for simplicity. Generalizing Eq. (3.21) for the dynamical susceptibility to any quantities $X$ and $Y$, we introduce the retarded (R) Green function as follows:

$$\langle [X, Y]\rangle(z) \equiv -{\rm i}\int_0^{\infty} {\rm d}t\ \langle [X(t), Y]\rangle e^{{\rm i}zt} \equiv \int_{-\infty}^{\infty} {\rm d}t\ D_{XY}^{\rm R}(t)e^{{\rm i}zt}, \tag{3.22}$$

where the time $t'$ for $Y$ is set to zero because the integrand in Eq. (3.21) depends only on the time difference $t - t'$. The complex frequency $z$ is in the upper half plane, which guarantees the convergent integral. The name "retarded" for $D_{XY}^{\rm R}(t)$ or its Fourier transform $\langle [X, Y]\rangle(z)$ comes from the fact that the time $t$ for $X$ is later than that for $Y$. On the other hand, the name "Green function" originates from the special solution of a differential equation where the RHS is the delta function. Actually this feature applies also to the present case. Namely, by comparing the integration ranges in Eq. (3.22), we obtain

$$D_{XY}^{\rm R}(t) = -{\rm i}\theta(t)\langle [X(t), Y]\rangle. \tag{3.23}$$

Since the expression includes the step function $\theta(t)$, the $t$-derivative of this part gives rise to the delta function. It is obvious from Eq. (3.22) that the function $\langle [X, Y]\rangle(z)$ has finite derivatives of any order $n$ in the upper half plane of $z$, because $t^n \exp({\rm i}zt)$ goes to zero exponentially with $t \to \infty$. This means that the Green function is analytic in the upper half plane. Note that the convergent integrals originate from the positive integration range of $t$, which reflects the causality.

In the many-body theory, the Green function also describes propagation of particles. The simplest is the single-particle Green function of bosons, which corresponds to $\langle [X, Y] \rangle (z)$ provided $X, Y$ are related to the Bose operators as in $X = b, Y = b^\dagger$. The fermion Green function will be discussed later. In the case of lattice vibration, the lattice displacement $\phi$ is related to phonon operators as $\phi \propto b + b^\dagger$ in the symbolic notation. Hence the response function with $X = Y = \phi$ is regarded as the (single-particle) phonon Green function.

We now take the space and time-dependent magnetic field $\boldsymbol{B}(\boldsymbol{r}, t)$. The corresponding Hamiltonian is given by

$$\begin{aligned} H_{\text{ext}}(t) &= -\int \mathrm{d}\boldsymbol{r} \boldsymbol{M}(\boldsymbol{r}, t) \cdot \boldsymbol{B}(\boldsymbol{r}, t) \\ &= -\sum_{\boldsymbol{q}} \boldsymbol{M}(\boldsymbol{q}, t) \cdot \boldsymbol{B}(-\boldsymbol{q}) \exp(-\mathrm{i}\omega t), \end{aligned} \tag{3.24}$$

where the second line is a result of spatial Fourier transform. The response $\langle \boldsymbol{M}(\boldsymbol{q}, t) \rangle_{\text{ext}}$ depends on time according to $\exp(-\mathrm{i}\omega t)$. Setting the complex frequency as $z = \omega + \mathrm{i}0_+$ with positive but infinitesimal imaginary part, we obtain the generalized susceptibility tensor

$$\chi_{\mu\nu}(\boldsymbol{q}, \omega) = -\langle [M_\mu(\boldsymbol{q}), M_\nu(-\boldsymbol{q})] \rangle (\omega + \mathrm{i}0_+), \tag{3.25}$$

which describes the response of $\mu$-component of magnetization against the $\nu$-component of magnetic field. The quantity is often called the dynamical susceptibility, which depends on both wave number and frequency. The negative sign in the RHS is due to the definition (3.22). In general, the response function is determined in terms of the retarded Green function.

In many fermion systems, response functions correspond to two-particle Green functions of fermions. We consider an isotropic many-electron system where the susceptibility tensor is reduced to a scalar. The magnetic moment along the $z$-axis is written, in terms of creation and annihilation operators of electrons, as

$$M_z(\boldsymbol{q}) = -\mu_{\mathrm{B}} \frac{1}{\sqrt{V}} \sum_{\boldsymbol{k}\sigma} \sigma c_{\boldsymbol{k}\sigma}^\dagger c_{\boldsymbol{k}+\boldsymbol{q}\sigma}, \tag{3.26}$$

with $\sigma = \pm 1$, and $V$ being the volume of the system. In the case of non-interacting electrons, the dynamical susceptibility can be derived exactly as

$$\chi_0(\boldsymbol{q}, \omega) = 2\mu_{\mathrm{B}}^2 \frac{1}{V} \sum_{\boldsymbol{k}} \frac{f(\epsilon_{\boldsymbol{k}+\boldsymbol{q}}) - f(\epsilon_{\boldsymbol{k}})}{\omega - \epsilon_{\boldsymbol{k}+\boldsymbol{q}} - \epsilon_{\boldsymbol{k}}}. \tag{3.27}$$

The derivation exploits the reduction:

$$[c_1^\dagger c_2, c_2^\dagger c_1] = c_1^\dagger [c_2, c_2^\dagger c_1] + [c_1^\dagger, c_2^\dagger c_1] c_2 = c_1^\dagger c_1 - c_2^\dagger c_2, \tag{3.28}$$

where 1 and 2 symbolically represent quantum numbers. Further details of derivation are the subject of Problem 3.3.

The dynamical susceptibility reduces to the Pauli susceptibility in the limit of $\omega = 0$, $\boldsymbol{q} \to 0$, and $T \to 0$. This can be confirmed as

$$\lim_{\boldsymbol{q}\to 0} \chi_0(\boldsymbol{q}, 0) = 2\mu_{\mathrm{B}}^2 \frac{1}{V} \sum_{\boldsymbol{k}} \left( -\frac{\partial f(\epsilon_{\boldsymbol{k}})}{\partial \epsilon_{\boldsymbol{k}}} \right) \underset{T\to 0}{\Rightarrow} \mu_{\mathrm{B}}^2 \rho(\mu), \tag{3.29}$$

where the derivative of the Fermi function tends to the delta function peaked at the Fermi level $\mu$. For free electrons with $\epsilon_{\boldsymbol{k}} = \boldsymbol{k}^2/(2m)$, the density of states per spin is given by $\rho(\mu) = mk_F/\pi^2$ with $k_F$ the Fermi momentum. It is necessary to keep response functions independent of $V$ so as to have a meaningful thermodynamic limit. For this purpose we always take finite $\boldsymbol{q}$ in the calculation. The homogeneous response is derived in the limit $\boldsymbol{q} \to 0$, as demonstrated in Eq. (3.29).

Let us investigate analytic properties of the Green function. It is most convenient for this purpose to take matrix elements of $X, Y$ in terms of the exact many-body eigenstates $|n\rangle$ and $|m\rangle$ with the eigenenergies $E_n$ and $E_m$. Then integration in Eq. (3.22) can be performed explicitly, with the result

$$\begin{aligned} \langle [X, Y] \rangle(z) &= \mathrm{Av}_n \sum_m \frac{X_{nm} Y_{mn}}{z - \omega_{mn}} [1 - \exp(-\beta\omega_{mn})] \\ &= \sum_n \sum_m \frac{\rho_n - \rho_m}{z - \omega_{mn}} X_{nm} Y_{mn}. \end{aligned} \tag{3.30}$$

Here $\mathrm{Av}_n$ means the statistical average with respect to $|n\rangle$, and the explicit use of the weight $\rho_n$ gives the second line. We have used the notations $X_{nm} = \langle n|X|m\rangle$ and $\omega_{mn} \equiv E_m - E_n$.

Equation (3.30) shows that the function $\langle [X, Y] \rangle(z)$ has the cut singularity only along the real axis of $z$, and is analytic in both upper and lower half planes of $z$. The part with $\mathrm{Im}\, z < 0$, however, cannot be represented by the same time integral as the retarded Green function. Instead we introduce the alternative representation for $\mathrm{Im}\, z < 0$:

$$\langle [X, Y] \rangle(z) = \mathrm{i} \int_{-\infty}^{0} \mathrm{d}t \langle [X(t), Y] \rangle e^{\mathrm{i}zt} = \int_{-\infty}^{\infty} \mathrm{d}t\, D_{XY}^{\mathrm{A}}(t) e^{\mathrm{i}zt}, \tag{3.31}$$

which is called the advanced Green function since the integration is in the negative range of $t$. Note that the integral in Eq. (3.31) converges only in the lower half plane of $z$. Summarizing the argument above, the function $\langle [X, Y] \rangle(z)$ is analytic except for the real axis of $z$, and reduces to retarded and advanced Green functions in the upper and lower half planes, respectively. For response functions, the advanced Green function is merely a mathematical device to construct the analytic function. However, the advanced Green function for fermions describes physical (causal) propagation of a hole (antiparticle) as explained in Sect. 3.7.

As a very useful quantity, we introduce the spectral function $I_{XY}(\omega)$ by

$$I_{XY}(\omega) = \underset{n}{\mathrm{Av}} \sum_{m} X_{nm} Y_{mn} [1 - \exp(-\beta\omega)] \delta(\omega - \omega_{mn})$$
$$= \int_{-\infty}^{\infty} \frac{\mathrm{d}t}{2\pi} e^{\mathrm{i}\omega t} \langle [X(t), Y] \rangle, \tag{3.32}$$

with the relation to retarded and advanced Green functions:

$$\langle [X(t), Y] \rangle = \mathrm{i}[D^{\mathrm{R}}_{XY}(t) - D^{\mathrm{A}}_{XY}(t)]. \tag{3.33}$$

In terms of $I_{XY}(\omega)$, the Green function is represented by

$$\langle [X, Y] \rangle (z) = \int_{-\infty}^{\infty} \mathrm{d}\omega \frac{I_{XY}(\omega)}{z - \omega}, \tag{3.34}$$

which is called the spectral representation, or the Lehmann representation. The most important is the case $Y = X^\dagger$ for practical purpose, where $I_{XX^\dagger}(\omega)$ is real according to Eq. (3.32). In this case, the spectral function is an odd function of $\omega$, being positive for $\omega > 0$. It is related to the imaginary part of the response function as

$$I_{XX^\dagger}(\omega) = -I_{XX^\dagger}(-\omega) = -\frac{1}{\pi} \mathrm{Im}\, \langle [X, X^\dagger] \rangle (\omega + \mathrm{i}0_+), \tag{3.35}$$

which is verified by use of the identity $\mathrm{Im}\ (x + \mathrm{i}0_+)^{-1} = -\pi\delta(x)$. We shall frequently use the odd function property in discussing various kinds of Green functions.

The spectral function describes dissipation of energy. To exemplify this, we consider the simplest case where the AC electric field along the $z$ direction is coupled to the polarization density $P(\boldsymbol{r})$ in the isotropic system. The corresponding Hamiltonian is given by

$$H_{\mathrm{ext}}(t) = -\int \mathrm{d}\boldsymbol{r}\, P(\boldsymbol{r}) E(\boldsymbol{r}) \exp(-\mathrm{i}\omega t). \tag{3.36}$$

According to the linear response theory, the dynamical polarizability $\alpha(\boldsymbol{q}, \omega)$ for the electric field with wave vector $\boldsymbol{q}$ is given by

$$\alpha(\boldsymbol{q}, \omega) = -\langle [P_{\boldsymbol{q}}, P_{-\boldsymbol{q}}] \rangle (\omega + \mathrm{i}0_+), \tag{3.37}$$

where $P_q$ is the Fourier transform defined by

$$P_q = \frac{1}{\sqrt{V}} \int \mathrm{d}\boldsymbol{r}\, P(\boldsymbol{r}) \exp(-\mathrm{i}\boldsymbol{q} \cdot \boldsymbol{r}). \tag{3.38}$$

The corresponding spectral function satisfies the relation

$$I_{PP}(\boldsymbol{q}, \omega) = \frac{1}{\pi} \mathrm{Im}\, \alpha(\boldsymbol{q}, \omega). \tag{3.39}$$

The polarizability is related to the dielectric function $\varepsilon(\boldsymbol{q}, \omega)$ and the conductivity $\sigma(\boldsymbol{q}, \omega)$ by

$$\varepsilon(\boldsymbol{q}, \omega) = 1 + 4\pi\alpha(\boldsymbol{q}, \omega) = 1 + \frac{4\pi i}{\omega}\sigma(\boldsymbol{q}, \omega). \tag{3.40}$$

Hence from Eqs. (3.39) and (3.40) we obtain

$$I_{PP}(\boldsymbol{q}, \omega) = \frac{1}{\pi\omega} \mathrm{Re}\, \sigma(\boldsymbol{q}, \omega), \tag{3.41}$$

where the RHS represents the Joule heat for unit magnitude of electric field. Thus we understand that the spectral function $I_{PP}(\boldsymbol{q}, \omega)$ represents the dissipation of energy.

The spectral function for $X, Y$ in general is related to the correlation function of $X$ and $Y$. In quantum statistical physics, a convenient quantity is the symmetrized correlation function defined by

$$\langle \{X(t), Y\} \rangle \equiv \langle X(t)Y + YX(t) \rangle. \tag{3.42}$$

In the classical limit, or if $X, Y$ commute with each other, the symmetrization is simply equivalent to multiplication by two. By taking the matrix element $X_{nm}Y_{mn}$ in (3.42) as in Eq. (3.32), we can derive the relation

$$\langle \{X(t), Y\} \rangle = \int_{-\infty}^{\infty} \mathrm{d}\omega \coth\left(\frac{\beta\omega}{2}\right) I_{XY}(\omega) \exp(-\mathrm{i}\omega t). \tag{3.43}$$

In the special case of $Y = X^\dagger$ with $t = 0$, the LHS of Eq. (3.43) represents the fluctuation of $X$. On the other hand, the spectral function $I_{XX^\dagger}(\omega)$ on the RHS represents the dissipation of energy as we have seen for $X = P$. Namely, Eq. (3.43) with $Y = X^\dagger$ relates the fluctuation of the quantity $X$ to the dissipation of energy. Hence Eq. (3.43) is often called the fluctuation–dissipation theorem.

## 3.4 Green Functions with Imaginary Time

In the linear response theory, both the statistical operator $\exp(-\beta H)$ and the time-boost operator $\exp(-\mathrm{i}tH)$ play important roles. The two operators take similar exponential forms, and identification of $\beta$ as an imaginary time leads to the Matsubara picture defined by Eq. (3.7). The use of imaginary time provides an extremely powerful framework in the Green function, which is often called the thermal (or Matsubara) Green function. This Section explains the basics of the thermal Green function [2] which are related to response functions. The fermion Green function will be explained later in Sect. 3.7. In the Matsubara picture, a physical observable $X$, which behave as a bosonic operator, is represented by

$$e^{\tau H} X e^{-\tau H} \equiv X^{\mathrm{M}}(\tau) \tag{3.44}$$

for imaginary time $\tau$. Assuming $-\beta < \tau < \beta$, we introduce the thermal Green function $D_{XY}[\tau]$ by

$$D_{XY}[\tau] = -\langle T_\tau X^{\mathrm{M}}(\tau) Y\rangle = \begin{cases} -\langle X^{\mathrm{M}}(\tau) Y\rangle, & (\tau > 0) \\ -\langle Y X^{\mathrm{M}}(\tau)\rangle, & (\tau < 0), \end{cases} \tag{3.45}$$

where $T_\tau$ arranges the operators $X, Y$ in descending order of the imaginary time, and is called the time-ordering operator. The angular bracket used in $D_{XY}[\tau]$ distinguishes the quantity from its Fourier decomposition to be defined by Eq. (3.47). As we shall discuss later, if $X$ and $Y$ are fermionic operators, the sign is reversed for $\tau < 0$, leading to $+\langle Y X^{\mathrm{M}}(\tau)\rangle$.

By direct evaluation of Eq. (3.45), we recognize that $D_{XY}[\tau]$ is a periodic function of $\tau$ with periodicity $\beta$. Namely, using the relation $\rho_n = \rho_m \exp(\beta\omega_{mn})$ with eigenstates $|n\rangle$ and $|m\rangle$, we obtain for $-\beta < \tau < 0$:

$$\begin{aligned} D_{XY}[\tau] &= -\sum_{nm} \rho_n Y_{nm} X_{mn} \exp(\tau\omega_{mn}) \\ &= -\sum_{nm} \rho_m Y_{nm} X_{mn} \exp[(\tau+\beta)\omega_{mn}] = D_{XY}[\tau+\beta]. \end{aligned} \tag{3.46}$$

Hence, the Fourier decomposition is carried out in the range $-\beta < \tau < \beta$ as

$$D_{XY}[\tau] = T \sum_n D_{XY}(\mathrm{i}\nu_n) \exp(-\mathrm{i}\nu_n \tau), \tag{3.47}$$

where the imaginary frequency $\mathrm{i}\nu_n = 2\pi \mathrm{i} nT$ is called the (even) Matsubara frequency with $n$ an integer. The even Matsubara frequency is associated with bosonic operators.

The expression for $0 < \tau < \beta$ is given with use of the spectral function as

$$D_{XY}[\tau] = -\int_{-\infty}^{\infty} d\omega I_{XY}(\omega)\frac{e^{-\tau\omega}}{1-e^{-\beta\omega}}, \tag{3.48}$$

which leads to the Fourier component $D_{XY}(i\nu_n)$ as

$$D_{XY}(i\nu_n) = \int_0^{\beta} d\tau D_{XY}[\tau]\exp(i\nu_n\tau) = \int_{-\infty}^{\infty} d\omega\frac{I_{XY}(\omega)}{i\nu_n - \omega}. \tag{3.49}$$

The rightmost form demonstrates a surprising relation

$$D_{XY}(i\nu_n) = \langle [X, Y]\rangle(i\nu_n), \tag{3.50}$$

by comparison with Eq. (3.34). Namely, the thermal Green function is related to retarded and advanced Green functions by analytic continuation [3]. This relation is practically most useful since the thermal Green function can be analyzed in terms of diagrammatic perturbation theory, as we discuss in later chapters. Thus a physical response function can also be analyzed diagrammatically, which greatly facilitate a systematic approximation.

In the special case of $X = Y = M$ in Eq. (3.50), the thermal Green function corresponds to the dynamical susceptibility with imaginary frequency. In the case of free electrons, $D_{XX^\dagger}(i\nu_n)$ with $X = c_1^\dagger c_2$ can be derived exactly as

$$D_{XX^\dagger}(i\nu_n) = \frac{f_2 - f_1}{i\nu_n - \epsilon_2 + \epsilon_1}. \tag{3.51}$$

This result will be repeatedly used in the rest of this chapter. The derivation is the subject of Problem 3.4.

In concluding this section, we give a special case where the analytic continuation in Eq. (3.50) does not work. Consider a system in which $M$ is conserved and Eq. (3.10) gives the Curic law. If the analytic continuation is valid, the Matsubara Green function $D_{MM}(i\nu_n = 0)$ and the retarded Green function $\langle [M, M]\rangle(z = 0)$ should be the same. However, since $[M(t), M] = 0$, Eq. (3.21) leads to $\chi(\omega) = 0$ for any $\omega$, including $\omega = 0$. The vanishing dynamic susceptibility originates from the lack of ergodicity in the motion of isolated magnetic moment. Without ergodicity, the ensemble average gives the Curie law, while the time average gives $\chi(0) = 0$. Therefore, due care is necessary in dealing with conserved quantities. Problem 3.5 discusses a procedure to recover the ergodicity, and derive the static limit properly.

## 3.5 Relaxation Function

Among various modes of response to external perturbations, we consider the situation where a constant external field was present for $t < 0$, but it is suddenly switched off at $t = 0$. For example, if a small external magnetic field is given by $B(t) = B\theta(-t)$, the magnetization $M$ at $t < 0$ is given by $\chi B$. For $t > 0$, the magnetization remains nonzero for a while. The behavior of magnetization for $t > 0$ is referred to as relaxation.

Generalizing this example to arbitrary observable, we consider a quantity $\langle X(t)\rangle_{\text{ext}}$ for $t > 0$, which was coupled to an external field via $Y$ up to $t = 0$. As we shall prove in the end of this section, the behavior is determined by

$$\langle X(t); Y\rangle \equiv \frac{1}{\beta}\int_0^\beta \mathrm{d}\lambda \langle X(t - \mathrm{i}\lambda) Y\rangle, \tag{3.52}$$

which is called the canonical correlation function. This name implies a quantum version of the correlation function associated with the average over $\lambda$. The quantity $\beta\langle X(t); Y\rangle$ is called the relaxation function. By definition, the initial value $\langle X(t = 0); Y\rangle$ corresponds to the static susceptibility since the statistical operator for $\langle X(t \le 0)\rangle_{\text{ext}}$ is given by $\exp[-\beta(H + H_{\text{ext}})]/Z_{\text{tot}}$.

The Laplace transform of the canonical correlation function is given by

$$\langle X; Y\rangle(z) \equiv \int_0^\infty \mathrm{d}t \langle X(t); Y\rangle e^{\mathrm{i}zt}, \tag{3.53}$$

which is related to the response function via the spectral function. Namely, by taking matrix elements such as $X_{nm}$ in terms of eigenbases $n, m$ of $H$, we obtain

$$\langle X(t); Y\rangle = \int_{-\infty}^\infty \mathrm{d}\omega \frac{I_{XY}(\omega)}{\beta\omega} e^{-\mathrm{i}\omega t}. \tag{3.54}$$

Hence the Laplace transform of $\exp(-\mathrm{i}\omega t)$ and the use of Eq. (3.34) leads to the relation

$$\langle X; Y\rangle(z) = (\mathrm{i}z\beta)^{-1}\{\langle [X, Y]\rangle(z) - \langle [X, Y]\rangle(0)\}. \tag{3.55}$$

On the other hand, the time derivative $\dot{Y}$ of the operator $Y$ has the spectral function $I_{X\dot{Y}}(\omega)$ satisfying $I_{X\dot{Y}}(\omega) = \mathrm{i}\omega I_{XY}(\omega)$, as can be checked with use of Eq. (3.32). Hence the following relations hold:

$$\beta\langle X; \dot{Y}\rangle(z) = -\langle [X, Y]\rangle(z), \tag{3.56}$$

$$\beta\langle X(t); \dot{Y}\rangle = \mathrm{i}\langle [X(t), Y]\rangle. \tag{3.57}$$

We now show that Eq. (3.56) gives the electrical conductivity in terms of a relaxation function. The perturbation by electric field is given by Eq. (3.36), and the electric current $J$ and the polarization $P$ are related by $\dot{P} = J$. Thus the Kubo formula together with Eq. (3.56) gives

$$\sigma(z) = -\langle [J, P] \rangle(z) = \beta \langle J; J \rangle(z). \tag{3.58}$$

where the rightmost side is precisely the relaxation function.

Finally we return to the case of magnetic relaxation with $X = Y = M$, and explain how Eq. (3.52) describes the relaxation. We obtain

$$\beta \langle M(t); M \rangle = \beta \int_{-\infty}^{0} \mathrm{d}t' \langle M(t); \dot{M}(t') \rangle = \mathrm{i} \int_{-\infty}^{0} \mathrm{d}t' \langle [M(t), M(t')] \rangle, \tag{3.59}$$

where we have used Eq. (3.57) in the second equality. The rightmost quantity gives $\langle M(t) \rangle / B$ for $t > 0$ according to Eq. (3.19). Hence the relaxation function indeed describes the decay of the induced magnetization after the constant external field is switched off.

Figure 3.1 illustrates an example of $\langle M(t) \rangle$ where the dynamical susceptibility takes the form

$$\chi(\omega) = \frac{\chi \Gamma}{-\mathrm{i}\omega + \Gamma}. \tag{3.60}$$

In this case, we obtain the relaxation function from Eq. (3.54) as

$$\beta \langle M(t); M \rangle = \chi \exp(-\Gamma t), \tag{3.61}$$

for $t > 0$.

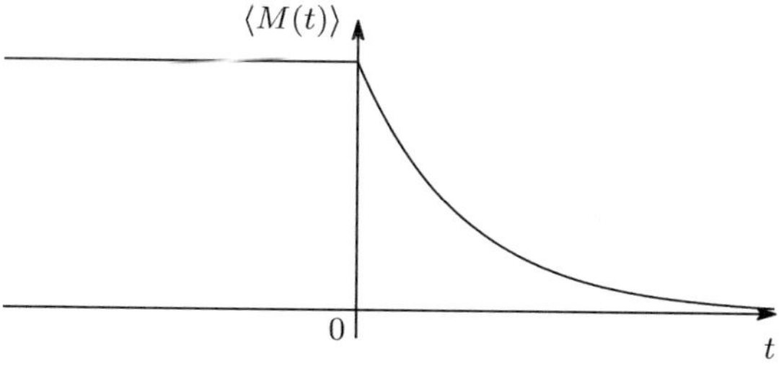


**Fig. 3.1** An example of magnetic relaxation where the external field is given by $B(t) = B\theta(-t)$. The magnetization $\langle M(t) \rangle$ for $t > 0$ decays in accordance with the relaxation function $\beta \langle M(t); M \rangle$

## 3.6 Liouville Operator Formalism

In more advanced treatment of the nonequilibrium property, it is convenient to regard the relaxation function as a kind of inner product between vectors $|X\rangle$ and $|Y\rangle$ associated with the Hermitian operators (observables) $X$ and $Y$. For example, the static susceptibility $\chi_{XY}$ is regarded as the inner product $\langle X|Y\rangle$ by the relation

$$\langle X|Y\rangle \equiv \int_0^\beta \mathrm{d}\tau \langle \exp(\tau H) X \exp(-\tau H) Y\rangle = \beta\langle X; Y\rangle, \tag{3.62}$$

where the canonical correlation function $\langle X; Y\rangle$ has been defined by Eq. (3.52). It can be checked that the inner product satisfies $\langle X|Y\rangle = \langle Y|X\rangle$ in addition to the standard linear relation such as

$$\langle X|a_1 Y_1 + a_2 Y_2\rangle = a_1\langle X|Y_1\rangle + a_2\langle X|Y_2\rangle.$$

We now introduce the Liouville operator $\mathcal{L}$ by the relation $\mathcal{L}X \equiv [H, X]$ with the Hamiltonian $H$. Simple calculation shows

$$\langle X|\mathcal{L}|Y\rangle \equiv \langle X|\mathcal{L}Y\rangle = \langle \mathcal{L}X|Y\rangle, \tag{3.63}$$

which means that $\mathcal{L}$ is a Hermitian operator for the present inner product. The Liouville operator together with the inner product provides the compact expression for the relaxation function. Namely, using $X(t) = \exp(\mathrm{i}\mathcal{L}t)X$ in Eq. (3.53) and the Hermiticity of $\mathcal{L}$, we obtain

$$\langle X; Y\rangle(z) = \langle X|\frac{\mathrm{i}}{z - \mathcal{L}}|Y\rangle \equiv C_{XY}(z). \tag{3.64}$$

Furthermore, the dynamical susceptibility defined by Eq. (3.22) is represented by the use of Eq. (3.55) as

$$\chi_{XY}(z) = \langle X|\frac{\mathcal{L}}{\mathcal{L} - z}|Y\rangle. \tag{3.65}$$

These compact expressions lead to a framework which enables one to focus on slow observables such as magnetization or electric current that is relevant to relaxation. We take the dynamical magnetic susceptibility $\chi_M(z)$ as a representative example. Corresponding to the observable $M$, a projection operator $\mathcal{P}$ is introduced by

$$\mathcal{P} \equiv |M\rangle\chi_M^{-1}\langle M|, \tag{3.66}$$

where $\chi_M = \langle M|M\rangle$ is the static susceptibility. The property $\mathcal{P}^2 = \mathcal{P} = \mathcal{P}^\dagger$ is easily checked, which is a necessary condition for the projection operator. Another projection operator $\mathcal{Q}$ is defined by $\mathcal{Q} = 1 - \mathcal{P}$ as the complement of $\mathcal{P}$. These projection operators are useful to deal with $(z - \mathcal{L})^{-1}$, which appears in the relaxation function $C_M(z)$.

We now recall the following relation for a $2 \times 2$ matrix:

$$\left[\begin{pmatrix} a & b^* \\ b & c \end{pmatrix}^{-1}\right]_{11} = \frac{1}{a - b^* c^{-1} b}. \tag{3.67}$$

The relation is generalized to matrices $a, b, c$ with dimensions $n \times n$, $m \times n$, and $m \times m$, respectively. Then the RHS of Eq. (3.67) should be regarded as the inverse matrix of the denominator with the dimension $n \times n$, and $b^*$ should be modified to $b^\dagger$.

Using the projection operators, we make the following substitution:

$$a = \mathcal{P}(z - \mathcal{L})\mathcal{P}, \quad b = -\mathcal{Q}\mathcal{L}\mathcal{P}, \quad c = \mathcal{Q}(z - \mathcal{L})\mathcal{Q}. \tag{3.68}$$

Then corresponding to Eq. (3.67), we obtain the expression [4]:

$$C_M(z) = \mathrm{i}\chi_M[z - \Omega_M + \mathrm{i}\Gamma(z)]^{-1}, \tag{3.69}$$

or, equivalently,

$$\chi_M(z) = \chi_M \Gamma(z)/[-\mathrm{i}(z - \Omega_M) + \Gamma(z)], \tag{3.70}$$

where

$$\Omega_M = \chi_M^{-1}\langle M|\mathcal{L}|M\rangle, \tag{3.71}$$

describes the frequency of averaged motion. We obtain using Eq. (3.54)

$$\mathrm{i}\langle M|\mathcal{L}|M\rangle = \int_{-\infty}^{\infty} \mathrm{d}\omega I_{MM}(\omega) = 0, \tag{3.72}$$

where the spectral function $I_{MM}(\omega) = I_{M\dot{M}}(\omega)/(\mathrm{i}\omega)$ is an odd function. The result $\Omega = 0$ holds in most cases of our interest. In the presence of finite magnetic field, however, the precession of magnetic moment is described by $\Omega$ which is extended to matrix. Problem 3.6 deals with such extension where multiple variables $X_i$ $(i = 1, 2, \ldots)$ become relevant.

On the other hand, $\Gamma(z)$ is given by

$$\Gamma(z) = \chi_M^{-1}\langle \mathcal{Q}\dot{M}|\frac{\mathrm{i}}{z - \mathcal{Q}\mathcal{L}\mathcal{Q}}|\mathcal{Q}\dot{M}\rangle, \tag{3.73}$$

with $\dot{M} = \mathrm{i}\mathcal{L}M$. As will be explained soon, $\Gamma(z)$ describes the dissipation associated with dynamics of $M$ in terms of the canonical correlation function of the random force $\mathcal{Q}\dot{M}$. Hence Eq. (3.73) is regarded as another form of the fluctuation-dissipation theorem [4, 5], which is in fact closer to the original form for the Brownian motion of a classical particle in terms of random force. In the classical case, the correlation function of the random force gives decay of the particle velocity by dissipation. The velocity is replaced by $M$ in the present case.

So far the discussion is exact, but tautological in the sense of playing with identical expressions. Equation (3.69) provides physical intuition on relaxation if $\mathcal{Q}\dot{M}$ is characterized by dynamics faster than that of $M$. More generally the choice of $\mathcal{P}$ should be made for slow variables. The framework leading to Eqs. (3.69)–(3.73) is often called the Mori (or Mori–Zwanzig) formalism [6], which is also practical for approximate treatment. As an example, we derive the magnetic relaxation rate of a magnetic impurity with spin $\boldsymbol{S}$ in metallic matrix. The perturbation Hamiltonian is given by

$$H_1 = J\boldsymbol{S}\cdot\boldsymbol{s} \equiv \frac{J}{2N}\sum_i\sum_{\boldsymbol{k},\boldsymbol{k}'}\sum_{\alpha\beta} S_i\,(\sigma_i)_{\alpha\beta}\,c^{\dagger}_{\boldsymbol{k}\alpha}c_{\boldsymbol{k}'\beta}, \tag{3.74}$$

where $\boldsymbol{s}$ is the spin of conduction electrons at the impurity site $\boldsymbol{R} = 0$, and $N$ the total number of lattice sites. The unperturbed Hamiltonian $H_0$, which is the kinetic energy of conduction electrons, conserves $\boldsymbol{S}$ and hence $M \equiv g\mu_{\mathrm{B}}S_z$. Then it follows that $\dot{M} = \mathrm{i}\mathcal{L}M = \mathrm{i}[H_1, M]$. Since this is already first order in $H_1$, $\Gamma(z)$ in the lowest order $O(H_1^2)$ is derived by replacing $H$ by $H_0$ in time development. Furthermore, we may put $\mathcal{Q} = 1$ since $\langle M|\dot{M}\rangle = 0$. Thus we obtain from Eq. (3.73),

$$\begin{aligned}\Gamma(z) &= \chi_M^{-1}\langle\dot{M}|i(z-\mathcal{L}_0)^{-1}|\dot{M}\rangle_0\\ &= \frac{1}{iz\chi_M}\left\{\langle[\dot{M},\dot{M}]\rangle_0(z) - \langle[\dot{M},\dot{M}]\rangle_0(0)\right\},\end{aligned} \tag{3.75}$$

with use of Eq. (3.55). The suffix 0 indicates the average with respect to $H_0$.

The retarded Green function is most easily derived from the Matsubara Green function which is computed as

$$\langle[\dot{M},\dot{M}]\rangle_0(\mathrm{i}\nu_n) = \frac{1}{2}J^2C_M\frac{1}{N^2}\sum_{\boldsymbol{k},\boldsymbol{k}'}\frac{f(\epsilon_{\boldsymbol{k}}) - f(\epsilon_{\boldsymbol{k}'})}{\mathrm{i}\nu_n - \epsilon_{\boldsymbol{k}} + \epsilon_{\boldsymbol{k}'}}, \tag{3.76}$$

where $C_M = \mu_{\mathrm{B}}^2 S(S+1)/3$ is the Curie constant. By analytic continuation to $z = \omega + \mathrm{i}0_+$, and taking the limit $\omega \to 0$, we obtain

$$\Gamma(0) = \frac{1}{2}(J\rho_c)^2T, \tag{3.77}$$

with $\rho_c$ being the density of states. As Eq. (3.76) indicates, the $\omega$ dependence in $\Gamma(\omega)$ becomes relevant only for $\omega$ of the order of the bandwidth. Thus the $T$-linear relaxation is observed widely as characteristic of the itinerant fermion reservoir. This is often called the Korringa relaxation, the name of which comes from relaxation phenomena of the nuclear spin responsible for $M$ [7]. In this case $J$ corresponds to the hyperfine interaction. The Korringa relaxation is also observed with a magnetic impurity where localized electrons form $M$. Details of derivation for Eqs. (3.76) and (3.77) are the subject of Problem 3.7.

In a similar manner, we can derive the optical conductivity $\sigma(\omega)$ from the current relaxation function. For simplicity, we assume the spectrum $\epsilon_{\boldsymbol{k}} = k^2/(2m)$ for electrons and neglect the spin. Electrons are scattered by impurities which are located randomly. The interaction with the impurity at the origin is given by

$$H_{10} = \int \mathrm{d}\boldsymbol{r} u(\boldsymbol{r}) \Psi^\dagger(\boldsymbol{r}) \Psi(\boldsymbol{r}), \tag{3.78}$$

where $u(\boldsymbol{r}) = u(r)$ is the spherically symmetric impurity potential and

$$\Psi(\boldsymbol{r}) = \frac{1}{\sqrt{V}} \sum_{\boldsymbol{k}} c_{\boldsymbol{k}} \exp(\mathrm{i}\boldsymbol{k}\cdot\boldsymbol{r}), \tag{3.79}$$

with $V$ being the volume of the system. We take the volume of the unit cell as unity so that $V = N$. The perturbation Hamiltonian $H_1$ is the sum of $H_{10}$ and similar ones from other impurities with the density $n_i$. The Fourier transform $J_{\boldsymbol{q}}$ of the electric current along the $z$-direction is given by

$$J_{\boldsymbol{q}} = -\frac{e}{m\sqrt{V}} \sum_{\boldsymbol{k}} k_z c^\dagger_{\boldsymbol{k}-\boldsymbol{q}/2} c_{\boldsymbol{k}+\boldsymbol{q}/2}, \tag{3.80}$$

which in the $\boldsymbol{q} \to 0$ limit commutes with the unperturbed Hamiltonian $H_0$ describing the kinetic energy. We use Eqs. (3.58) and (3.69) to obtain

$$\sigma(\boldsymbol{q}, z) = \langle J_{\boldsymbol{q}} | J_{-\boldsymbol{q}} \rangle / [-\mathrm{i}z + \Gamma(\boldsymbol{q}, z)]. \tag{3.81}$$

In the lowest order in the impurity potential, $\Gamma(\boldsymbol{q}, z)$ is given by the form of Eq. (3.75) in which $M$ is replaced by $J_{\pm\boldsymbol{q}}$. After some algebra, which is the subject of Problem 3.7, we obtain for $\boldsymbol{q} \to 0$

$$\langle J_{\boldsymbol{q}} | J_{-\boldsymbol{q}} \rangle_0 = \left(\frac{e}{m}\right)^2 \frac{1}{V} \sum_{\boldsymbol{k},\boldsymbol{k}'} \delta(\boldsymbol{k} - \boldsymbol{k}' + \boldsymbol{q}, 0) k_z k'_z \frac{f(\epsilon_{\boldsymbol{k}}) - f(\epsilon_{\boldsymbol{k}'})}{\epsilon_{\boldsymbol{k}'} - \epsilon_{\boldsymbol{k}}} \to \frac{ne^2}{m}, \tag{3.82}$$

$$\langle [\dot{J}_{\boldsymbol{q}}, \dot{J}_{-\boldsymbol{q}}] \rangle_0 (\mathrm{i}\nu_n) = \left(\frac{e}{m}\right)^2 \frac{n_i}{V^2} \sum_{\boldsymbol{k},\boldsymbol{k}'} |u[\boldsymbol{k} - \boldsymbol{k}']|^2 (k_z - k'_z)^2 \frac{f(\epsilon_{\boldsymbol{k}}) - f(\epsilon_{\boldsymbol{k}'})}{\mathrm{i}\nu_n - \epsilon_{\boldsymbol{k}} + \epsilon_{\boldsymbol{k}'}}, \tag{3.83}$$

where $n$ and $n_i$ are the electron and impurity densities, respectively, $\delta(\boldsymbol{p}, 0)$ is the Kronecker delta, and $u[\boldsymbol{k} - \boldsymbol{k}']$ is the Fourier transform of $u(\boldsymbol{r})$. Proceeding in the same manner as in the case of the Korringa relaxation, $\Gamma(\boldsymbol{q} = 0, \omega = +\mathrm{i}0_+)$ is derived as

$$\Gamma(\boldsymbol{0}, \mathrm{i}0_+) = 2\int_0^{\pi} \mathrm{d}\theta (1 - \cos\theta) \left|u[k_\mathrm{F}(\boldsymbol{e}_0 - \boldsymbol{e}_\theta)]\right|^2 \rho_c(\epsilon_\mathrm{F}) \equiv \frac{1}{\tau_\mathrm{tr}}, \tag{3.84}$$

where $\boldsymbol{e}_\theta$ is the unit vector along the polar angle $\theta$ from the $z$-axis of $\boldsymbol{k}$. The $\omega$-dependence of $\Gamma$ is negligible as long as $\omega \ll \epsilon_\mathrm{F}$. The temperature dependence is also negligible as long as $T \ll \epsilon_\mathrm{F}$. More details of the derivation is the subject of Problem 3.7. The result for the transport relaxation time $1/\tau_\mathrm{tr}$ is consistent with the one obtained by the kinetic (Boltzmann) equation [6]. In contrast to the latter approach, the derivation here is automatic, relying only on the lowest-order perturbation theory for the relaxation rate. In summary, the optical conductivity takes the Drude form:

$$\sigma(\omega) = \frac{ne^2}{m} \Big/ \left(-\mathrm{i}\omega + \frac{1}{\tau_\mathrm{tr}}\right), \tag{3.85}$$

which in the static limit reduces to the dc resistivity

$$\rho = m/(ne^2\tau_\mathrm{tr}), \tag{3.86}$$

with $1/\tau_\mathrm{tr}$ given microscopically by Eq. (3.84).

We have learned that the Lorentzian form of the response function is valid if the focused frequency $\omega$ is much smaller than the characteristic frequency in $\Gamma(\omega)$ so that its $\omega$-dependence can be neglected. In order to study the spectral shape more generally, we turn to another approach that deals with time $t$ instead of frequency. As a preliminary, we represent the relaxation function as

$$\langle M(t)|M\rangle = \langle M|\exp(-\mathrm{i}\mathcal{L}t)|M\rangle \equiv \langle\langle \exp(-\mathrm{i}\mathcal{L}t)\rangle\rangle \langle M|M\rangle, \tag{3.87}$$

where the average $\langle\langle \cdots \rangle\rangle$ has been introduced. Then expanding $\exp(-\mathrm{i}\mathcal{L}t)$ and comparing the coefficient of $t^n$ with that obtained from the expression Eq. (3.54), we obtain the spectral moments $\langle \omega^n \rangle$ $(n = 0, 1, 2 \ldots)$ as

$$\langle \omega^n \rangle \equiv \langle\langle \mathcal{L}^n \rangle\rangle = \int_{-\infty}^{\infty} \frac{\mathrm{d}\omega}{\pi} \omega^{n-1} \mathrm{Im} \frac{\chi_M(\omega)}{\chi_M}, \tag{3.88}$$

in terms of the dynamical susceptibility.

In addition to the moments, the related quantity called the cumulant is commonly used in characterizing the statistical distribution. In the case of relaxation phenomena, we manipulate as follows:

$$\langle\langle \exp(-\mathrm{i}\mathcal{L}t)\rangle\rangle = \exp[\langle\langle \exp(-\mathrm{i}\mathcal{L}t)\rangle\rangle_c - 1] \equiv \exp[-\mathrm{i}\langle\omega\rangle t + X(t)], \qquad (3.89)$$

where $X(t)$ generates the cumulants $\langle\omega^n\rangle_c \equiv \langle\langle\mathcal{L}^n\rangle\rangle_c$ by

$$X(t) = \sum_{n=2}^{\infty} \frac{(-\mathrm{i}t)^n}{n!}\langle\omega^n\rangle_c. \qquad (3.90)$$

Of particular importance is the second cumulant

$$\langle\omega^2\rangle_c = \langle\langle\mathcal{L}^2\rangle\rangle - \langle\langle\mathcal{L}\rangle\rangle^2. \qquad (3.91)$$

The moments and cumulants are derived by expansion in terms of $t$, which assumes smooth continuation from the initial stage of the relaxation. In the Gaussian distribution, the cumulants higher than the third all vanish, the proof of which is the subject of Problem 3.8. On the other hand, in the case of the Lorentzian spectrum, all moments with $n \geq 2$ are divergent. As we shall discuss below, the Lorentzian spectrum is relevant only to the the long-time behavior, which corresponds to the small frequency region of the relaxation spectrum.

In order to combine the complimentary aspects of long and short times, we appeal to less rigorous but more intuitive argument. The function $X(t)$ in Eqs. (3.90) starts from 0 at $t = 0$, and should decrease as the time passes, as long as the relaxation takes place in Eq. (3.89). We define the characteristic time $\tau_R$ in $X(t)$ so that $\mathrm{Re}\, X(\tau_R) = -1$. Then the relaxation function should be negligibly small for $t \gg \tau_R$. In such large time $t$, however, the expansion form Eq. (3.90) may not be valid, since the dissipation effect breaks the analyticity in time. In fact, the Lorentzian spectrum is equivalent to the following behavior for $t \gg \tau_R$:

$$X(t) \to -\mathrm{i}(\delta\omega - \mathrm{i}\Gamma)t, \qquad (3.92)$$

where $\delta\omega$ is the shift in resonance frequency. Note that the $t$-linear term with $\Gamma$ is absent in Eq. (3.90). The term with $\Gamma$ breaks the time reversal, and emerges only with the boundary condition imposing the causality. The simplest form with account of the causality is given by

$$X(t) = -\int_0^t dt_1 \int_0^{t_1} dt_2 g(t_1 - t_2) = -\int_0^t d\tau (t-\tau) g(\tau), \qquad (3.93)$$

where $g(\tau)$ is the cumulant correlation function of $Q\dot{M}$. In order to understand the qualitative feature, we take a phenomenological model

$$g(\tau) = g(0)\exp(-\tau/\tau_C), \qquad (3.94)$$

with another characteristic time $\tau_C$. For simplicity we assume that $g(0)$ is real, which implies $\langle\omega\rangle = \delta\omega = 0$. In this case we obtain

$$X(t) = -g(0)\tau_C^2[1 - t/\tau_C + \exp(-t/\tau_C)] \to \begin{cases} -\frac{1}{2}g(0)t^2 & (t \ll \tau_C), \\ -g(0)\tau_C t & (t \gg \tau_C). \end{cases} \tag{3.95}$$

As extreme cases, we consider (i) $\tau_C \gg \tau_R$ and (ii) $\tau_C \ll \tau_R$, with $g(0) = D^2 \sim \tau_R^{-2}$. Namely, $\tau_C$ separates the short-time behavior described by Eq. (3.90), and the long-time behavior described by Eq. (3.92). In the case (i), the short-time behavior is dominant in $X(t)$. The spectrum becomes the Gaussian with the explicit form

$$\mathrm{Im}\frac{\chi_M(\omega)}{\omega} = \sqrt{\frac{\pi}{2}}\frac{\chi_M}{D}\exp\left[-\frac{1}{2}\left(\frac{\omega}{D}\right)^2\right], \tag{3.96}$$

where

$$D^2 = \chi_M^{-1}\langle\dot{M}|\dot{M}\rangle. \tag{3.97}$$

It is possible to derive $D$ by perturbation theory.

In the case (ii), Eq. (3.95) shows that the relaxation is dominated by the Lorentzian form as described by Eq. (3.92). The relaxation rate is given by Eq. (3.75) with $z = 0$, or by $\Gamma = D^2\tau_C \sim \tau_C/\tau_R^2$ according to Eq. (3.95). Hence with given $D$, the relaxation rate in the case (ii) is smaller than that in (i) by the factor $\tau_C/\tau_R \ll 1$. The reduction associated with the Lorentzian spectrum is often referred to as the motional narrowing [8].

Figure 3.2 illustrates the typical spectra for both cases (i) and (ii). The ratio $\tau_C/\tau_R$ in the figure should not be taken seriously, which is shown just for illustration.

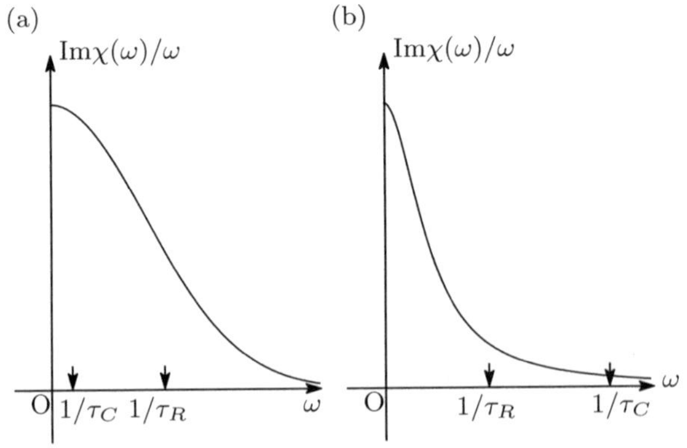


**Fig. 3.2** Spectral shapes of dynamical susceptibility for (**a**) the Gaussian case with $\tau_C \gg \tau_R$ and (**b**) the Lorentzian case with $\tau_R \gg \tau_C$. The time $\tau_C$ characterizes the switching scale in the cumulant function $X(t)$ as given by Eqs. (3.89) and (3.92), while $\tau_R$ controls the initial behavior of $X(t)$ as $X(t) \sim -\frac{1}{2}(t/\tau_R)^2$. See text for details

## 3.7 Green Function for Fermions

The single-particle Green function gives a building block for response functions, which corresponds to the two-particle Green function. Hence it is appropriate to discuss the Green function for fermions here. We use indices $i, j$ to represent symbolically spatial coordinates, momentum, and other degrees of freedom of fermions. With the fermionic creation $\psi_j^\dagger$ and annihilation $\psi_i$ operators, the thermal Green function $G_{ij}[\tau]$ is defined by

$$G_{ij}[\tau] = -\langle T_\tau \psi_i^{\mathrm{M}}(\tau)\psi_j^\dagger\rangle = \begin{cases} -\langle \psi_i^{\mathrm{M}}(\tau)\psi_j^\dagger\rangle & (\tau > 0) \\ \langle \psi_j^\dagger \psi_i^{\mathrm{M}}(\tau)\rangle & (\tau < 0). \end{cases} \tag{3.98}$$

with $-\beta < \tau < \beta$. It is straightforward to show, following the bosonic case given by Eq. (3.46), that the Green function is anti-periodic in $\tau$ with the period $\beta$. Namely, for $-\beta < \tau < 0$ we obtain

$$G_{ij}[\tau] = -G_{ij}[\tau + \beta]. \tag{3.99}$$

Hence the Green function can be Fourier decomposed into $G_{ij}(\mathrm{i}\epsilon_n)$ with odd Matsubara frequencies $\mathrm{i}\epsilon_n = \mathrm{i}\pi(2n+1)T$.

As in the case of response functions, a useful quantity is the spectral function $\rho_{ij}(\epsilon)$ defined by

$$\begin{aligned} \rho_{ij}(\epsilon) &= \underset{n}{\mathrm{Av}} \sum_m (\psi_i)_{nm}(\psi_j^\dagger)_{mn} \left[1 + \exp(-\beta\epsilon)\right] \delta(\epsilon - \epsilon_{mn}) \\ &= \int_{-\infty}^{\infty} \frac{\mathrm{d}t}{2\pi} e^{\mathrm{i}\epsilon t} \langle\{\psi_i(t), \psi_j^\dagger\}\rangle. \end{aligned} \tag{3.100}$$

Note that the anticommutator appears in Eq. (3.100) in contrast with the commutator in Eq. (3.32). Using the spectral function, we obtain the representation for $0 < \tau < \beta$:

$$G_{ij}[\tau] = -\int_{-\infty}^{\infty} \mathrm{d}\epsilon \rho_{ij}(\epsilon)[1 - f(\epsilon)] \exp(-\epsilon\tau) \tag{3.101}$$

which leads to

$$G_{ij}(\mathrm{i}\epsilon_n) = \int_0^\beta \mathrm{d}\tau G_{ij}[\tau] \exp(\mathrm{i}\epsilon_n\tau) = \int_{-\infty}^{\infty} \mathrm{d}\epsilon \frac{\rho_{ij}(\epsilon)}{\mathrm{i}\epsilon_n - \epsilon}. \tag{3.102}$$

The rightmost form is the spectral (or Lehmann) representation for the fermionic case. The definition (3.100) immediately leads to the sum rule

$$\int_{-\infty}^{\infty} \mathrm{d}\epsilon\, \rho_{ij}(\epsilon) = \delta_{ij}. \tag{3.103}$$

Hence Eq. (3.102) can be regarded as superposition of the Green functions $1/(\mathrm{i}\epsilon_n - \epsilon)$ for free particles with energy $\epsilon$. The effect of interactions appears in the weight $\rho_{ij}(\epsilon)$.

In performing the inverse transform of Eq. (3.102), we use the following formula for summation over Matsubara frequencies:

$$T\sum_n \frac{\exp(\mathrm{i}\epsilon_n\tau)}{\mathrm{i}\epsilon_n - \epsilon} = \begin{cases} f(\epsilon)\exp(\epsilon\tau), & (\tau < 0) \\ [f(\epsilon) - 1]\exp(\epsilon\tau), & (\tau > 0), \end{cases} \tag{3.104}$$

where $f(\epsilon) = 1/(e^{\beta\epsilon} + 1)$ is the Fermi distribution function. In the case of $\tau = 0$, the slow asymptotic decay $1/(\mathrm{i}\epsilon_n)$ of the summand makes the summation logarithmically divergent. However, with use of the pair-wise combination:

$$\frac{1}{\mathrm{i}\epsilon_n - \epsilon} + \frac{1}{-\mathrm{i}\epsilon_n - \epsilon} = -\frac{\epsilon}{\epsilon_n^2 + \epsilon^2}, \tag{3.105}$$

the asymptotic behavior $1/\epsilon_n^2$ leads to convergence. This conditional convergence of the summation is the origin of discontinuity at $\tau = 0$, which is consistent with the definition given by Eq. (3.98). The proof of Eq. (3.104) is the subject of Problem 3.9.

For free particles with the Hamiltonian, $H = \sum_k \epsilon_k \psi_k^\dagger \psi_k$, the time dependence can be derived explicitly as

$$\psi_k(t) = \exp(\mathrm{i}Ht)\psi_k \exp(-\mathrm{i}Ht) = \exp(-\mathrm{i}\epsilon_k t)\psi_k. \tag{3.106}$$

Therefore the spectral function defined by Eq. (3.100) is derived as

$$\rho_k(\epsilon) = \delta(\epsilon - \epsilon_k). \tag{3.107}$$

The thermal Green function follows from Eq. (3.102) as

$$G_k(\mathrm{i}\epsilon_n) = \frac{1}{\mathrm{i}\epsilon_n - \epsilon_k}. \tag{3.108}$$

For a general case with the Hamiltonian matrix $h_{ij}$, we obtain

$$G_{ij}(\mathrm{i}\epsilon_n) = \left(\frac{1}{\mathrm{i}\epsilon_n - h}\right)_{ij}, \tag{3.109}$$

which can be confirmed by diagonalization of $h$ with a proper unitary transformation.

By analytic continuation of Matsubara frequencies to complex frequency $z$ in the upper half plane, we obtain $G_{ij}(z)$, which corresponds to the Fourier transform

$$G_{ij}(z) = \int_{-\infty}^{\infty} dt e^{izt} G_{ij}^{R}(t), \tag{3.110}$$

of the retarded Green function

$$G_{ij}^{R}(t) = -i\theta(t)\langle\{\psi_i(t), \psi_j^\dagger\}\rangle. \tag{3.111}$$

Note that Eq. (3.111) involves the anticommutator in contrast with the commutator in Eq. (3.22) for the bosonic Green function. On the other hand, the analytic continuation to $z$ in the lower half plane gives the Fourier transform of the advanced Green function

$$G_{ij}^{A}(t) = i\theta(-t)\langle\{\psi_i(t), \psi_j^\dagger\}\rangle. \tag{3.112}$$

Namely, we obtain with $\text{Im}\, z < 0$:

$$G_{ij}(z) = \int_{-\infty}^{\infty} dt e^{izt} G_{ij}^{A}(t). \tag{3.113}$$

The retarded and advanced Green functions in the time domain are reproduced from Eq. (3.108) which is the subject of Problem 3.10. The difference of retarded and advanced Green functions amounts to

$$\langle\{\psi_i(t), \psi_j^\dagger\}\rangle = i\left[G_{ij}^{R}(t) - G_{ij}^{A}(t)\right], \tag{3.114}$$

which leads to the spectral function by Fourier transform, as given by Eq. (3.100). It is instructive to compare with the bosonic case in Eq. (3.33).

In many fermion systems, we may interpret the retarded Green function as describing the propagation of particles with positive energy measured from the chemical potential $\mu$. On the other hand, the advanced Green function describes the propagation of holes, which is equivalent to backward propagation of particles with negative energy. To recognize the hole propagation, we take the non-interacting system at $T = 0$, and choose the energy $\epsilon_k < 0$ measured from $\mu$. We write the advanced Green function as

$$G_k^{A}(t) = i\theta(-t) f(\epsilon_k) \exp(-i\epsilon_k t) = i\theta(t')[1 - f(\epsilon_k')] \exp(-i\epsilon_k' t'), \tag{3.115}$$

where the backward propagation ($t < 0$) is translated into the forward propagation ($t' = -t > 0$) of the hole with positive energy $\epsilon_k' = -\epsilon_k$.

We now consider another non-interacting system where the basis set has both localized (impurity) state and itinerant states. The latter is characterized by the crystal momentum $\boldsymbol{k}$. This system is convenient in introducing concept of the self-energy in an explicit manner. The Hamiltonian is written as

$$H = \epsilon_f f^\dagger f + \sum_{\boldsymbol{k}} \epsilon_{\boldsymbol{k}} c_{\boldsymbol{k}}^\dagger c_{\boldsymbol{k}} + \frac{V}{\sqrt{N}} \sum_{\boldsymbol{k}} (c_{\boldsymbol{k}}^\dagger f + f^\dagger c_{\boldsymbol{k}}), \tag{3.116}$$

where $f$ and $c_{\boldsymbol{k}}$ denote annihilation operators of localized and itinerant states, respectively, and $N$ denotes the number of $\boldsymbol{k}$ states. These states hybridize with the local state by the strength $V$. We use the matrix $h$ to represent the Hamiltonian in the present basis set, and obtain

$$z - h = \begin{pmatrix} z - \epsilon_f & -\boldsymbol{V}^\dagger \\ -\boldsymbol{V} & z - h_c \end{pmatrix}, \tag{3.117}$$

where all elements of the vector $\boldsymbol{V}$ and its conjugate $\boldsymbol{V}^\dagger$ consist of $V/\sqrt{N}$, and $h_c$ is an $N \times N$ diagonal matrix with elements $\epsilon_{\boldsymbol{k}}$. In such a case, the matrix identity Eq. (3.67) gives the Green function $G_f(z)$ of the localized state as

$$\begin{aligned} G_f(z) &= [(z-h)^{-1}]_{11} = \left(z - \epsilon_f - \boldsymbol{V}^\dagger (z - h_c)^{-1} \boldsymbol{V}\right)^{-1} \\ &\equiv \left(z - \epsilon_f - \Sigma_f(z)\right)^{-1}, \end{aligned} \tag{3.118}$$

where $\Sigma_f(z)$ is the simplest example of the self-energy, which modifies the bare spectrum characterized by $\epsilon_f$. Namely, with $z = \epsilon + \mathrm{i}0_+$, the real part Re $\Sigma_f(\epsilon)$ describes the shift $\Delta\epsilon$ from the bare energy $\epsilon_f$. Here $\mathrm{i}0_+$ can be neglected. The shift satisfies the self-consistent relation,

$$\Delta\epsilon = \mathrm{Re}\,\Sigma_f(\epsilon_f + \Delta\epsilon), \tag{3.119}$$

which resembles the Brillouin–Wigner perturbation theory as given by Eq. (1.6). On the other hand, $-\mathrm{Im}\,\Sigma_f(\epsilon + \mathrm{i}0_+) \geq 0$ describes the damping of the excitation. The meaning of real and imaginary parts of the self-energy is general and applicable also to homogeneous systems with mutual interactions. We shall discuss typical examples later, especially in Chap. 10.

It is instructive to derive $\Sigma_f(z)$ explicitly in the special case of constant density of states $\rho_c$ inside the energy band: $-D < \epsilon < D$ of itinerant states. We can immediately obtain

$$\Sigma_f(z) = V^2 \rho_c \int_{-D}^{D} \mathrm{d}\epsilon \frac{1}{z - \epsilon} = V^2 \rho_c \ln \frac{z + D}{z - D}. \tag{3.120}$$

Due care is necessary in evaluating the logarithm with complex argument. With $z = \epsilon \pm i0_+$, the logarithm is evaluated as

$$\ln \frac{\epsilon + D \pm i0_+}{\epsilon - D \pm i0_+} = \ln \left| \frac{\epsilon + D}{\epsilon - D} \right| \mp i\pi\theta(D^2 - \epsilon^2). \tag{3.121}$$

There is a discontinuity of $2\pi$ above and below the branch cut $\epsilon \in [-D, D]$. As long as $|z| \ll D$, the real part of the logarithm has much smaller absolute value of $O(|z|/D)$ than the imaginary part $\pi$. Hence we obtain for small $|z|$,

$$\Sigma_f(z) = -\mathrm{sgn}(\mathrm{Im}\, z)\mathrm{i}\Delta + O(z/D), \tag{3.122}$$

with $\Delta \equiv \pi V^2 \rho_c$. Note that we have the finite damping even though all the eigenvalues of the Hamiltonian matrix $h$ are real. The complex value of the self-energy is the result of projection onto the localized state, for which itinerant states behave like the reservoir. We have met analogous situation in the linear response theory where the total Hamiltonian, which is Hermitian, can bring about damping of certain degrees of freedom observed.

It is remarkable that the sign of the imaginary part is consistent with analytic properties of the Green functions. Namely, the retarded Green function is given, with neglect of $\mathrm{Re}\Sigma_f$, by

$$G_f(z) = (z - \epsilon_f + \mathrm{i}\Delta)^{-1}, \tag{3.123}$$

which has no singularity in the upper half plane of $z$. The spectral density is given by

$$\rho_f(\epsilon) = -\frac{1}{\pi} \,\mathrm{Im}\, G_f(\epsilon + \mathrm{i}0_+) = \frac{1}{\pi} \cdot \frac{\Delta}{(\epsilon - \epsilon_f)^2 + \Delta^2}. \tag{3.124}$$

The Lorentzian density of states is typical of the resonant state. In the time domain, we obtain from the inverse Fourier transform of Eq. (3.123),

$$G_f^{\mathrm{R}}(t) = -\mathrm{i}\theta(t) \exp\left[-\mathrm{i}(\epsilon_f - \mathrm{i}\Delta)t\right], \tag{3.125}$$

which has the damping $\Delta$, or the lifetime $1/\Delta$. Thus the imaginary part of the self-energy is analogous to the relaxation rate of magnetic moment, or that of current as described in Eq. (3.85).

## Problems

**3.1** Show that in the magnetization given by Eq. (3.9), correction terms begin from $O(B^3)$.

**3.2** Derive the Kubo formula Eq. (3.19) by using the density matrix.

**3.3** Derive the dynamical magnetic susceptibility for free electrons, and show that the static and long-wavelength limit reproduces the Pauli spin susceptibility.

**3.4** Derive the dynamical susceptibility with even Matsubara frequencies for free electrons, and confirm the property given by Eq. (3.50).

**3.5** Assume a finite magnetic field $B_z$ along the $z$-direction, and derive the transverse dynamical susceptibility $\chi_\perp(\omega)$ for an isolated spin. Show that the Curie law is obtained in the limit $B_z \to 0$.

**3.6** Generalize Eq. (3.69) to the case where multiple variables $X_i$ $(i = 1, 2, \ldots)$ become relevant for slow relaxation.

**3.7** * Derive the Korringa relaxation rate given by Eq. (3.77), and the current relaxation rate given by Eq. (3.84)

**3.8** Show that cumulants in the Gaussian distribution vanish in all orders higher than two.

**3.9** Prove Eq. (3.104) for summation over Matsubara frequencies.

**3.10** Derive retarded and advanced Green functions with real time using the thermal Green function for free fermions,

## Solutions to Problems

### Problem 3.1

The first line of the RHS in Eq. (3.9) is expanded in terms of $B$ in $H_{\text{ext}}$ for both numerator and denominator as follows:

$$\langle M \rangle_{\text{ext}} = (a_1 B + a_3 B^3 + \ldots)/(a_0 + a_2 B^2 + \ldots), \tag{3.126}$$

where the nonzero coefficients $a_i$ are constrained by the time reversal; only odd terms in the numerator and only even terms in the denominator. Hence the correction to $\chi B$ begins from $B^3$. Up to $B^3$ we may write

$$\langle M \rangle_{\text{ext}} = \chi B + \chi_3 B^3. \tag{3.127}$$

The linear response dominates for weak magnetic field $B \ll B_c$, where $B_c$ is estimated as

$$B_c^2 \sim \chi/\chi_3. \tag{3.128}$$

**Problem 3.2**
The density matrix $\rho_{\rm tot}(t)$ in external field obeys the equation of motion

$$\frac{\partial \rho_{\rm tot}(t)}{\partial t} = {\rm i}\left[\rho_{\rm tot}(t), H + H_{\rm ext}(t)\right], \tag{3.129}$$

which follows from the Schrödinger equation. Instead of solving Eq. (3.129) directly, we start from the formal solution

$$\rho_{\rm tot}(t) = \exp(-{\rm i}Ht)U(t)\rho U(t)^{\dagger} \exp({\rm i}Ht), \tag{3.130}$$

as obtained from Eq. (3.13). Putting the lowest-order result of $U(t)$ given by Eq. (3.16), we obtain

$$\rho_{\rm tot}(t) = \rho + {\rm i}\int_{-\infty}^{t} {\rm d}t' \exp(-{\rm i}Ht)[\rho, H_{\rm ext}^{\rm H}(t')] \exp({\rm i}Ht), \tag{3.131}$$

up to $O(H_{\rm ext})$. Using the identity

$$\mathrm{Tr} A\, [\rho, H_{\rm ext}^{\rm H}] = \mathrm{Tr}\rho\, [A, H_{\rm ext}^{\rm H}], \tag{3.132}$$

for any given observable $A$, we obtain Eq. (3.19). The derivation is of course equivalent to the one employed in the text.

**Problem 3.3**
We evaluate the Kubo formula involving the $z$-component of the magnetic moment with wave number $\boldsymbol{q}$:

$$\chi_0(\boldsymbol{q}, \omega) = {\rm i}\int_0^{\infty} {\rm d}t \langle [M_{\boldsymbol{q}}^{\rm H}(t), M_{-\boldsymbol{q}}^{\rm H}(0)]\rangle \exp({\rm i}\omega t), \tag{3.133}$$

where

$$M_{\boldsymbol{q}} = -\mu_{\rm B}\frac{1}{\sqrt{V}}\sum_{\boldsymbol{k}\sigma} \sigma c_{\boldsymbol{k}\sigma}^{\dagger} c_{\boldsymbol{k}+\boldsymbol{q}\sigma}. \tag{3.134}$$

The time dependence for free electrons is made explicit as

$$c_1^{\dagger}(t)c_2(t) = \exp\left[{\rm i}(\epsilon_1 - \epsilon_2)t\right] c_1^{\dagger}c_2, \tag{3.135}$$

where use has been made of $c_1^{\dagger}(t) = c_1^{\dagger}{\rm e}^{{\rm i}\epsilon_1 t}$ with eigenenergy $\epsilon_1$, and its Hermitian conjugate. Furthermore, the commutator is decomposed as

$$[c_1^{\dagger}c_2, c_2^{\dagger}c_1] = c_1^{\dagger}[c_2, c_2^{\dagger}c_1] + [c_1^{\dagger}, c_2^{\dagger}c_1]c_2 = c_1^{\dagger}c_1 - c_2^{\dagger}c_2. \tag{3.136}$$

At this stage, the statistical average is obtained with use of the Fermi distribution function $f(\epsilon_1) \equiv f_1$ as $\langle c_1^\dagger c_1 \rangle = f_1$. Hence we obtain as a part in $\chi(\boldsymbol{q}, \omega)$,

$$\chi_{12}(z) = \mathrm{i}\int_0^\infty \mathrm{d}t \exp[\mathrm{i}(\epsilon_1 - \epsilon_2)t](f_1 - f_2) = \frac{f_2 - f_1}{z - \epsilon_2 + \epsilon_1}. \tag{3.137}$$

By summation over spin and momentum, it is straightforward to obtain the dynamical susceptibility given by Eq. (3.27).

**Problem 3.4**
With imaginary time, the quantity that corresponds to Eq. (3.135) is evaluated as

$$\langle c_1^\dagger(\tau) c_2(\tau) c_2^\dagger c_1 \rangle = \exp[\tau(\epsilon_1 - \epsilon_2)] f_1(1 - f_2), \tag{3.138}$$

where the Matsubara index M has been omitted for creation and annihilation operators. We have used the fact that the average with indices 1,2 can be taken independently, and that the operator for an eigenstate depends on imaginary time as

$$c_1^\dagger(\tau) = \exp(\tau H) c_1^\dagger \exp(-\tau H) = c_1^\dagger \exp(\tau\epsilon_1). \tag{3.139}$$

Corresponding to Eq. (3.137), the Matsubara version is given by

$$\begin{aligned} \chi_{12}(\mathrm{i}\nu_n) &= \int_0^\beta \mathrm{d}\tau \exp[(\epsilon_1 - \epsilon_2 + \mathrm{i}\nu_n)\tau] f_1(1 - f_2) \\ &= \frac{\exp[\beta(\epsilon_1 - \epsilon_2)] - 1}{\mathrm{i}\nu_n - \epsilon_2 + \epsilon_1} f_1(1 - f_2) = \frac{f_2 - f_1}{\mathrm{i}\nu_n - \epsilon_2 + \epsilon_1}, \end{aligned} \tag{3.140}$$

where the last equality is due to the property $\exp(\beta\epsilon) f(\epsilon) = 1 - f(\epsilon)$. Equation (3.140) corresponds to analytic continuation of $z$ in Eq. (3.137) to even Matsubara frequencies $\mathrm{i}\nu_n$. Hence the result gives an example to show the validity of general formula given by Eq. (3.50).

**Problem 3.5**
In the spectral function

$$I_{XY}(\omega) = \mathop{\mathrm{Av}}_n \sum_m X_{nm} Y_{mn} [1 - \exp(-\beta\omega)] \delta(\omega - \omega_{mn}), \tag{3.141}$$

$X, Y$ are taken as $X = S_+ = S_x + iS_y$ and $Y = S_- = S_x - iS_y$ with $\boldsymbol{S}$ being the spin 1/2 operator. Then $|n\rangle = |\uparrow\rangle$ and $|m\rangle = |\downarrow\rangle$ are the only states that give finite matrix elements ($= 1$) for $X$ and $Y$. Provided the temperature $T = 1/\beta$ is much larger than the Zeeman splitting $|\omega_{mn}|$, we may take $1 - \exp(-\beta\omega) \sim \beta\omega$ and obtain

$$\chi_{+-}(z) = \int_{-\infty}^{\infty} \mathrm{d}\omega \frac{I_{+-}(\omega)}{\omega - z} \xrightarrow[z=0]{} \frac{1}{2}\beta, \tag{3.142}$$

which implies the Curie law $\chi_\perp = \chi_{xx} = \chi_{yy} = \beta/4$ since $\chi_{+-} = \chi_{xx} + \chi_{yy}$. We interpret the result as due to the ergodicity recovered by the small but nonzero precession frequency $\omega_{mn}$.

**Problem 3.6**

The projection operator is modified as

$$\mathcal{P} \equiv \sum_{ij} |X_i\rangle (\chi^{-1})_{ij} \langle X_j|, \tag{3.143}$$

where the matrix $\chi$ is composed of elements $\chi_{ij} = \langle X_i|X_j\rangle$. Then the matrix $C(z)$ for the relaxation function is obtained as

$$C(z) = \mathrm{i}\chi[z - \Omega + \mathrm{i}\Gamma(z)]^{-1}, \tag{3.144}$$

where the matrices $\Omega$ and $\Gamma(z)$ are composed of elements:

$$\Omega_{ij} = \sum_l (\chi^{-1})_{il} \langle X_l|\mathcal{L}|X_j\rangle, \tag{3.145}$$

$$\Gamma_{ij}(z) = \sum_l (\chi^{-1})_{il} \langle \mathcal{Q}\dot{X}_l| \frac{\mathrm{i}}{z - \mathcal{QLQ}} |\mathcal{Q}\dot{X}_j\rangle. \tag{3.146}$$

If the Hamiltonian is invariant under time reversal, we obtain $\Omega_{ij} = 0$ as long as all $X_i$'s have the same parity under time reversal. Namely, the inner product of $|X_i\rangle$ and $\mathcal{L}|X_j\rangle$ with different parities vanishes. If the external magnetic field $B_z$ along the $z$-axis is finite and fixed, the Hamiltonian $H_{\text{ext}} = -M_z B_z$ breaks the time reversal. The magnetic moment $\boldsymbol{M} = g\mu_{\text{B}}\boldsymbol{S}$ has the time derivative proportional to $[S_z, \boldsymbol{S}]$. Considering the component $S_j$, we obtain as a part of $\Omega_{ij}$:

$$\langle S_i|[S_z, S_j]\rangle = \mathrm{i}\epsilon_{zjk}\langle S_i|S_k\rangle, \tag{3.147}$$

with $\epsilon_{zjk}$ being the completely antisymmetric unit tensor. The RHS is finite with $i = x$, $j = y$ and $i = y$, $j = x$. The finite inner product contributes to $\Omega_{xy} = -\Omega_{yx}$ which describes the precession frequency.

**Problem 3.7***

With the perturbation $H_1 = J\boldsymbol{S}\cdot\boldsymbol{s}$, we obtain

$$[S_i, H_1] = \mathrm{i}J\sum_{jk} \epsilon_{ijk} S_k s_j. \tag{3.148}$$

To derive $\Gamma(z)$ we need to evaluate

$$\langle S_k(\tau)s_j(\tau)S_k s_j\rangle_0 = \frac{1}{3}S(S+1)\langle s_j(\tau)s_j\rangle_0. \tag{3.149}$$

The $s_j$ part is Fourier transformed as

$$\langle [s_j, s_j] \rangle_0(\mathrm{i}\nu_n) = \frac{1}{4N^2}\sum_{\boldsymbol{k},\boldsymbol{k}'}\frac{f(\epsilon_{\boldsymbol{k}}) - f(\epsilon_{\boldsymbol{k}'})}{\mathrm{i}\nu_n - \epsilon_{\boldsymbol{k}} + \epsilon_{\boldsymbol{k}'}}, \tag{3.150}$$

referring to Eq. (3.140). Then in $\Gamma(z)$, as given by Eq. (3.75), we deal with the following quantity

$$\frac{1}{N^2}\sum_{\boldsymbol{k},\boldsymbol{k}'}\frac{f(\epsilon_{\boldsymbol{k}}) - f(\epsilon_{\boldsymbol{k}'})}{-\epsilon_{\boldsymbol{k}} + \epsilon_{\boldsymbol{k}'}} \cdot \frac{1}{z - \epsilon_{\boldsymbol{k}} + \epsilon_{\boldsymbol{k}'}}. \tag{3.151}$$

With $z = \mathrm{i}0_+$, the real part vanishes by double summation and the imaginary part becomes

$$-\frac{\pi}{N^2}\sum_{\boldsymbol{k},\boldsymbol{k}'}\delta(-\epsilon_{\boldsymbol{k}} + \epsilon_{\boldsymbol{k}'})\left(-\frac{\partial f(\epsilon_{\boldsymbol{k}})}{\partial \epsilon_{\boldsymbol{k}}}\right) = -\pi\rho_c(\mu)^2. \tag{3.152}$$

Substituting $\chi_M = C/T$, we obtain Eq. (3.77).

We proceed to the conductivity $\sigma(\boldsymbol{q}, z)$ in the small $|\boldsymbol{q}|$ limit. The static current susceptibility is derived by the formula

$$\langle J_{\boldsymbol{q}} | J_{-\boldsymbol{q}} \rangle = \langle [J_{\boldsymbol{q}}, J_{-\boldsymbol{q}}] \rangle(z = 0), \tag{3.153}$$

which leads to the equality in Eq. (3.82). In the limit $\boldsymbol{q} \to 0$, we obtain

$$\frac{1}{V^2}\sum_{\boldsymbol{k}}\Delta(\boldsymbol{k} - \boldsymbol{k}' + \boldsymbol{q})k_z k_z' \frac{f(\epsilon_{\boldsymbol{k}}) - f(\epsilon_{\boldsymbol{k}'})}{\epsilon_{\boldsymbol{k}'} - \epsilon_{\boldsymbol{k}}} = \frac{1}{3}k_{\mathrm{F}}^2\rho_c(\epsilon_{\mathrm{F}}), \tag{3.154}$$

with $\rho_c(\epsilon_{\mathrm{F}}) = 3n/(2\epsilon_{\mathrm{F}})$. Then we obtain the rightmost result in Eq. (3.82). Note if one works with $\boldsymbol{q} = 0$ from the beginning, singular contribution due to $[J_0, H_0] = 0$ leads to wrong result instead of Eq. (3.154).

On the other hand, we can set $\boldsymbol{q} = 0$ in dealing with $\dot{J}_{\boldsymbol{q}}$ since the correction is negligible for $\boldsymbol{q} \to 0$. Noting $\dot{J}_0$ is contributed by $H_1$ only, we obtain

$$[J_0, H_1] = n_i \sum_{\boldsymbol{k},\boldsymbol{k}'}(k_z - k_z')u(\boldsymbol{k} - \boldsymbol{k}')c_{\boldsymbol{k}}^\dagger c_{\boldsymbol{k}'}. \tag{3.155}$$

Proceeding in the same way as in Eq. (3.150) we obtain Eq. (3.83). The next step is also the same as in Eq. (3.151). Instead of Eq. (3.152), however, we now have to average over the solid angles $\Omega_{\boldsymbol{k}}$ and $\Omega_{\boldsymbol{k}'}$. In the case of spherically symmetric potential $u(\boldsymbol{r}) = u(r)$, only the direction of $\boldsymbol{k}'$ relative to that of $\boldsymbol{k}$ matters. Hence we may take the $z$-direction along $\boldsymbol{k}$, and obtain

$$\frac{\partial}{\mathrm{i}\partial\omega}\langle[\dot{J}_0, \dot{J}_0]\rangle_0(\omega+\mathrm{i}0_+)\Big|_{\omega=0}$$
$$= \frac{2\pi e^2 n_i}{3m}\rho_c(\epsilon_\mathrm{F})^2 k_\mathrm{F}^2 \int \frac{\mathrm{d}\Omega_{\boldsymbol{k}'}}{4\pi}|u[k_\mathrm{F}(\hat{z}-\hat{\Omega}_{\boldsymbol{k}'})]|^2(1-\cos\theta_{\boldsymbol{k}'}), \tag{3.156}$$

where $\hat{z}$ and $\hat{\Omega}_{\boldsymbol{k}'}$ are the unit vector along the $z$-axis and $\boldsymbol{k}'$, respectively, and $\theta_{\boldsymbol{k}'}$ is the polar angle along $\boldsymbol{k}'$. Combining with Eq. (3.82), we obtain Eq. (3.84).

**Problem 3.8**

We take a (classical) stochastic variable $x$, and consider the moments $\langle x^n\rangle$ and cumulants $\langle x^n\rangle_c$ with $n = 0, 1, \ldots$. As a convenient device, we employ the generating function $G(q)$ defined by

$$G(q) = \langle\exp(\mathrm{i}qx)\rangle = \sum_{n=0}^{\infty}\frac{(\mathrm{i}q)^n}{n!}\langle x^n\rangle. \tag{3.157}$$

Correspondingly, the $n$-th cumulant $\langle x^n\rangle_c$ is defined by

$$\ln G(q) = \sum_{n=1}^{\infty}\frac{(\mathrm{i}q)^n}{n!}\langle x^n\rangle_c. \tag{3.158}$$

If the variable $x$ has the Gaussian distribution with the average $\bar{x}$ and the variance $\sigma^2$, the generating function is derived in the closed form

$$G(q) = \int_{-\infty}^{\infty}\frac{\mathrm{d}x}{\sqrt{2\pi}\sigma}\exp\left(-\frac{1}{2\sigma^2}(x-\bar{x})^2+\mathrm{i}qx\right) = e^{-\sigma^2 q^2/2+\mathrm{i}q\bar{x}}. \tag{3.159}$$

By taking the logarithm of both sides in Eq. (3.159), and compare with Eq. (3.158), we obtain immediately $\langle x\rangle_c = \bar{x}$, $\langle x^2\rangle_c = \sigma^2$, and $\langle x^n\rangle_c = 0$ for $n > 2$. On the other hand, all moments $\langle x^n\rangle$ remain finite for the Gaussian distribution. These moments are derived explicitly by expansion of the rightmost side of Eq. (3.159) in terms of $q$, and compare with Eq. (3.157).

**Problem 3.9**

Let us consider first the case $\tau > 0$. We replace the summation by integral over the complex energy $z$ as

$$T\sum_n\frac{\exp(\mathrm{i}\epsilon_n\tau)}{\mathrm{i}\epsilon_n-\epsilon} = \int_C\frac{\mathrm{d}z}{2\pi\mathrm{i}}\frac{\exp(z\tau)}{z-\epsilon}f(z), \tag{3.160}$$

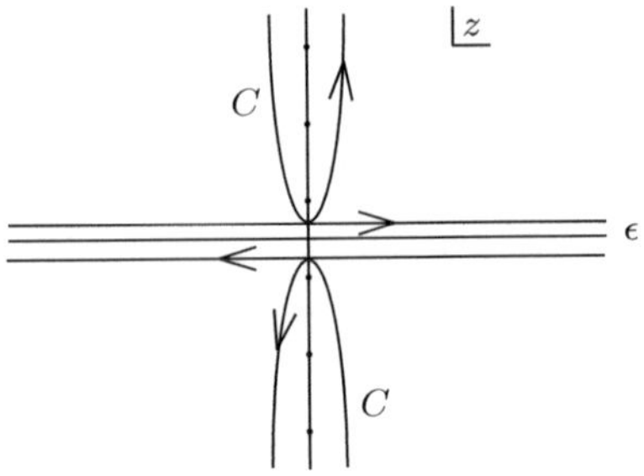


**Fig. 3.3** The integration path $C$ around the imaginary axis picks up the poles of $f(z)$. It is deformed along both directions of the real axis surrounding a pole at $z = \epsilon$

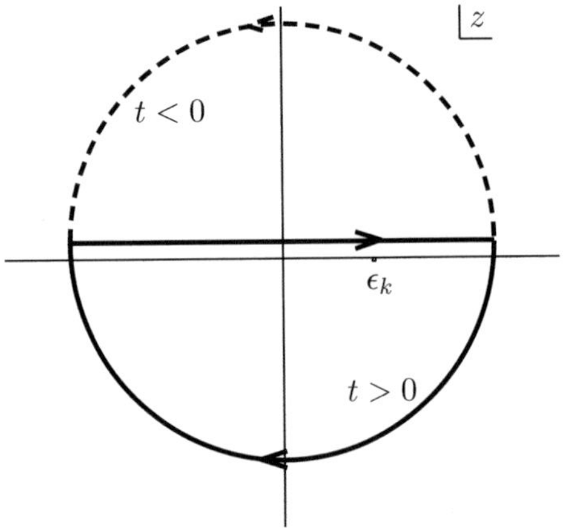


**Fig. 3.4** Integration path for the retarded Green function. The dashed semicircle is for the case $t < 0$, while the solid semicircle for $t > 0$

where $f(z) = 1/(e^{\beta z} + 1)$ has first-order poles at Matsubara frequencies $z = i\epsilon_n$ on the imaginary axis of $z$. The integration contour $C$ is shown in Fig. 3.3, which runs originally around the imaginary axis to pick up all poles of $f(z)$, but is continuously deformed so that it runs parallel to the real axis in both directions. The deformed contour picks up a pole at $z = \epsilon$, and the integral is convergent since $f(z)\exp(z\tau)$ goes to zero exponentially at $z \to \pm\infty$ as long as $0 < \tau < \beta$.

In the second case of $\tau < 0$, we use $1 - f(z) = e^{\beta z}/(e^{\beta z} + 1)$ instead of $f(z)$ in Eq. (3.160). We have the same poles as the first case, and the integral is again convergent since $(1 - f(z))\exp(z\tau)$ goes to zero exponentially at $z \to \pm\infty$ as long as $-\beta < \tau < 0$. Hence we obtain Eq. (3.104) by picking up the pole on the real axis in both cases.

**Problem 3.10**

The analytic continuation $i\epsilon_n \to z$ gives the retarded Green function in the upper half plane, and the advanced Green function in the lower half plane. The inverse Fourier transform of $(z - \epsilon_k)^{-1}$ is performed by the integral over $z$ along the horizontal line slightly above the real axis. The integration path is illustrated in Fig. 3.4. If $t < 0$, the integration path along the real axis can be supplemented by a large semicircle in the upper half plane. The contribution from the semicircle tends to zero as the radius goes to infinity, and the integral along the closed loop also gives zero because there is no singularity inside. On the other hand, for $t > 0$, the large semicircle must be taken in the lower half plane to have the convergent (zero) contribution from the semicircle. The closed loop in this case includes a pole at $z = \epsilon_k$, and gives a finite result. Thus we obtain

$$\int_{-\infty+i0_+}^{\infty+i0_+} \frac{dz}{2\pi} \frac{e^{-izt+\delta t}}{z-\epsilon_k} = -i\theta(t)\exp(-i\epsilon_k t) = G_k^R(t). \tag{3.161}$$

This result from Eq. (3.161) is of course the same as the direct calculation using Eq. (3.111).

## References


1. Kubo, R.: J. Phys. Soc. Jpn. **12**, 570 (1957)
2. Matsubara, T.: Prog. Theor. Phys. **14**, 351 (1955)
3. Abrikosov, A.A., Gorkov, L.P., Dzyaloshinskii, I.E.: Methods of Quantum Field Theory in Statistical Physics. Dover, New York (1975)
4. Mori, H.: Prog. Theor. Phys. **34**, 399 (1965)
5. Kubo, R.: Rep. Prog. Phys. **29**, 255 (1966)
6. Zwanzig, R.: Nonequilibrium Statistical Mechanics. Oxford University Press, New York (2001)
7. Korringa, J.: Physica **16**, 601 (1950)
8. Anderson, P.W.: J. Phys. Soc. Jpn. **9**, 316 (1954)

# Chapter 4
# Fermi Liquid Theory

**Abstract** The Fermi liquid theory constructed by Landau efficiently describes strongly correlated fermions in terms of small number of parameters. The core of the theory is the concept of quasi-particles. This chapter explains how the Fermi liquid theory accounts for interaction effects taking examples such as specific heat and magnetic susceptibility. Also discussed is the dynamical response of Fermi liquid against slowly varying external fields.

## 4.1 Quasi-Particles and Their Distributions

Let $H_0$ be the Hamiltonian for $N$ electrons without electron–electron interactions, and $\phi_n$ be the $n$-th eigenstate with the eigenenergy $E_n$. The Schrödinger equation reads

$$H_0\phi_n = E_n\phi_n. \tag{4.1}$$

Each many-electron state $\phi_n$ of free electrons is constructed by the Slater determinant of one-electron states. We assume the translational and rotational symmetries in the system. Then each one-electron state is specified by the momentum $\boldsymbol{p}$ and spin $\sigma$. Hence the distribution $n_{p\sigma}$ (=0, 1) of one-electron states, determines the many-electron state.

For discussing the interaction effect, the electron–electron interaction Hamiltonian $\lambda H_1$ is scaled by the parameter $\lambda$ which controls the strength in the range $0 \le \lambda \le 1$. The many-body state $\Phi_n$ corresponding to $\lambda = 1$ develops continuously from $\phi_n$. If there is no level crossing in between, this process is called adiabatic by analogy with thermodynamic process. The eigenenergy $\tilde{E}_n$ for $H = H_0 + H_1$ is also connected continuously with $E_n$ as a function of $\lambda$.

For a small system, the interval between neighboring $E_n$ can be much larger than the characteristic magnitude of $H_1$. Then it is obvious that no level crossing takes place in the course of increasing $\lambda$. The bold assumption taken by Landau [1–3] is that even though the interval of $E_n$ is much smaller than $H_1$, the adiabatic continuity holds from $E_n$ to $\tilde{E}_n$ and from $\phi_n$ to $\Phi_n$. For example, if there is a phase



Y. Kuramoto, *Quantum Many-Body Physics*, Lecture Notes in Physics 934,
https://doi.org/10.1007/978-4-431-55393-9_4

transition at zero temperature for certain $\lambda < 1$, this means a level crossing of the ground states. Then the adiabatic continuity certainly breaks down. Even without a phase transition, the level crossing occurs in many realistic systems. For example, the Fermi surface in metals is generally not spherical, and the shape depends on the strength of the interaction. This means that level crossings actually take place as a function of $\lambda$. In order to avoid such complexity in anisotropic systems, we first confine the discussion to the isotropic system with the spherical Fermi surface as is realized in liquid $^3$He.

If Landau's assumption is valid for the one-to-one correspondence between $\Phi_n$ and $\phi_n$, the density matrix $\rho$, or the weight of each $\Phi_n$, is completely specified by the distribution function $\{n_{p\sigma}\}$ of the non-interacting system. Hence the exact excitation energy $\delta E$ is regarded as a functional of $\{n_{p\sigma}\}$. In the interacting system, the excitation that corresponds to the change of particular $n_{p\sigma}$ is called the quasi-particle. The concept is valid for all excitation energies as long as the adiabatic continuation holds. However, the actual utility is confined to the range of low excitation energies.[1] Namely, if the change $\delta n_{p\sigma}$ from the ground state is small, $\delta E$ can be expanded in terms of $\delta n_{p\sigma}$ as follows:

$$\delta E = \sum_{p\sigma} \epsilon_{p\sigma} \delta n_{p\sigma} + \frac{1}{2} \sum_{pp'} \sum_{\sigma\sigma'} f(p\sigma, p'\sigma') \delta n_{p\sigma} \delta n_{p'\sigma'} + \cdots , \tag{4.2}$$

where $\epsilon_{p\sigma}$ is interpreted as energy of a quasi-particle measured from the Fermi level $\mu$. This energy is in general different from that of the non-interacting particle with the same quantum numbers. Without magnetic field, $\epsilon_{p\sigma}$ is independent of spin $\sigma$. The function $f(p\sigma, p'\sigma')$ describes interaction between quasi-particles.

Note that two terms in the RHS of Eq. (4.2) are both second order quantities with respect to $\sum_{p\sigma} |\delta n_{p\sigma}|/N$, since $\epsilon_{p\sigma}$ is also a small quantity measured from the Fermi level. Hence, in the limit of low excitation energies, those terms which are higher than the second order in $\delta n_{p\sigma}$ can be neglected. Conversely, with increasing excitation energy, the higher order terms become relevant, and the quasi-particle description is no longer useful.

The Fermi surface of free fermions is a sphere with radius $p_F$. Because of the one-to-one correspondence between $\Phi_n$ and $\phi_n$, the Fermi surface of quasi-particles does not change by the interaction, and the Fermi momentum remains the same as $p_F$. The one-to-one correspondence has a further consequence that the entropy of the system is determined only by distribution of quasi-particles. The maximum entropy under the constraint of given energy $E$ and the number $N$ of quasi-particles is equivalent to the minimum thermodynamic potential $\Omega = E - \mu N - TS$, as in

[1] In the original Fermi liquid theory, the quasi-particle is defined only near the Fermi level. This is because the lifetime becomes shorter as the energy of an added or removed fermion goes off from the Fermi level. In later development, the concept was modified so as to be consistent with real value of $\epsilon_{p\sigma}$ in Eq. (4.2) for arbitrary $\boldsymbol{p}$, and sometimes called the "statistical" quasi-particle [4]. We have mainly followed the latter description here.

the case of free fermions. Throughout this book we take the origin of energy so as to set the chemical potential $\mu = 0$, unless stated otherwise. Then the stationary condition $\delta\Omega/\delta n_{p\sigma} = 0$ gives the quasi-particle distribution as

$$n_{p\sigma} = \left[\exp(\beta\epsilon_{p\sigma}) + 1\right]^{-1}, \tag{4.3}$$

which takes the form analogous to free fermions. However, $\epsilon_{p\sigma}$ here is not a constant but depends on temperature and interactions. We emphasize that all interaction effects, including fluctuations of spin and charge, are implicitly taken into account in the quasi-particle energy.

It is very convenient to parameterize $f(\boldsymbol{p}\sigma, \boldsymbol{p}'\sigma')$ in terms of spherical harmonics or, equivalently, Legendre polynomials. Namely, we consider two quasi-particles with momenta $\boldsymbol{p}$ and $\boldsymbol{p}'$ both of which have magnitude equal to $p_{\mathrm{F}}$. Then taking the angle $\theta_{\boldsymbol{pp}'}$ between $\boldsymbol{p}$ and $\boldsymbol{p}'$ we make the decomposition

$$\rho^*(\mu) f(\boldsymbol{p}\sigma, \boldsymbol{p}'\sigma') = \sum_{l=0}^{\infty} (F_l + \sigma\sigma' Z_l) P_l(\cos\theta_{\boldsymbol{pp}'}), \tag{4.4}$$

with $\sigma, \sigma' = \pm 1$. Here $\rho^*(\mu)$ is the density of states for quasi-particles at the Fermi level, as given by

$$\rho^*(\mu) = \sum_{p\sigma} \delta(\epsilon_{\boldsymbol{p}}) = \frac{V}{\pi^2} m^* p_{\mathrm{F}}, \tag{4.5}$$

where $V$ is the volume of the system, taken to be unity, and $m^*$ the effective mass. The dimensionless quantities $F_l$, $Z_l$ in Eq. (4.4) are called the Landau parameters.

Similarly, the distribution function $\delta n_{p\sigma}$ is decomposed by spherical harmonics as

$$\delta n_{p\sigma} = \frac{4\pi}{\rho^*(\mu)} \delta(c_{\boldsymbol{p}}) \sum_{lm} \delta n_{lm\sigma} Y_{lm}(\hat{p}), \tag{4.6}$$

where $\hat{p}$ represents the solid angle of $\boldsymbol{p}$. In taking summation over a function that is sharply peaked near the Fermi level, we can make the approximation

$$\sum_{p\sigma} \sim \rho^*(\mu) \int \mathrm{d}\epsilon_{\boldsymbol{p}} \int \frac{\mathrm{d}\Omega}{4\pi}, \tag{4.7}$$

where $\Omega$ is the solid angle. Furthermore using the addition theorem

$$P_l(\cos\theta_{\boldsymbol{pp}'}) = \frac{4\pi}{2l+1} \sum_m Y_{lm}(\hat{p}) Y_{lm}(\hat{p}'), \tag{4.8}$$

the interaction part of the excitation energy, which is the second term in the RHS of Eq. (4.2) is written as

$$\begin{aligned}\delta E_{\text{int}} &= \frac{4\pi}{2\rho^*(\mu)}\sum_{lm}\sum_{\sigma\sigma'}\frac{1}{2l+1}(F_l+\sigma\sigma' Z_l)\delta n_{lm\sigma}\delta n_{lm\sigma'} \\ &= \frac{4\pi}{2\rho^*(\mu)}\sum_{lm}\frac{1}{2l+1}\left(F_l\delta n_{lm+}^2 + Z_l\delta n_{lm-}^2\right),\end{aligned} \tag{4.9}$$

where we have introduced the symmetric and antisymmetric components

$$\delta n_{lm+} = \sum_{\sigma}\delta n_{lm\sigma}, \quad \delta n_{lm-} = \sum_{\sigma}\sigma\delta n_{lm\sigma}, \tag{4.10}$$

with respect to spins $\sigma = \pm 1$.

In the presence of other quasi-particles, the energy $\tilde{\epsilon}_{p\sigma}$ of a quasi-particle is modified from $\epsilon_{p\sigma}$ as

$$\frac{\delta E}{\delta n_{p\sigma}} \equiv \tilde{\epsilon}_{p\sigma} = \epsilon_{p\sigma} + \sum_{p'\sigma'} f(p\sigma, p'\sigma')\delta n_{p'\sigma'}, \tag{4.11}$$

where the rightmost term is interpreted as a molecular field due to interactions. Since the excited states are not translationally invariant in general, $\delta n_{p\sigma}$ may slowly depend on $\boldsymbol{r}$. In this case $\tilde{\epsilon}_{p\sigma}(\boldsymbol{r})$ depends also on $\boldsymbol{r}$, and is called the local quasi-particle energy.

The distribution function $\delta n_{p\sigma}(\boldsymbol{r})$, which depends both on $\boldsymbol{p}$ and $\boldsymbol{r}$, is an example of the Wigner distribution. The latter is extension of the classical distribution function in the phase space $(\boldsymbol{p}, \boldsymbol{r})$. For free fermions, it is given explicitly by

$$\delta n_{p\sigma}(\boldsymbol{r}) = \sum_{\boldsymbol{q}}\langle c^\dagger_{\boldsymbol{p}-\boldsymbol{q}/2,\sigma} c_{\boldsymbol{p}+\boldsymbol{q}/2,\sigma}\rangle \exp(\mathrm{i}\boldsymbol{q}\cdot\boldsymbol{r}), \tag{4.12}$$

where the RHS vanishes in the ground state, but can remain finite in excited states with momentum $\boldsymbol{q}$. Taking account of other quasi-particle excitations, we introduce the local distribution function [2]

$$\bar{n}_{p\sigma} = \left[\exp(\beta\tilde{\epsilon}_{p\sigma}) + 1\right]^{-1}. \tag{4.13}$$

In terms of the deviation $\delta\bar{n}_{p\sigma} = n_{p\sigma} - \bar{n}_{p\sigma}$ from the local equilibrium, a useful relation

$$\delta\bar{n}_{p\sigma} = \delta n_{p\sigma} + \delta(\epsilon_{p})\sum_{p'\sigma'} f(p\sigma, p'\sigma')\delta n_{p'\sigma'}, \tag{4.14}$$

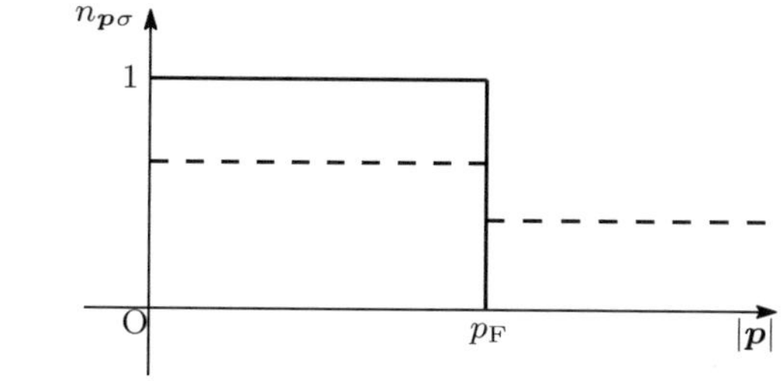


**Fig. 4.1** Distribution $\{n_{p\sigma}\}$ of quasi-particles in the ground state (solid line), and an example of $\langle c^\dagger_{p\sigma} c_{p\sigma}\rangle$ of fermions with strong repulsion (dashed lines). Both distributions have a discontinuity at the Fermi momentum $p_F$

follows because of the relation between $\tilde{\epsilon}_{p\sigma}$ and $\epsilon_{p\sigma}$ as specified in Eq. (4.11). If the Landau parameter is positive (repulsive), the presence of other quasi-particles leads to $\tilde{\epsilon}_{p\sigma} > \epsilon_{p\sigma}$. This entails $|\delta \bar{n}_{p\sigma}| > |\delta n_{p\sigma}|$, which means suppression of the response $\delta n_{p\sigma}$ by the repulsion between quasi-particles. Problem 4.1 addresses the deviation from local equilibrium for each component of the angular momentum.

It is possible to derive the Fermi liquid theory microscopically using the many-body perturbation theory, as first carried out by Landau himself [5]. The microscopic theory can also deal with anisotropic systems with deformable Fermi surface [6]. The energy shift $\delta E$ in Eq. (4.11) is regarded as a quasi-classical form of the effective Hamiltonian, which is valid for particle-hole excitations with long wavelength. Furthermore, the spectrum of quasi-particles is interpreted as the pole of the single-particle Green function introduced in Sect. 3.7. In this context it is instructive to compare the momentum distribution of quasi-particles and the average $\langle c^\dagger_{p\sigma} c_{p\sigma}\rangle$ of interacting fermions. Figure 4.1 illustrates both distributions. Suppose a case where the interaction between fermions is strongly repulsive at short distance. Then the particles try to avoid each other at the expense of larger kinetic energy. As a result, the momentum distribution extends beyond $p_F$. The actual form of distribution depends on interactions and the single-particle spectrum. In the Hubbard model with large $U$, for example, $\langle c^\dagger_{p\sigma} c_{p\sigma}\rangle$ becomes almost a constant ($\sim 1/2$) if the number of electrons is close to unity per site. As shown schematically by dashed lines in Fig. 4.1, a signature of the Fermi liquid is the finite discontinuity at $p_F$. The reduction of discontinuity measures the degree of renormalization in forming the quasi-particle. We shall discuss microscopic details of renormalization later in Chaps. 6 and 10.

## 4.2 Specific Heat and Magnetic Susceptibility

As a simple application of the Fermi liquid theory, we first discuss the specific heat in a metal which takes the form

$$C = \gamma T + \beta T^3, \tag{4.15}$$

at low temperatures. The first term originates from quasi-particles and the $T^3$ term from lattice vibrations.[2] In the Fermi liquid theory, the coefficient $\gamma$ is given by

$$\gamma = \frac{\pi^2}{3} k_B^2 \rho^*(\mu), \tag{4.16}$$

with $k_\mathrm{B}$ the Boltzmann constant. The $T$-linear specific heat comes from the fact that the entropy in the Fermi liquid, including interaction effects, is exhausted by quasi-particle excitations in the low temperature limit. The effective mass is in general different from the vacuum mass $m_0 = 9.1 \times 10^{-28}$ g of an electron.

We next consider the spin susceptibility $\chi$. The magnetization $M$ of the whole volume is given by

$$M = -\mu_\mathrm{B} \sum_{p\sigma} \sigma \delta n_{p\sigma}, \tag{4.17}$$

where $\mu_\mathrm{B} = e\hbar/(2m_0 c)$ is the Bohr magneton with the $g$-factor taken to be 2. In the magnetic field $B$, the quasi-particle energy is shifted by $\mu_\mathrm{B} B\sigma$ that is called the Zeeman term. Then the deviation from the local equilibrium is simply given by

$$\delta \bar{n}_{p\sigma} = \mu_\mathrm{B} \sigma B \delta(\epsilon_p). \tag{4.18}$$

The response of the system is, however, described by $\delta n_{p\sigma}$ instead of $\delta \bar{n}_{p\sigma}$. The relation between the two kinds of deviation has been given by Eq. (4.14). By decomposition into angular momentum component $(lm)$, the relation is described in terms of the Fermi liquid parameters as discussed in Problem 4.1. Since the homogeneous field only affects the $l = 0$ component of the angular momentum, the magnetic susceptibility is obtained as

$$\chi = \frac{\mu_\mathrm{B}^2 \rho^*(\mu)}{1 + Z_0} \equiv \frac{\chi_0}{1 + Z_0}, \tag{4.19}$$

the derivation of which is the subject of Problem 4.2.

The Landau parameter $Z_0$ describes the isotropic exchange interaction between quasi-particles. With negative sign, $Z_0$ corresponds to ferromagnetic interaction. In particular the susceptibility is divergent as $Z_0 \to -1$. This means the instability of the paramagnetic state, which is called the normal Fermi liquid, toward a ferromagnetic state. It may happen that spin fluctuations become very soft near the instability toward a magnetically ordered state. In such a case, the specific heat is influenced by spin fluctuations through the enhanced effective mass of quasi-particles. For magnetic impurity systems, the modified Fermi liquid theory gives

[2] It is known that interaction between quasi-particles gives rise to a term of the form $T^3 \ln T$, which has been observed in liquid $^3$He [3], but is unobservable in most metals.

the exact relation between the specific heat and susceptibilities for spin and charge. The details will be discussed in Chap. 6 in relation to the Kondo effect.

## 4.3 Dynamical Response of Fermi Liquid

The Fermi liquid theory can describe the dynamics of the system with small momentum $q$ ($\ll p_{\rm F}$) and frequency $\omega$ ($\ll \epsilon_F$) where $\epsilon_F$ is the Fermi energy. We introduce a slowly varying fictitious external field $\varphi_{p\sigma}(\boldsymbol{q})$ that acts on a quasi-particle with momentum $\boldsymbol{p}$ and spin $\sigma$. Here $\boldsymbol{q}$ represents the wave vector of the external field. The deviation $\delta\bar{n}_{p\sigma}(\boldsymbol{q})$ from the local equilibrium, which is proportional to $\varphi_{p\sigma}(\boldsymbol{q})$, is derived along the line of linear response theory as

$$\delta\bar{n}_{p\sigma}(\boldsymbol{q}) = \frac{f(\epsilon_-) - f(\epsilon_+)}{\omega - \epsilon_+ + \epsilon_-}\varphi_{p\sigma}(\boldsymbol{q}) \sim \frac{\boldsymbol{v}\cdot\boldsymbol{q}}{\omega - \boldsymbol{v}\cdot\boldsymbol{q}}\delta(\epsilon_p)\varphi_{p\sigma}(\boldsymbol{q})$$
$$\equiv \Pi_p(\boldsymbol{q},\omega)\varphi_{p\sigma}(\boldsymbol{q}), \tag{4.20}$$

where $\boldsymbol{v} = \partial\epsilon_p/\partial\boldsymbol{p}$, $\epsilon_\pm = \epsilon_{p\pm q/2}$, and we have defined $\Pi_p(\boldsymbol{q},\omega)$ in the second line. Equation (4.20) is a dynamical generalization of Eq. (4.18). On the other hand, the deviation $\delta n_{p\sigma}(\boldsymbol{q})$ is the response against the external field plus the molecular field as given by

$$\delta n_{p\sigma}(\boldsymbol{q}) = \Pi_p(\boldsymbol{q},\omega)\left(\varphi(\boldsymbol{p},\boldsymbol{q}) + \sum_{p'\sigma'} f(\boldsymbol{p}\sigma,\boldsymbol{p}'\sigma')\delta n_{p'\sigma'}(\boldsymbol{q})\right), \tag{4.21}$$

which forms the integral equation for $\delta n_{p\sigma}(\boldsymbol{q})$. The equation is also written symbolically in the matrix form:

$$\left(1 - \Pi(\boldsymbol{q},\omega)\hat{f}\right)\delta\boldsymbol{n} = \Pi(\boldsymbol{q},\omega)\boldsymbol{\varphi}, \tag{4.22}$$

where $\boldsymbol{n}$ and $\boldsymbol{\varphi}$ are vectors, and $\hat{f}$ is the matrix with momentum and spin indices. In the static limit $\omega = 0$, $\Pi(\boldsymbol{q},\omega)\hat{f}$ reduces to Landau parameters through decomposition into components of spherical harmonics. Derivation of Eq. (4.22) is the subject of Problem 4.3.

We now discuss collective excitation modes which appear as the solution of Eq. (4.22) with the RHS set to zero. The spectrum corresponds to zero of the determinant of the matrix in the LHS. Then finite $\delta\boldsymbol{n}$ is allowed as the solution even in the absence of external fields. As the simplest example, we derive the spectrum of a collective mode called the zero sound in a neutral Fermi liquid such as $^3$He. In the case of electronic Fermi liquid, on the other hand, we have to include the long-range Coulomb interaction in addition to the Landau parameters. We take a hypothetical case where only $F_0$ is finite as the Landau parameters. By angular average of the matrix elements, the $\ell = 0$ component is easily extracted. With the dimensionless parameter $s = \omega/vq$, the spectrum of the zero sound is determined by the condition

$$\frac{1}{F_0} + \frac{1}{2}\int_{-1}^{1} \mathrm{d}\mu \frac{\mu}{\mu - s} = \frac{1}{F_0} + 1 + \frac{s}{2} \ln \frac{s-1}{s+1} = 0, \tag{4.23}$$

where the term with $\mu = \cos\theta$ comes from $\Pi(\boldsymbol{q}, \omega)$. Derivation of the solution to Eq. (4.23) is the subject of Problem 4.4.

In actual metals with a non-spherical Fermi surface, the adiabatic continuity with respect to the strength $\lambda$ does not hold, since an occupied state for small $\lambda$ may become empty for larger $\lambda$, and vice versa. Nevertheless, many metals are described reasonably well in terms of quasi-particles. Because of the crystalline anisotropy, the Landau parameters can no longer be classified by the spherical harmonics. Concerning the volume inside the Fermi surface, more general argument can show that the volume is independent of interactions. We shall discuss this aspect in Sect. 9.4.

## Problems

**4.1** By decomposing into components of angular momentum, derive the proportionality relation between $\delta\bar{n}_{p\sigma}$ and $\delta n_{p\sigma}$.

**4.2** Derive the magnetic susceptibility of the Fermi liquid as given by Eq. (4.19).

**4.3** Derive the dynamical response in the Fermi liquid in the matrix form as given by Eq. (4.22).

**4.4** Discuss how the spectrum of zero sound is influenced by the Landau parameter with use of Eq. (4.23).

## Solutions to Problems

### Problem 4.1

Equation (4.14) is decomposed into components $(lm)$ of the angular momentum as

$$\delta\bar{n}_{lm\sigma} = \delta n_{lm\sigma} + \frac{1}{2l+1}\sum_{\sigma'}(F_l + \sigma\sigma' Z_l)\delta n_{lm\sigma'}, \tag{4.24}$$

where no mixing occurs between different angular momenta. Each component is further decomposed into spin symmetric and antisymmetric components as

$$\delta\bar{n}_{lm+} = \sum_\sigma \delta\bar{n}_{lm\sigma}, \quad \delta\bar{n}_{lm-} = \sum_\sigma \sigma\delta\bar{n}_{lm\sigma}, \tag{4.25}$$

with $\sigma = \pm 1$. Then Eq. (4.24) is written as

$$\delta\bar{n}_{lm+} = \left(1 + \frac{F_l}{2l+1}\right)\delta n_{lm+}, \quad \delta\bar{n}_{lm-} = \left(1 + \frac{Z_l}{2l+1}\right)\delta n_{lm-}. \tag{4.26}$$

If the Landau parameter for certain $l$ is negative, the deviation $\delta\bar{n}_{lm\sigma}$ from local equilibrium induces larger (in absolute magnitude) response $\delta n_{lm\sigma}$. In particular, a multipolar fluctuation with angular momentum $(l, m)$ is divergent if the condition $\delta\bar{n}_{lm\sigma}/\delta n_{lm\sigma} = 0$ is met. Namely, finite $\delta n_{lm\sigma}$ is generated spontaneously even by negligible $\delta\bar{n}_{lm\sigma}$, which means negligible external field. Hence for the electric multipole, the stability of the normal Fermi liquid requires $F_l > -(2l+1)$. Similarly for the magnetic multipole, the stability requires $Z_l > -(2l+1)$.

In actual liquid helium $^3$He, the Landau parameter $F_0$ becomes as large as 10 to 100 near the solidification by pressure. This means that the system becomes hard against compression. On the other hand, an estimate $Z_0 \sim -0.75$ is made against the magnetic response. This means that the system is near the ferromagnetic instability, and the response is four times larger than that of independent quasi-particles [7].

**Problem 4.2**
The magnetization corresponds to the homogeneous component $\delta\bar{n}_{00-}$ in Eq. (4.26). Hence we obtain $\delta\bar{n}_{00-} = (1 + Z_0)\,\delta n_{00-}$. Since $\delta\bar{n}_{00-}/B$ is proportional to the density of states $\rho^*(\mu)$ of quasi-particles, we obtain Eq. (4.19) as representing $\delta n_{00-}/B$. By similar argument, we obtain the compressibility of the Fermi liquid as

$$\chi_{00+} = \frac{\rho^*(\mu)}{1 + F_0}. \tag{4.27}$$

**Problem 4.3**
With use of $\Pi_{\boldsymbol{p}}$, Eq. (4.20) is symbolically written as $\delta n_{\boldsymbol{p}} = -\Pi_{\boldsymbol{p}}\varphi_{\boldsymbol{p}}$. According to Eq. (4.21), $\varphi_{\boldsymbol{p}}$ in the RHS is modified to add $f(\boldsymbol{p}, \boldsymbol{k})$, which is then moved to the LHS. Then we obtain Eq. (4.22).

**Problem 4.4**
Equation (4.23) is written as $1/F_0 = f(s)$, in terms of the function $f(s)$ defined by

$$f(s) = (s/2)\ln[(s+1)/(s-1)] - 1. \tag{4.28}$$

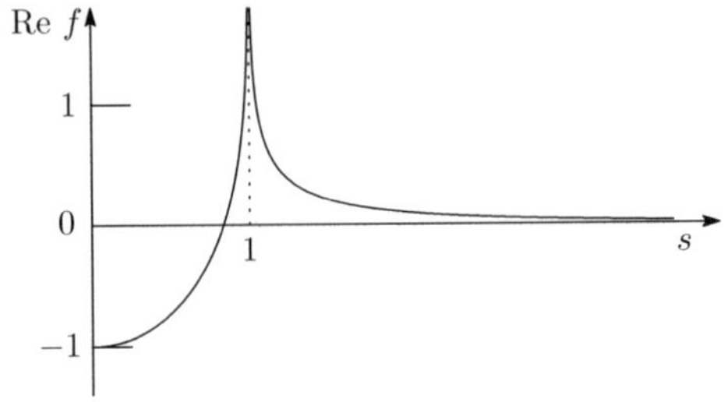


**Fig. 4.2** The real part of the function $f(s)$

Figure 4.2 shows the real part of $f(s)$. The imaginary part $\mathrm{Im} f(s) = \pi s/2$, which is not shown in the figure, is obviously finite for $0 < s < 1$. If we have $F_0 < -1$, the Fermi liquid becomes unstable because the compressibility becomes negative as shown in Eq. (4.27).

As evident from the figure, two real solutions for $s$ exist for $F_0 > 0$. The solution with $s > 1$ corresponds to larger energy than the upper limit of particle-hole excitation continuum, resulting in infinite lifetime. This collective excitation is called the zero sound. In the limit of large $F_0 \gg 1$, we may expand the logarithm assuming $s \gg 1$. Then we obtain $s \sim \sqrt{F_0/3}$. In this way the velocity of the zero sound increases with increasing $F_0$. As $F_0$ decreases to $F_0 \to 0$, the solution corresponds to the infinitely large logarithm. This gives $s \to 1$, which connects smoothly to individual quasi-particle excitations.

On the other hand, the solution for $-1 < F_0 < 0$ corresponds to complex $s$. This means a collective mode in the continuum of individual particle-hole excitations. The resultant finite lifetime is called the Landau damping. It is impossible to derive the solution from Fig. 4.2. Finally, with $F_0 \to -1$, the solution corresponds to $s \to 0$. This means that the velocity of the collective mode becomes zero just before the collapse against pressure takes place.

## References

1. Landau, L.D.: Soviet Phys. JETP **3**, 920 (1957)
2. Pines, D., Nozières, P.: The Theory of Quantum Liquids: Normal Fermi Liquids. Perseus Books Group, New York, (1994)
3. Baym, G., Pethick, C.: Landau Fermi-Liquid Theory: Concepts and Applications. Wiley, New York (2008)
4. Balian, R., de Dominicis, C.: Ann. Phys. **62**, 229 (1971)
5. Landau, L.D. Soviet Phys. JETP **8**, 70 (1959)
6. Noziéreres, P.: Theory of Interacting Fermi Systems. CRC Press, Boca Raton (1997)
7. Vollhardt, D., Wölfle, P.: The Superfluid Phases of Helium 3. Dover, New York (2013)

# Chapter 5
# Superconductivity

**Abstract** Superconductivity derives from a kind of two-electron bound state called the Cooper pair. Many Cooper pairs are multiply overlapping with one another, in contrast with an ordinary bound state such as hydrogen molecule. Then at any spatial point one cannot pinpoint a pair. This feature is related to the tiny binding energy as compared with the Fermi energy. This chapter explains basic mechanisms leading to formation of the Cooper pair. The discussion includes some recent development, concerning new types of superconductors with high transition temperatures.

## 5.1 Breakdown of Gauge Symmetry

We shall deal with Cooper pairs which are regarded as coherent superposition of many bound states with a definite quantum mechanical phase. The phase is a variable conjugate to the number of particles in the system, which is usually conserved by the gauge symmetry. Hence the definite phase means spontaneous breakdown of the gauge symmetry, accompanied by a phase transition. In the superconducting state, the number of electrons, and hence the number of Cooper pairs, are fluctuating. On the other hand, in the hypothetical strong-coupling limit where the binding energy is larger than the Fermi energy, the Cooper pair goes over to an ordinary molecule made up of two electrons. These molecules do not necessarily realize superconductivity, unless the gauge symmetry is broken spontaneously. Namely, presence or absence of the coherent phase separates the Cooper pair from the two-electron molecule. For simple understanding of the gauge symmetry, we start from the coherent states in the harmonic oscillator and then in Bose particles. It is straightforward to construct the Cooper pair with these preliminaries.

The Hamiltonian of a harmonic oscillator is given by

$$H = \omega \left( a^\dagger a + \frac{1}{2} \right), \tag{5.1}$$



Y. Kuramoto, *Quantum Many-Body Physics*, Lecture Notes in Physics 934,
https://doi.org/10.1007/978-4-431-55393-9_5

where $a^\dagger$ is the creation operator which changes an eigenstate $|n\rangle$ to another one as

$$a^\dagger|n\rangle = \sqrt{n+1}|n+1\rangle, \tag{5.2}$$

with $n = 0, 1, 2, \ldots$ and $|0\rangle$ is the ground state. Hence the operators $a, a^\dagger$ obey the bosonic commutation rule $[a, a^\dagger] = 1$. In terms of arbitrary complex number $z$, a coherent state $|z)$ is constructed by superposition of $|n\rangle$ as

$$|z) = e^{-|z|^2/2} \sum_{n=0}^{\infty} \frac{z^n}{\sqrt{n!}}|n\rangle = e^{-|z|^2/2} \exp\left(za^\dagger\right)|0\rangle, \tag{5.3}$$

with the normalization property $(z|z) = 1$. The coherent state is an eigenstate of $a$ such that

$$a|z) = z|z), \tag{5.4}$$

as will be demonstrated in Eq. (7.11). This leads to the property $(z|a|z) = z$.

We proceed to the system of identical Bose particles. Let $a_0^\dagger, a_0$ represent the creation and annihilation operators for the wave number $\boldsymbol{k} = 0$. On the analogy of harmonic oscillators, superposition of states with different numbers of particles having $\boldsymbol{k} = 0$ gives a coherent state $|z)$. Specifying the occupation numbers with different wave vectors by $\{n_{\boldsymbol{k}\neq 0}\}$, the state $\Psi$ for the whole system is given by

$$\Psi = |z) \otimes |\{n_{\boldsymbol{k}\neq 0}\}\rangle = e^{-|z|^2/2} \exp\left(za_0^\dagger\right)|0, \{n_{\boldsymbol{k}\neq 0}\}\rangle, \tag{5.5}$$

where the details of $\{n_{\boldsymbol{k}\neq 0}\}$ are not relevant. By choosing $z = \sqrt{N}\exp(\mathrm{i}\theta)$, we obtain

$$\langle\Psi|a_0|\Psi\rangle = \sqrt{N}\exp(\mathrm{i}\theta), \tag{5.6}$$

$$\langle\Psi|a_0^\dagger a_0|\Psi\rangle = N, \tag{5.7}$$

which show that the phase of $\Psi$ is fixed at $\theta$, and that the expectation value of the particles with wave number 0 is given by $N$. The case with $N \gg 1$ is referred to the Bose condensed state. The state with wave number 0 has the lowest kinetic energy, and macroscopic number of Bose particles can occupy this state. The condensation does not need interactions between bosons, and occurs even for ideal bosons.

In the Hamiltonian, ordinary terms such as kinetic energy and two-body interaction do not change the total number of particles in the system. Namely, the number of particles is a conserved quantity, and the relevant symmetry is the gauge symmetry. The simplest example of the gauge invariance is to take an arbitrary state $i$ and perform the unitary transformation

$$a_i \to a_i \exp(\mathrm{i}\phi),\ a_i^\dagger \to a_i^\dagger \exp(-\mathrm{i}\phi).$$

The change of phase leaves the operators such as $a_i^\dagger a_j$ and $a_i^\dagger a_j a_l^\dagger a_m$ remain the same, provided $\phi$ is common to all states $i$. On the other hand, the state $\Psi$ constructed by Eq. (5.5) is not invariant under the phase change, since different number of particles are contributing to the state, with $N$ being their average. Hence a state specified by the phase $\phi$ as in Eq. (5.6) breaks the gauge symmetry. If the ground state corresponds to such state, the gauge symmetry in the Hamiltonian is not reflected in the state. This is an example of the spontaneous symmetry breaking.

Suppose that a Bose condensed state, regarded as a liquid, is moving as a whole with a velocity smaller than the minimum velocity $v$ of excitations in the liquid. Then the moving liquid cannot slow down while conserving the total energy and momentum, and becomes a superfluid. This was the original argument by Landau [1] for the superfluidity. In the case of free bosons with finite mass, the kinetic energy is proportional to $q^2$, and the minimum velocity goes to zero as $q \to 0$. In the presence of repulsive interaction between the bosons, on the other hand, the lowest velocity becomes finite because a collective excitation called the phonon dominates the lowest excitations. It can be shown that the excitation energy $\omega$ starts as $\omega = v|\boldsymbol{q}|$ for small wave vector $\boldsymbol{q}$ [2].

According to the Bardeen–Cooper–Schrieffer (BCS) theory [3], the superconducting ground state is a coherent state of electrons. Since each electron obeys the Fermi statistics, the coherent state is formed by bound electron pairs which behave as a kind of bosons. The explicit form of the BCS state is given in terms of real parameters $u_{\boldsymbol{k}}$, $v_{\boldsymbol{k}}$, and $\phi$ as

$$\Psi_{\rm BCS} = \prod_{\boldsymbol{k}} \left[ u_{\boldsymbol{k}} + v_{\boldsymbol{k}} e^{i\phi} c_\uparrow^\dagger(\boldsymbol{k}) c_\downarrow^\dagger(-\boldsymbol{k}) \right] |0\rangle, \tag{5.8}$$

where $c_\uparrow^\dagger(\boldsymbol{k})$ $(\equiv c_{\boldsymbol{k}\uparrow}^\dagger)$ represents the electron creation operator, and $|0\rangle$ is the vacuum with no electrons. The angle $\phi$ is independent of $\boldsymbol{k}$, and corresponds to a phase of the many-electron coherent state. Derivation of the norm of $\Psi_{\rm BCS}$ is the subject of Problem 5.1. As shown in Eq. (5.94), the normalization requires $u_{\boldsymbol{k}}^2 + v_{\boldsymbol{k}}^2 = 1$. As a special case, the ground state of the free fermions is described by taking

$$u_{\boldsymbol{k}} = \theta(|\boldsymbol{k}| - k_F), \quad v_{\boldsymbol{k}} = \theta(k_F - |\boldsymbol{k}|), \tag{5.9}$$

with $u_{\boldsymbol{k}} v_{\boldsymbol{k}} = 0$. In this case, the total number is fixed and $\phi$ is irrelevant.

In the case of $u_{\boldsymbol{k}} v_{\boldsymbol{k}} \neq 0$, on the other hand, the total number of electrons is fluctuating. Namely, we obtain by straightforward calculation

$$\langle \Psi_{\rm BCS} | c_\downarrow(-\boldsymbol{k}) c_\uparrow(\boldsymbol{k}) | \Psi_{\rm BCS} \rangle = u_{\boldsymbol{k}} v_{\boldsymbol{k}} e^{i\phi}, \tag{5.10}$$

$$\langle \Psi_{\rm BCS} | c_\uparrow(\boldsymbol{k})^\dagger c_\uparrow(\boldsymbol{k}) | \Psi_{\rm BCS} \rangle = \langle \Psi_{\rm BCS} | c_\downarrow(-\boldsymbol{k})^\dagger c_\downarrow(-\boldsymbol{k}) | \Psi_{\rm BCS} \rangle = v_{\boldsymbol{k}}^2. \tag{5.11}$$

Equation (5.10) shows that $u_{\boldsymbol{k}} v_{\boldsymbol{k}} \neq 0$ is necessary for the coherent state. For each $(\boldsymbol{k}, \sigma)$, $\Psi_{\rm BCS}$ is a superposition of occupied and empty states, with the weight $v_{\boldsymbol{k}}^2$

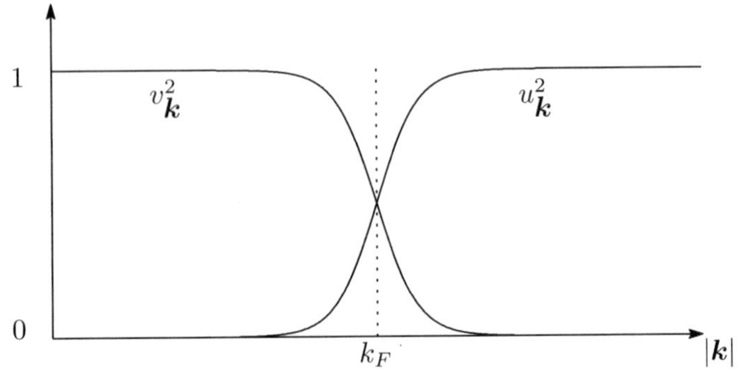


**Fig. 5.1** Illustration of the parameters $u_k^2$ and $v_k^2$ corresponding to the BCS state of electrons

for the occupied state. As long as $u_k v_k \neq 0$, the average occupation number $v_k^2$ is neither 0 nor 1. The kinetic energy increases by the superposition since the momentum distribution deviates from the step function that gives the minimum kinetic energy. If the interaction energy gains more by the superposition, however, the bound pair of electrons is favored. The origin of such interaction energy will be discussed in the next section. Figure 5.1 illustrates an example of $u_k^2$ and $v_k^2$ in the BCS state. The momentum range of deviation from the step function becomes narrower as the size of the pair wave function becomes larger. The corresponding energy gain gives a measure of transition temperature, which will be quantified later in Sect. 5.3.

## 5.2 Attractive Interaction Mediated by Phonons

Since electrons are negatively charged, they feel the mutual Coulomb repulsion. Then it may appear unlikely that two electrons make a pair. There are several mechanisms to overcome the repulsion, leading finally to effective attraction. In the BCS theory, the origin of the attraction is ascribed to the electron–phonon interaction, which is the most important and well-established mechanism. Inspired by the success of the BCS theory, other mechanisms have been searched for, and prove to be effective in superfluid $^3$He with strong hard-core repulsion among $^3$He atoms. We shall discuss an alternative mechanism from Sect. 5.4, and especially in Sect. 5.9 in relation to the high-temperature superconductivity in Cu oxides and Fe pnictides/chalcogenides. Here pnictides mean a family of compounds including group V elements such as As, and chalcogenides mean those including group VI elements such as Se.

Let us first explain the phonon mechanism of attraction. We use the effective Hamiltonian formalism, which has been discussed in Chap. 1. Because of the screening effect in metals, the electron–phonon interaction can be regarded as a local interaction. At position $\boldsymbol{r}$, the interaction is given by $C\rho(\boldsymbol{r})\nabla \cdot \boldsymbol{u}(\boldsymbol{r})$ where $C$ is the coupling constant, $\boldsymbol{u}(\boldsymbol{r})$ is the lattice displacement, and $\rho(\boldsymbol{r})$ the density of electrons. Here we have considered only the longitudinal phonons for simplicity, which emerge by quantization of $\boldsymbol{u}(\boldsymbol{r})$. The quantized form is given by Eq. (7.6) in Chap. 7. In terms of creation and annihilation operators $b_{\boldsymbol{q}}^\dagger, b_{-\boldsymbol{q}}$ of phonons, and $c_\sigma^\dagger(\boldsymbol{k}), c_\sigma(\boldsymbol{k}+\boldsymbol{q})$ of electrons which compose the density fluctuation, we obtain the interaction Hamiltonian

$$H' = \frac{1}{\sqrt{N}} \sum_{\boldsymbol{q}} \alpha_q \left( b_{\boldsymbol{q}}^\dagger + b_{-\boldsymbol{q}} \right) \sum_{\boldsymbol{k}\sigma} c_\sigma^\dagger(\boldsymbol{k}) c_\sigma(\boldsymbol{k}+\boldsymbol{q}), \tag{5.12}$$

where $N$ is the number of lattice sites in the system and

$$\alpha_q = Cq/\sqrt{2M\omega_q} \tag{5.13}$$

is the interaction coefficient with $q = |\boldsymbol{q}|$. Here $M$ is the mass of the unit cell of the lattice, and $\omega_q$ is the frequency of the phonon, which is assumed to be isotropic. In more general cases including transverse or optical phonons, the interaction $\alpha_q$ should be modified. However, the form Eq. (5.12) remains the same.

As the model space for the effective Hamiltonian, we take the one where the electron–phonon interaction is absent. By proper renormalization procedure, the low-energy levels for the original system are reproduced by inclusion of an emergent effective interaction between electrons, which we now derive. The unperturbed Hamiltonian is taken as $H_0 = H_\mathrm{e} + H_\mathrm{ph}$ where $H_\mathrm{e}$ and $H_\mathrm{ph}$ describe the non-interacting electrons and phonons, respectively. According to the general framework in Chap. 1, the effective Hamiltonian is constructed as

$$H_\mathrm{eff} = PH_0P + PH' \frac{1}{E_0 - H_0} QH'P + O(H'^3), \tag{5.14}$$

where $P$ is the projection operator of our model space, and $E_0$ is the ground state energy of $H_0$. We consider the weak-coupling case where terms of $O(H'^3)$ are safely neglected. The second term in the RHS represents the effective interaction $V_\mathrm{eff}$, where the denominator is the difference of energies between ground and intermediate excited states. With simplified notation of the indices, we obtain

$$V_\mathrm{eff} = \frac{1}{2N} \sum_{1,2,3,4} \delta(1+3, 2+4) |\alpha_q|^2 \left( \frac{1}{\epsilon_2 - \epsilon_1 - \omega_q} + \frac{1}{\epsilon_4 - \epsilon_3 - \omega_q} \right) c_1^\dagger c_2 c_3^\dagger c_4, \tag{5.15}$$

which is symmetric with respect to $(1, 2)$ and $(3, 4)$. The delta function represents conservation of momentum and spin of the electron pair.

Let us consider an electron pair with $(2, 4) \sim (\boldsymbol{k} \uparrow, -\boldsymbol{k} \downarrow)$ near the Fermi surface which is scattered into $(1, 3) \sim (\boldsymbol{p} \uparrow, -\boldsymbol{p} \downarrow)$, which involves the momentum transfer $\boldsymbol{q} = \boldsymbol{p} - \boldsymbol{k}$. If the electron momenta are all at the Fermi surface, the positive energy $\omega_q$ gives rise to the attractive (negative) effective interaction. The negative sign corresponds to the general consequence of second-order perturbation theory. The situation remains qualitatively the same as long as the electronic energy is close enough to the Fermi level with deviation smaller than $\omega_q$. Hence in Eq. (5.15), we keep only electron pairs each of which has the zero total momentum, and put all the individual energy $\epsilon_i = 0$ in the energy denominator. Then we obtain the approximate electron–electron interaction as

$$V_{\text{ee}} = -\frac{1}{N}\sum_{1\sim 4}{}' V(1-2)c_3^\dagger c_1^\dagger c_2 c_4$$
$$= -\frac{1}{N}\sum_{\boldsymbol{k},\boldsymbol{p}}{}' V(\boldsymbol{p}-\boldsymbol{k})c_\downarrow^\dagger(-\boldsymbol{p})c_\uparrow^\dagger(\boldsymbol{p})c_\uparrow(\boldsymbol{k})c_\downarrow(-\boldsymbol{k}), \tag{5.16}$$

where $V(\boldsymbol{q}) = |\alpha_q|^2/\omega_q$, and the prime (′) in summation represents the restriction that the energy associated with $\boldsymbol{k}$ and $\boldsymbol{p}$ should be smaller than the Debye frequency $\omega_{\text{D}}$, which is of the order of average of $\omega_q$. In the following, we omit writing the prime in the summation by modifying the definition of $V(\boldsymbol{p}-\boldsymbol{k})$ so that it vanishes out of the restricted range.

## 5.3 Mean Field Approximation

The mean field approximation becomes more accurate as the number $N_c$ of particles forming the mean field becomes larger, since the fluctuation relative to the mean field is proportional to $1/\sqrt{N_c}$. The BCS superconductor is the best example of the accurate case with $N_c \gg 1$, which will be demonstrated at the end of this section. The mean field by these Cooper pairs is called the pairing field. The mean field approximation consists of the following replacement in Eq. (5.16):

$$c_3^\dagger c_1^\dagger c_2 c_4 \to \langle c_3^\dagger c_1^\dagger\rangle c_2 c_4 + c_3^\dagger c_1^\dagger \langle c_2 c_4\rangle - \langle c_3^\dagger c_1^\dagger\rangle\langle c_2 c_4\rangle. \tag{5.17}$$

The bracketed quantities represent the pairing field, which is zero if the gauge symmetry is preserved. In other words, the finite average requires the coherent state such as the BCS state in Eq. (5.10). On the other hand, we neglect terms such as $\langle c_1^\dagger c_2\rangle$ which play the major role in the Hartree–Fock approximation in Sect. 2.2. Then $V_{\text{ee}}$ in Eq. (5.16) is modified in the mean field approximation as

$$V_{\text{BCS}} = \sum_{\boldsymbol{k}}\Big[\Delta(\boldsymbol{k})^* c_\downarrow(-\boldsymbol{k})c_\uparrow(\boldsymbol{k}) + \Delta(\boldsymbol{k})c_\uparrow^\dagger(\boldsymbol{k})c_\downarrow^\dagger(-\boldsymbol{k}) - \Delta(\boldsymbol{k})\Big\langle c_\uparrow^\dagger(\boldsymbol{k})c_\downarrow^\dagger(-\boldsymbol{k})\Big\rangle\Big], \tag{5.18}$$

where the gap function $\Delta(\boldsymbol{k})$ is defined by

$$\Delta(\boldsymbol{k}) = \frac{1}{N}\sum_{\boldsymbol{p}} V(\boldsymbol{k}-\boldsymbol{p})\langle c_{\uparrow}(\boldsymbol{p})c_{\downarrow}(-\boldsymbol{p})\rangle. \tag{5.19}$$

Together with the Hamiltonian for the kinetic energy

$$H_{\mathrm{e}} = \sum_{\boldsymbol{k}\sigma} \epsilon_{\boldsymbol{k}} c_{\sigma}^{\dagger}(\boldsymbol{k}) c_{\sigma}(\boldsymbol{k}), \tag{5.20}$$

the BCS Hamiltonian is given by $H_{\mathrm{BCS}} = H_{\mathrm{e}} + V_{\mathrm{BCS}}$. Since each momentum pair $(\boldsymbol{k}, -\boldsymbol{k})$ is independent of other pairs, $H_{\mathrm{BCS}}$ can be diagonalized for each momentum pair. The diagonalization can be performed conveniently in terms of the two-component field $\boldsymbol{\psi}(\boldsymbol{k})$ defined by

$$\boldsymbol{\psi}(\boldsymbol{k}) \equiv \left(c_{\uparrow}(\boldsymbol{k}), c_{\downarrow}^{\dagger}(-\boldsymbol{k})\right)^{T} \equiv \left(c_{\uparrow}(\boldsymbol{k}), h_{\uparrow}(\boldsymbol{k})\right)^{T}, \tag{5.21}$$

where the upper index $T$ denotes the transpose of a vector or matrix, and $h_{\uparrow}(\boldsymbol{k})$ represents the annihilation operator of a hole. Each component in $\boldsymbol{\psi}$ obeys the fermionic anticommutation relation. Since the kinetic energy of the hole is given by $-\epsilon_{-\boldsymbol{k}}$, we obtain the compact expression[1]

$$H_{\mathrm{BCS}} = \sum_{\boldsymbol{k}} \boldsymbol{\psi}^{\dagger}(\boldsymbol{k}) \begin{pmatrix} \epsilon_{\boldsymbol{k}} & \Delta(\boldsymbol{k}) \\ \Delta(\boldsymbol{k})^{*} & -\epsilon_{-\boldsymbol{k}} \end{pmatrix} \boldsymbol{\psi}(\boldsymbol{k}) \equiv \sum_{\boldsymbol{k}} \boldsymbol{\psi}^{\dagger}(\boldsymbol{k}) h_{\mathrm{BCS}}(\boldsymbol{k}) \boldsymbol{\psi}(\boldsymbol{k}). \tag{5.22}$$

Diagonalization of $h_{\mathrm{BCS}}(\boldsymbol{k})$ in Eq. (5.22) is the subject of Problem 5.2. According to Eq. (5.101), $h_{\mathrm{BCS}}$ becomes diagonal in terms of the new two-component field

$$\tilde{\boldsymbol{\psi}}(\boldsymbol{k}) = (\alpha(\boldsymbol{k}), \beta(\boldsymbol{k}))^{T}$$

which is related to $\boldsymbol{\psi}(\boldsymbol{k})$ by the unitary transformation

$$\tilde{\boldsymbol{\psi}}(\boldsymbol{k}) = \begin{pmatrix} u_{\boldsymbol{k}} & -e^{-i\phi} v_{\boldsymbol{k}} \\ e^{i\phi} v_{\boldsymbol{k}} & u_{\boldsymbol{k}} \end{pmatrix} \boldsymbol{\psi}(\boldsymbol{k}). \tag{5.23}$$

[1] The two-component operator $\boldsymbol{\psi}(\boldsymbol{k})$ annihilates either an electron or a hole with spin up, the latter of which means creation of a spin-down electron. On the other hand, the hole creation operator in $\boldsymbol{\psi}^{\dagger}(\boldsymbol{k})$ actually annihilates an electron with spin down. Hence bilinear combinations of $\boldsymbol{\psi}^{\dagger}(\boldsymbol{k})$ and $\boldsymbol{\psi}(\boldsymbol{k})$ as in Eq. (5.22) are sufficient for describing the singlet pairing with up and down spins. For other types of pairing, however, the two-component operator is not sufficient. A convenient device for a general case is the four-component field to be defined by Eq. (5.62).

The eigenvalues of $h_{\mathrm{BCS}}$ are given by $E_{\pm}(\boldsymbol{k}) = \pm\sqrt{\epsilon_{\boldsymbol{k}}^2 + |\Delta(\boldsymbol{k})|^2}$ for $\alpha(\boldsymbol{k})$ with the $+$ sign, and $\beta(\boldsymbol{k})$ for the $-$ sign.

The operators $\alpha(\boldsymbol{k}), \beta(\boldsymbol{k})$ represent annihilation of quasi-particles, which are superposition of electrons and holes. The transformation (5.23) is called the Bogoliubov transformation. Note that the quasi-particle here has a different meaning from that in Fermi liquids, and is sometimes referred to as the Bogoliubov quasi-particle.

At zero temperature, all $\beta$ states are occupied and all $\alpha$ states are empty. It is possible to reconstruct $\Psi_{\mathrm{BCS}}$ by successive operation of $\beta(\boldsymbol{k})^{\dagger}\alpha(\boldsymbol{k})$ to vacuum. Explicit construction is the subject of Problem 5.3. At finite temperature, each quasi-particle with energy $E$ populates according to the Fermi distribution function $f(E)$. Namely, we obtain

$$\langle \alpha^{\dagger}(\boldsymbol{k})\alpha(\boldsymbol{k})\rangle = f(E_{+}(\boldsymbol{k})), \tag{5.24}$$

$$\langle \beta^{\dagger}(\boldsymbol{k})\beta(\boldsymbol{k})\rangle = f(E_{-}(\boldsymbol{k})) = 1 - f(E_{+}(\boldsymbol{k})). \tag{5.25}$$

These populations determine the pairing amplitude as

$$\langle c_{\uparrow}(\boldsymbol{p})c_{\downarrow}(-\boldsymbol{p})\rangle = \frac{\Delta(\boldsymbol{p})}{2E_{+}(\boldsymbol{p})}\left[\langle \beta^{\dagger}(\boldsymbol{p})\beta(\boldsymbol{p})\rangle - \langle \alpha^{\dagger}(\boldsymbol{p})\alpha(\boldsymbol{p})\rangle\right]. \tag{5.26}$$

Using the relation $1 - 2f(E) = \tanh(\frac{1}{2}E/T)$ between the Fermi distribution function and the hyperbolic function, we obtain from Eqs. (5.19) and (5.26)

$$\Delta(\boldsymbol{k}) = \frac{1}{N}\sum_{\boldsymbol{p}} V(\boldsymbol{k}-\boldsymbol{p})\frac{\Delta(\boldsymbol{p})}{2E_{+}(\boldsymbol{p})}\tanh\frac{E_{+}(\boldsymbol{p})}{2T}, \tag{5.27}$$

which is called the gap equation. The superconducting state has a self-consistent solution with $\Delta(\boldsymbol{p}) \neq 0$.

The non-linear integral equation (5.27) can be solved only numerically in general. However, in the special cases at $T = 0$ and at the transition temperature $T = T_c$, it can be solved analytically. As temperature increases, $|\Delta(\boldsymbol{p})|$ becomes smaller and eventually vanishes at $T_c$. Since we have $E_{+}(\boldsymbol{p}) \to |\epsilon(\boldsymbol{p})|$ at $T_c$, the gap equation becomes a linear integral equation for $\Delta(\boldsymbol{k})$. In terms of the function

$$\Pi(\boldsymbol{p}) = \frac{1}{2\epsilon(\boldsymbol{p})}\tanh\frac{\epsilon(\boldsymbol{p})}{2T}, \tag{5.28}$$

the gap equation (5.27) reduces to

$$\Delta(\boldsymbol{k}) = \frac{1}{N}\sum_{\boldsymbol{p}} V(\boldsymbol{k}-\boldsymbol{p})\Pi(\boldsymbol{p})\Delta(\boldsymbol{p}). \tag{5.29}$$

At $T = T_c$, there is a nontrivial solution for $\Delta(\boldsymbol{k})$, provided $T_c \neq 0$. Note that $\Delta(\boldsymbol{k})$ is infinitesimal at $T_c$, which is not explicit in Eq. (5.29).

Let us focus on the special solution corresponding to the $s$-wave pair. Then $\Delta(\boldsymbol{p})$ is independent of the direction $\hat{p}$, and the summation over $\boldsymbol{p}$ in Eq. (5.29) is converted to the energy integration with use of the density of states $\rho(\epsilon)$. If the density of states varies only slightly within the range of $\pm\omega_{\mathrm{D}}$ around the Fermi level, one may use the constant $\rho(\mu)$ in the energy integration. The $|\boldsymbol{p}|$-dependence in $\Delta(\boldsymbol{p})$ can also be neglected in this energy range. Let us first derive the density of states $\rho_{\mathrm{BCS}}(E)$ of quasi-particles. With $E_{\pm}(\boldsymbol{k}) = \pm\sqrt{\epsilon_{\boldsymbol{k}}^2 + \Delta^2}$ we obtain

$$\rho_{\mathrm{BCS}}(E) = \frac{1}{2}\sum_{\boldsymbol{k},\pm}\delta(E - E_{\pm}(\boldsymbol{k})) = \rho(\mu)\frac{|E|}{\sqrt{E^2 - \Delta^2}}, \tag{5.30}$$

for $|E| > \Delta$, and zero otherwise. Here $\Delta = \Delta(T)$ is the energy gap at temperature $T$. Note that $\rho_{\mathrm{BCS}}(E)$ is divergent at $E = \pm\Delta$, and tends to $\rho(\mu)$ as $|E| \gg \Delta$.

Next we derive the BCS formula for $T_c$. Discarding the common factor $\Delta$ on both sides of Eq. (5.29), we obtain a transcendental equation for $T_c$. With the dimensionless variable $x = \epsilon/(2T_c)$, the integral in question is evaluated as

$$\begin{aligned}\int_0^w \mathrm{d}x\frac{\tanh x}{x} &= \ln x \tanh x\big|_0^w - \int_0^w \mathrm{d}x \ln x (\tanh x)' \\ &\sim \ln w + \ln\frac{4}{\pi} + \gamma,\end{aligned} \tag{5.31}$$

with $w = \omega_{\mathrm{D}}/(2T_c)$ and $\gamma \sim 0.577$ being the Euler's constant. In the first line of the RHS, the upper limit $w$ of the integral including $(\tanh x)'$ can be safely replaced by $\infty$ provided $w \gg 1$. The resultant error is exponentially small. With Eq. (5.31) put into the gap equation, a little algebra gives

$$T_c \simeq 1.13\omega_{\mathrm{D}}\exp\left(-g^{-1}\right), \tag{5.32}$$

which is one of the most famous results in the BCS theory [3]. Here the dimensionless constant $g$ is given by $g = \langle V(\boldsymbol{k}-\boldsymbol{p})\rangle\rho(\mu)$ with $\langle\cdots\rangle$ being the angular average over $\boldsymbol{k}$ and $\boldsymbol{p}$ on the Fermi surface. Since the dependence on the coupling constant $g$ in Eq. (5.32) is singular around $g = 0$, the BCS state cannot be obtained by perturbation theory in $g$. In other words, the BCS theory accounts for the collective bound state of electrons. The bound state is possible only for attractive interaction, which corresponds to positive sign of $g$. As is clear from the derivation, Eq. (5.32) is accurate only for the weak-coupling case with $T_c/\omega_{\mathrm{D}} \ll 1$.

Equation (5.27) at $T = 0$ can also be solved for the $s$-wave pair. Discarding $\Delta_0 \equiv \Delta(T = 0)$ on both sides and using $\tanh(E_+/2T) \to 1$, we are left with the implicit equation

$$1 = g \int_{-\omega_D}^{\omega_D} d\epsilon \frac{1}{2\sqrt{\epsilon^2 + \Delta_0^2}} = g \ln \left[ x + \sqrt{x^2 + 1} \right] \Big|_0^{x_D}, \tag{5.33}$$

for $x_D = \omega_D/\Delta_0$. Assuming the weak-coupling condition $x_D \gg 1$, we obtain another famous result of the BCS theory:

$$\Delta_0 = 2\omega_D \exp(-g^{-1}) \simeq 1.76 T_c. \tag{5.34}$$

We can now estimate the number $N_c$ composing the mean field as $N_c \sim n\xi^3$ with $n$ being the electron density and $\xi$ the coherence length. The latter is estimated as $\xi/a \sim \varepsilon_F/\Delta_0$ in terms of the lattice constant $a$ and the Fermi energy $\varepsilon_F$. Since $na^3 \sim O(1)$, we obtain $N_c \gg 1$ in the weak-coupling case $\xi/a \gg 1$.

## 5.4 Multiband Model

We now proceed to discuss other mechanisms of effective attraction than mediated by phonons. Such study was first motivated by high transition temperatures in some transition metals even though the electronic Coulomb repulsion is large. As a possible origin for high transition temperature, the interband transfer of Cooper pairs has been considered [4]. Namely, in terms of creation and annihilation operators for bands A and B, the interaction is given by

$$H_{AB} = \frac{J_H}{N} \sum_{\boldsymbol{k},\boldsymbol{p}} A_\uparrow^\dagger(\boldsymbol{k}) A_\downarrow^\dagger(-\boldsymbol{k}) B_\downarrow(-\boldsymbol{p}) B_\uparrow(\boldsymbol{p}) + \text{h.c.}, \tag{5.35}$$

where the interaction constant $J_H$ is related to the exchange coupling due to the Hund's rule, as discussed in Sect. 1.4. Namely, the same constant $J_H > 0$ appears in

$$H_{\text{Hund}} = -2J_H \sum_i \boldsymbol{S}_A(i) \cdot \boldsymbol{S}_B(i). \tag{5.36}$$

In order to understand the relation between $H_{AB}$ and $H_{\text{Hund}}$, we take the Wannier functions $w_A(\boldsymbol{r})$, $w_B(\boldsymbol{r}')$ localized at the origin, and represent the exchange integral as

$$J_H = \int d\boldsymbol{r} \int d\boldsymbol{r}' w_B(\boldsymbol{r})^* w_A(\boldsymbol{r}')^* \frac{e^2}{|\boldsymbol{r} - \boldsymbol{r}'|} w_B(\boldsymbol{r}') w_A(\boldsymbol{r}). \tag{5.37}$$

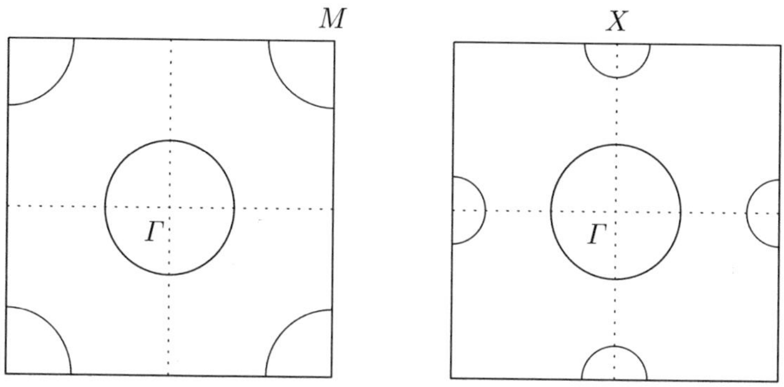


**Fig. 5.2** Examples of Fermi surfaces where the interband pair transfer can conserve the crystal momentum

Since the Wannier function can be taken real, we may write

$$w_{\mathrm{B}}(\boldsymbol{r})^* w_{\mathrm{A}}(\boldsymbol{r}) = w_{\mathrm{B}}(\boldsymbol{r}) w_{\mathrm{A}}(\boldsymbol{r})^*. \tag{5.38}$$

Then Eq. (5.37) assumes a form corresponding to the pair transfer from the band B to A.

The total momentum of each pair must be preserved in the interband transfer, which makes strong constraint on location of the Fermi surfaces. The transfer is possible if the bands A and B have the common center for the Fermi surfaces, or one of the Fermi surfaces has a center shifted from the other by half of the reciprocal lattice vector. In the latter case, the transfer involves a momentum corresponding to a reciprocal lattice vector, which is equivalent to zero as the crystal momentum. Figure 5.2 shows an example of the Brillouin zone for the square lattice. The right panel corresponds to Fe pnictides which have been intensively studied as a family showing high-temperature superconductivity [6, 7].

With band indices A, B written as $i, j$, the gap equation including multiple energy bands is given by

$$\Delta_i(\boldsymbol{k}) = \frac{1}{N}\sum_{\boldsymbol{p}}\sum_{j} V_{ij}(\boldsymbol{k}-\boldsymbol{p})\Pi_j(\boldsymbol{p})\Delta_j(\boldsymbol{p}), \tag{5.39}$$

where $V_{ij}(\boldsymbol{k}-\boldsymbol{p})$ describes the interband pair transfer for $i \neq j$, and corresponds to $J_{\mathrm{H}}$ in Eq. (5.35). For $i = j$, it reduces to the interaction within each band. We introduce the quantities $L_j$ and $g_i$ by

$$\frac{1}{N}\sum_{\boldsymbol{p}} \Pi_j(\boldsymbol{p}) = \rho_j L_j(T), \quad \rho_i V_{ii} \equiv g_i, \tag{5.40}$$

where $\rho_i$ is the density of states (per site) for each band A, B. Here we have neglected the momentum dependence in $V_{ii}$. In the case of two relevant bands, Eq. (5.39) takes the matrix form

$$\begin{pmatrix} 1 - g_A L_A & -J_H \rho_B L_B \\ -J_H \rho_A L_A & 1 - g_B L_B \end{pmatrix} \begin{pmatrix} \Delta_A \\ \Delta_B \end{pmatrix} = 0. \tag{5.41}$$

Nontrivial solution is possible if the determinant is zero:

$$(1 - g_A L_A)(1 - g_B L_B) - J_H^2 \rho_A \rho_B L_A L_B = 0. \tag{5.42}$$

With $J_H = 0$, the transition temperature $T_i$ for the band $i$ is determined by the condition $1 - g_i L_i(T_i) = 0$ with $g_i$ positive. Using the relation

$$L_i(T) \simeq \ln(\omega_D/T), \tag{5.43}$$

we obtain $T_i \sim \omega_D \exp(-1/g_i)$, which corresponds to Eq. (5.32) with the coefficient 1.13 approximated by 1. For finite $J_H$, the transition temperature is always larger than $T_A$ and $T_B$. The demonstration is the subject of Problem 5.4.

Let us take an extreme case where the interaction consists only of the pair transfer. Namely, $g_A = g_B = 0$. The superconductivity appears in this case independent of the sign of $J_H$. The transition temperature $T_c$ is determined from Eq. (5.42) by

$$1 - J_H^2 \rho_A \rho_B L_A L_B = 1 - J_H^2 \rho_A \rho_B \left[\ln(\omega_D/T_c)\right]^2 = 0. \tag{5.44}$$

The magnitude of $T_c$ is the same as a virtual single-band model with

$$g = |J_H| \sqrt{\rho_A \rho_B}$$

as the dimensionless coupling constant.

It is likely that the interband transfer plays an important role in realizing the high transition temperature ($T_c \sim 50\,\mathrm{K}$) in Fe pnictides. On the other hand, another aspect of the interband transfer seems important in $MgB_2$, which also has a high transition temperature ($T_c = 38\,\mathrm{K}$) [8]. Namely, the strongly covalent band B, originating from boron states, has a strong electron–phonon interaction, and is responsible for the high $T_c$ in $MgB_2$. The other conduction band A, which originates from Mg states, has another Fermi surface. Both Fermi surfaces A and B are centered at the $\Gamma$ point of the Brillouin zone, and the pair transfer is possible between the two bands. The A band alone does not have a high transition temperature, but the whole system becomes a high $T_c$ superconductor with the pair transfer. Because of rather different characters of A and B bands, superconductivity in $MgB_2$ is sensitive to a small amount of magnetic field and by finite temperature, which at first sight does not match the high transition temperature.

## 5.5 Renormalization of Coulomb Repulsion

The magnitude of the Coulomb repulsion between electrons is generally larger than that of the effective attraction mediated by phonons. Nevertheless, electron pairing can be realized because the energy range of their action is different from each other. As the simplest model of the Coulomb repulsion, we take the Hubbard type local interaction $U > 0$. In Eq. (5.39), the suffices $i, j$ are now interpreted as distinguishing the relevant energy range. Namely, instead of the band A, we take the range $\omega_D$ around the Fermi level. The other range up to the band edges is taken as B. The interaction in the range A is given by $g_A - U\rho_A$, while in the range B by $-U\rho_B$. As long as the temperature $T$ is low enough: $T \ll \omega_D$, we obtain

$$\frac{1}{N}\sum_{p} \Pi_B(p) = \rho_B L_B = \rho_B \ln \frac{D}{\omega_D}, \tag{5.45}$$

which is independent of $T$. Here the density of states per spin in the range B is approximated by the average $\rho_B$, and the half of the bandwidth is written as $D$. Figure 5.3 illustrates the energy dependence of the interactions.

The condition for $T_c$ is given, by analogy with Eq. (5.42) for the two-band model, as

$$\begin{aligned} 0 =& 1 - (g_A - U\rho_A)L_A - \frac{U^2\rho_A\rho_B L_A L_B}{1 + U\rho_B L_B} \\ =& 1 - \left(g_A - \frac{U\rho_A}{1 + U\rho_B L_B}\right) L_A \equiv 1 - g_{\text{eff}} L_A. \end{aligned} \tag{5.46}$$

The result implies that the effective Coulomb repulsion is renormalized to a smaller value. Namely, Eq. (5.46) shows

$$U \to U^* \equiv U/(1 + U\rho_B L_B) < U, \tag{5.47}$$

**Fig. 5.3** Energy dependence of the Coulomb repulsion and the phonon-mediated attraction

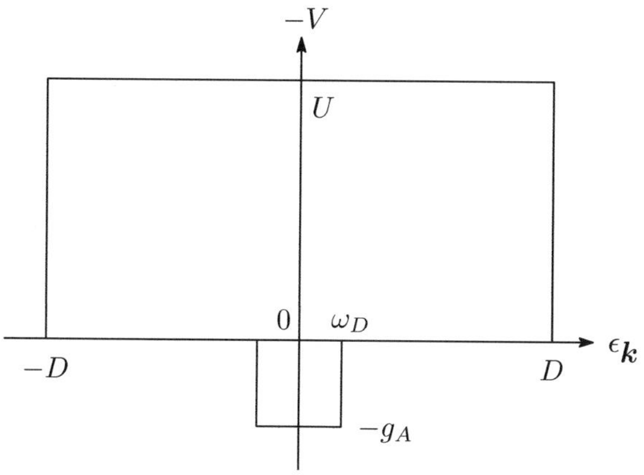

which gives much decreased $U^*$ in the case of $\omega_D \ll D$ with large $L_B$. Hence even though the bare Coulomb repulsion dominates over the attraction ($g_A < U\rho_A$), the effective interaction may become $g_A - U^*\rho_A > 0$ which realizes the pairing.

The reduction of the effective Coulomb repulsion is the typical example of the renormalization. Later in Chap. 6, we discuss the scaling which is one of the standard methods of renormalization. Problem 5.5 deals with derivation of Eq. (5.47) by the scaling method, which may better be appreciated after the reader finishes reading Chap. 6.

## 5.6 Isotope Effect

The transition temperature $T_c$ of a simple substance depends on its isotopes in a characteristic manner as given by

$$T_c \propto M^{-\alpha}, \tag{5.48}$$

where $M$ is the mass of the isotope. In the phonon-mediated pairing as in the BCS theory, the exponent is $\alpha = 1/2$ because of $\omega_D \propto M^{-1/2}$. This result is consistent with many metals including Hg and Zn, and gives strong support for the BCS theory. On the other hand, in transition metals with strong Coulomb repulsion, there are cases where $\alpha$ is much smaller than 1/2, or almost zero as in Ru [10].

This section discusses the isotope effect in the presence of the Coulomb repulsion. As $\omega_D$ decreases, $U^*$ also decreases according to Eq. (5.47). Then the effective attraction is enhanced, which competes with the ordinary isotope effect. Hence we expect $\alpha < 1/2$ in the presence of the Coulomb repulsion. To quantify the argument we take the logarithmic derivative of $T_c = \omega_D \exp(-1/g)$, and obtain

$$\delta T_c/T_c = \delta\omega_D/\omega_D + g^2\delta g. \tag{5.49}$$

In contrast with the original BCS theory with $\delta g = 0$, the situation for $g = g_{\text{eff}}$ in Eq. (5.46) is different. A little manipulation gives

$$\frac{\delta T_c}{T_c} = \frac{\delta\omega_D}{\omega_D}\left[1 - \left(\frac{U^*\rho_A}{g_A - U^*\rho_A}\right)^2 \frac{\rho_B}{\rho_A}\right], \tag{5.50}$$

which leads to the exponent in Eq. (5.48) as

$$\alpha = \frac{1}{2}\left[1 - \left(\frac{U^*\rho_A}{g_A - U^*\rho_A}\right)^2 \frac{\rho_B}{\rho_A}\right]. \tag{5.51}$$

Note that $\alpha$ is smaller than 1/2 in general, and can even be negative. Experimentally, negative isotope effect has been reported in $Sr_2RuO_4$ [11] for instance. It is important to recognize that the origin of attraction is phonons even if $\alpha$ deviates substantially from 1/2 as in Eq. (5.51). Thus the isotope effect alone is not sufficient to identify the origin of pairing.

We now mention an extreme case in the context of the isotope effect. With fixed force constant for the vibration, we obtain the relation $\omega_D \propto M^{-1/2}$ where $M$ is the mass of the atom relevant to the vibration. If conduction electrons interact with vibrations of hydrogen, the relevant $\omega_D$ can become an order of magnitude larger than the case of elements with mass number $O(10^2)$. Hence $T_c$ may also become an order of magnitude larger provided the coupling constant $g$ takes a reasonable magnitude. It has been suggested that such systems are realized in certain hydrides under high but presently attainable pressures [12]. The hydrogen bond formed by caged structures may contribute to realize larger $g$ [13]. Recently, there have been experimental reports on superconductivity with $T_c \sim 200\,\mathrm{K}$ in $SH_n$ $(n > 2)$ under 150 GPa [14], and $T_c \sim 260\,\mathrm{K}$ in $LaH_{10}$ under 190 GPa [15, 16]. Without application of pressure, the highest $T_c$ reported so far is 134 K in one of the Cu oxide superconductors [17]. In Cu oxides, electron–electron interaction plays a dominant role in the pairing as will be discussed later.

## 5.7 Spin Structure of Cooper Pairs

The singlet pairing in the BCS state can be extended to more general spin structures. This section discusses the symmetry of the Cooper pair including the triplet pairing. For simplicity, we first deal with the spherically symmetric system such as superfluid $^3$He. In solids, the crystalline symmetry is lower than spherical, and moreover spin–orbit interaction may be important. These complications will be discussed later in Sect. 5.8.

The pairing amplitude for a general spin structure is represented by

$$\Psi_{\alpha\beta}(\hat{k}) = \langle c_\alpha(\boldsymbol{k}) c_\beta(-\boldsymbol{k})\rangle, \tag{5.52}$$

where $\langle \cdots \rangle$ denotes the statistical average. The notation $\hat{k}$ means the unit vector in the direction of $\boldsymbol{k}$ on the Fermi surface. Since the pairing energy is much smaller than the Fermi energy, the magnitude of $|\boldsymbol{k}|$ in Eq. (5.52) is identified as the Fermi momentum in the corresponding direction. Before dealing with the symmetry of the pairing amplitude $\langle c_\alpha c_\beta\rangle$, we touch on the symmetry of $\langle c_\alpha^\dagger c_\beta\rangle$, which corresponds to the one-body density matrix. In the $k$-space, the density matrix $\rho_{\alpha\beta}(\boldsymbol{k}) = \langle c_\beta^\dagger(\boldsymbol{k}) c_\alpha(\boldsymbol{k})\rangle$ is parameterized as

$$2\rho_{\alpha\beta}(\boldsymbol{k}) = n(\boldsymbol{k})\delta_{\alpha\beta} + \boldsymbol{m}(\boldsymbol{k}) \cdot \boldsymbol{\sigma}_{\alpha\beta}(\boldsymbol{k}), \tag{5.53}$$

with use of scalar and vector components. Here the scalar $n(\boldsymbol{k})$ represents the occupation number, and the vector $\boldsymbol{m}(\boldsymbol{k})$ represents the spin polarization. The vector $\boldsymbol{\sigma}$ consists of three components of the Pauli matrices.

By introducing the concept of charge conjugation, we can deal with the pairing symmetry by analogy with Eq. (5.53). Namely, creation of a particle is equivalent to annihilation of a hole with opposite momentum, spin, and charge. Practically the annihilation operator of a spin-up hole in Eq. (5.21) is extended to spin-down case as

$$h_{\uparrow}(\boldsymbol{k}) \equiv c_{\downarrow}^{\dagger}(-\boldsymbol{k}), \quad h_{\downarrow}(\boldsymbol{k}) \equiv -c_{\uparrow}^{\dagger}(-\boldsymbol{k}), \tag{5.54}$$

where the negative sign in the second equation is analogous to the time reversal of spin down. The sign convention for hole operators leads to compact expression of the electron pair. With use of the Pauli matrix $\sigma_y$, the antisymmetric unit tensor is defined by $\varepsilon \equiv \mathrm{i}\sigma_y$. Then the two components of hole operators are written as

$$h_{\alpha}(\boldsymbol{k}) = \sum_{\beta} \varepsilon_{\alpha\beta} c_{\beta}^{\dagger}(-\boldsymbol{k}). \tag{5.55}$$

The creation operator of a singlet pair is written in the new notation as

$$c_{\uparrow}^{\dagger}(\boldsymbol{k})c_{\downarrow}^{\dagger}(-\boldsymbol{k}) - c_{\downarrow}^{\dagger}(\boldsymbol{k})c_{\uparrow}^{\dagger}(-\boldsymbol{k}) = \sum_{\alpha} c_{\alpha}^{\dagger}(\boldsymbol{k})h_{\alpha}(\boldsymbol{k}), \tag{5.56}$$

where the RHS is diagonal in spin indices, and behaves manifestly as a scalar under spin rotation.

On the other hand, the component $S_z = 1$ of the triplet pair is rewritten as

$$\begin{aligned} c_{\uparrow}^{\dagger}(\boldsymbol{k})c_{\uparrow}^{\dagger}(-\boldsymbol{k}) &= \frac{1}{2}\sum_{\alpha\beta} c_{\alpha}^{\dagger}(\boldsymbol{k})\,(\sigma_z + 1)_{\alpha\beta}\, c_{\beta}^{\dagger}(-\boldsymbol{k}) \\ &= -\frac{1}{2}\sum_{\alpha\beta} c_{\alpha}^{\dagger}(\boldsymbol{k})\left(\sigma_x + \mathrm{i}\sigma_y\right)_{\alpha\beta} h_{\beta}(\boldsymbol{k}), \end{aligned} \tag{5.57}$$

with use of the relation $\sigma_z + 1 = -\left(\sigma_x + i\sigma_y\right)\mathrm{i}\sigma_y$. Creation of a pair with $S_z = 1$ in the LHS is translated into the spin flip in the RHS. Similarly the pair with $S_z = -1$ is translated into the reverse spin flip by $(\sigma_x - \mathrm{i}\sigma_y)$, and the pair with $S_z = 0$ by $2\sigma_z$. Thus the type of a triplet pair in general is determined by three coefficients of Pauli matrices which form a vector $\boldsymbol{\Psi}_{\mathrm{t}}(\hat{k})$. For example, $\boldsymbol{\Psi}_{\mathrm{t}}(\hat{k}) \propto (0, 0, 1)^T$ gives $S_z = 0$. Including the singlet case by $\Psi_{\mathrm{s}}(\hat{k})$, the pairing amplitude in general is parameterized as

$$\Psi_{\alpha\beta}(\hat{k}) = \{[\Psi_{\mathrm{s}}(\hat{k}) + \boldsymbol{\Psi}_{\mathrm{t}}(\hat{k}) \cdot \boldsymbol{\sigma}]\varepsilon\}_{\alpha\beta}, \tag{5.58}$$

which assumes a form analogous to the density matrix given by Eq. (5.53). In the presence of inversion symmetry, the total spin of the pair is either 0 or 1, and $\Psi_s(\hat{k})$ and $\boldsymbol{\Psi}_t(\hat{k})$ are mutually exclusive. However, in crystalline solids without the inversion symmetry, spatially even and odd components can mix with each other. The spin–orbit interaction, on the other hand, can mix the spin-singlet and triplet components. Some actual systems show interesting behaviors by the mixing [18].

In the case of $s$-wave pairing, $\Psi_s(\hat{k})$ is a constant independent of the direction of $\hat{k}$. More generally the singlet pairing amplitude is decomposed as

$$\Psi_s(\hat{k}) = \sum_{lm} c_{lm} Y_{lm}(\hat{k}),$$

in terms of spherical harmonics. The Pauli principle requires an even value for $l$ since the spatial parity of the pair must be even for the singlet. For a triplet pair with $S_z = 0$, the pairing amplitude has the form

$$\boldsymbol{\Psi}_t(\hat{k}) = \hat{z} f(\hat{k}), \tag{5.59}$$

where $\hat{z}$ is the unit vector along $z$-axis, and $f(\hat{k})$ specifies the orbital state of the pair. As in the case of the singlet pair, $f(\hat{k})$ can be expanded in terms of spherical harmonics for isotropic systems. The Pauli principle requires now an odd value for $l$. In the case of $S_z = \pm 1$, $\boldsymbol{\Psi}_t(\hat{k})$ has the form $\hat{x} \pm \mathrm{i}\hat{y}$ in place of $\hat{z}$ in Eq. (5.59). .

If the spatial symmetry is lower than spherical, the angular momentum is not a good quantum number. The antisymmetry of the fermionic wave function still demands that $\Psi_s(\hat{k})$ should be an even function of $\hat{k}$, and $\boldsymbol{\Psi}_t(\hat{k})$ an odd function, provided the inversion symmetry is present. Hence singlet and triplet pairings are equivalently called even parity and odd parity, respectively.

We proceed to spin structure of the gap function in more detail. The matrix element of the pairing interaction is written as $\langle \alpha\beta | V(\boldsymbol{k}, \boldsymbol{p}) | \nu\mu \rangle$. Then the finite pairing amplitude gives a mean field $\Delta_{\alpha\beta}(\hat{k})$, which obeys the self-consistent equation:

$$\Delta_{\alpha\beta}(\hat{k}) = \frac{1}{N} \sum_{\mu\nu} \sum_{\boldsymbol{p}} \langle \alpha\beta | V(\boldsymbol{k}, \boldsymbol{p}) | \nu\mu \rangle \Psi_{\mu\nu}(\hat{p}). \tag{5.60}$$

Here we arrange the single-electron state as $|\alpha\beta\rangle^\dagger = \langle \beta\alpha|$ following the convention described by Sect. 2. The gap function including both singlet and triplet cases is parameterized as

$$\Delta_{\alpha\beta}(\hat{k}) = \{[D(\hat{k}) + \boldsymbol{d}(\hat{k}) \cdot \boldsymbol{\sigma}]\varepsilon\}_{\alpha\beta}, \tag{5.61}$$

where the scalar $D(\hat{k})$ and vector $\boldsymbol{d}(\hat{k})$ transform like $\Psi_s(\hat{k})$ and $\boldsymbol{\Psi}_t(\hat{k})$, respectively. Conventionally $\boldsymbol{d}(\hat{k})$ is called the $d$-vector. In contrast with the real parameters $n(\hat{k}), \boldsymbol{m}(\hat{k})$ in the density matrix, $D(\hat{k})$ and $\boldsymbol{d}(\hat{k})$ are complex numbers in general. This is a consequence of the spontaneous breakdown of the gauge symmetry.

In addition to the broken gauge symmetry, spontaneous breakdown may also happen to spatial symmetry and/or time-reversal symmetry. Such breakdown has been observed in superfluid $^3$He [9]. In order to deal with the pairing amplitude most generally, we introduce the four-component field $\psi_i(\boldsymbol{k})$ $(i = 1, 2, 3, 4)$ which forms a vector

$$\boldsymbol{\psi}(\boldsymbol{k}) = (c_\uparrow(\boldsymbol{k}), c_\downarrow(\boldsymbol{k}), h_\uparrow(\boldsymbol{k}), h_\downarrow(\boldsymbol{k}))^T, \tag{5.62}$$

with the hole annihilation operators defined by Eq. (5.55). It is obvious that the components with $i = 1, 3$ make the charge conjugation pair which appear in the two-component field in Eq. (5.21). The same applies to the components with $i = 2, 4$. In terms of the four-component field $\boldsymbol{\psi}(\boldsymbol{k})$, the Hamiltonian in the mean field theory is represented by

$$H = \frac{1}{2}\sum_{\boldsymbol{k}}\sum_{ij} h_{ij}(\boldsymbol{k})\psi_i^\dagger(\boldsymbol{k})\psi_j(\boldsymbol{k}) = \frac{1}{2}\sum_{\boldsymbol{k}} \boldsymbol{\psi}^\dagger(\boldsymbol{k})\hat{h}(\boldsymbol{k})\boldsymbol{\psi}(\boldsymbol{k}), \tag{5.63}$$

where the $4 \times 4$ matrix $\hat{h} = \{h_{ij}\}$ is given by

$$\hat{h}(\boldsymbol{k}) = \begin{pmatrix} \epsilon_{\boldsymbol{k}} & \bar{\Delta}(\hat{k}) \\ \bar{\Delta}^\dagger(\hat{k}) & -\epsilon_{-\boldsymbol{k}}, \end{pmatrix}. \tag{5.64}$$

with each component being a $2 \times 2$ matrix. We obtain

$$\bar{\Delta}(\hat{k}) = D(\hat{k}) + \boldsymbol{d}(\hat{k}) \cdot \boldsymbol{\sigma}, \tag{5.65}$$

which is to be compared with Eq. (5.61). With four components, double counting of each state occurs as particles and holes. The correction factor 1/2 is necessary to compensate the double counting.

We remark that definition of the four-component field in the literature is often different from Eq. (5.62), and is given by

$$(c_\uparrow(\boldsymbol{k}), c_\downarrow(\boldsymbol{k}), c_\uparrow^\dagger(-\boldsymbol{k}), c_\downarrow^\dagger(-\boldsymbol{k}))^T.$$

With this definition, $\bar{\Delta}(\hat{k})$ in Eq. (5.64) should be replaced by $\Delta(\hat{k})$ composed of elements in Eq. (5.61).

The eigenvalues of $\hat{h}(\boldsymbol{k})$ describe the spectrum of quasi-particles. The derivation is the subject of Problem 5.6. The result is given by

$$E_{\tau\sigma}(\boldsymbol{k}) = \tau\sqrt{\epsilon(\boldsymbol{k})^2 + |D(\hat{k})|^2 + |\boldsymbol{d}(\hat{k})|^2 + \sigma|\boldsymbol{w}(\hat{k})|},$$

where $\boldsymbol{w} \equiv \mathrm{i}\boldsymbol{d} \times \boldsymbol{d}^* = \boldsymbol{w}^*$. Here $\tau = \pm 1$ distinguishes the vacant and occupied states, and $\sigma = \pm 1$ distinguishes the spin of the triplet pair parallel or antiparallel to $\boldsymbol{w}$. The pairing is called unitary if $\boldsymbol{w} = 0$, which is the case with time reversal preserved. The case with $\boldsymbol{w} \neq 0$ is called the non-unitary pairing where the spin degeneracy is broken.

In the non-$s$-wave pairing, $E_{\tau\sigma}(\boldsymbol{k})$ can be zero for particular values of the momentum. The set of such momenta on the Fermi surface is called nodes. In the case of singlet pairing, the gap energy vanishes with $|D(\hat{k})|^2 = 0$, which may happen, for example, in a $d$-wave pairing. In the triplet case, the node occurs along $\boldsymbol{d}(\hat{k}) = 0$. It is possible for the non-unitary case that the node occurs only for a particular spin component. For example if we have $|\boldsymbol{d}(\hat{k})|^2 - |\boldsymbol{w}(\hat{k})| = 0$ for certain $\boldsymbol{k}$ with $\boldsymbol{d} \cdot \boldsymbol{d}^* = 0$, the node occurs only for $\sigma = -1$.

In superfluid $^3$He, various types of the $p$-wave pairing are realized with $d$-vectors given by

$$\boldsymbol{d}(\hat{k}) \propto \begin{cases} \hat{x}k_x + \hat{y}k_y + \hat{z}k_z, & \text{[BW]} \\ \hat{z}(k_x \pm i k_y), & \text{[chiral, ABM]} \\ \hat{z}k_z, & \text{[polar]}, \end{cases} \tag{5.66}$$

where, and in the next section, the angular component of $\hat{k}$ is simply written as $k_\alpha$ ($\alpha = x, y, z$). The first line in the RHS of Eq. (5.66) gives the BW (Balian–Werthamer) state which is isotropic. In other words, the total angular momentum $J$ of the pair is zero: $\boldsymbol{J} = \boldsymbol{L} + \boldsymbol{S} = 0$ where $\boldsymbol{L}$ and $\boldsymbol{S}$ are orbital and spin angular momenta, respectively [9]. The gap function has the spherical symmetry without nodes. The BW state is dominant in superfluid $^3$He and is called the B phase. The second line in Eq. (5.66) gives the pair with $S_z = 0$ and $L_z = \pm 1$, which breaks not only the spherical symmetry but the time reversal. However, it is a unitary state with $\boldsymbol{w} = 0$. This pairing is realized in the narrow region (A phase) of the phase diagram of superfluid $^3$He, and is called the ABM (Anderson–Brinkman–Morel) state [9], or the chiral $p$-state [11]. The ABM state or chiral $p$-state has point nodes at $k_x = k_y = 0$. Finally the third line in Eq. (5.66) has $S_z = L_z = 0$, and breaks the rotational symmetry. The pairing is called the polar state, and has nodes along the line $k_z = 0$.

## 5.8 Anisotropic Cooper Pairs

For more detailed discussion of the Cooper pair in solids, the spherical symmetry in the previous section must be replaced by the point-group symmetry. We take the tetragonal symmetry for a representative, which is relevant to high-temperature superconductivity in Cu oxides and Fe pnictides/chalcogenides where singlet pairs are formed. The tetragonal symmetry also applies to $CeCu_2Si_2$ which shows a singlet superconductivity by electrons with very large effective mass ($\sim$100 times the free electron mass), and $SrRu_2O_4$ where a triplet pairing seems responsible for superconductivity. Since these materials have a strong spin–orbit interaction,

**Table 5.1** Irreducible representations and examples of basis functions of an even parity pair in the point group $D_{4h}$

| Representation | Basis functions |
|---|---|
| $\Gamma_1^+$ $(A_{1g})$ | $1; k_x^2 + k_y^2 - 2k_z^2$ |
| $\Gamma_2^+$ $(A_{2g})$ | $\mathrm{Im}k_+^4$ |
| $\Gamma_3^+$ $(B_{1g})$ | $\mathrm{Re}k_+^2$ |
| $\Gamma_4^+$ $(B_{2g})$ | $\mathrm{Im}k_+^2$ |
| $\Gamma_5^+$ $(E_g)$ | $k_z k_\pm; k_z k_\pm^3$ |

Here $k_\pm \equiv k_x \pm ik_y$ and the semicolon separate different basis functions belonging to the same representation. Linear combinations of functions separated by semicolon can also be realized

the classification in terms of singlet and triplet of spins is inaccurate. However, classification by spatial parity is still valid.

The tetragonal point group is written as $D_{4h}$, and the irreducible representations, five in total, are listed in Table 5.1. Conventionally two main notations have been used for irreducible representations. The first one, after Bethe, uses $\Gamma_j^\pm$ where $j$ $(= 1, 2, \ldots)$ specifies a representation and $\pm$ indicates the parity. The identity representation has $j = 1$, and the number increases in accordance with increasing dimension of the representation. However, there is no systematic rule how to put the number within the same dimension. The second one, after Mulliken, uses $A_{1g}$, $B_{2g}$, $E_{2u}$, etc. and is popular particularly in chemistry. The one-dimensional representations are classified into $A$ and $B$ where $A$ shows invariance of the basis function against $\pi/2$ rotation around the $z$-axis, while $B$ undergoes sign change against the same rotation. The numbers 1,2 after $A$, $B$ indicate whether the basis function is even or odd against reflection about the $yz$- or $zx$-plane. This is equivalent to being even or odd against $\pi$ rotation around the $x$- or $y$-axis. The number is 1 for the even basis, and 2 for the odd one. After these numbers, attached is $g$ (gerade in German) if the basis function has even parity, and $u$ (ungerade) if it is odd. On the other hand, $E_{2u}$ means a two-dimensional (2D) irreducible representation with two basis functions, which are odd under space inversion. The suffix 2 is often omitted; the rule is not so simple as that in one-dimensional representations. Examples of basis functions are also given in Table 5.1. Note that the basis functions are not exhaustive, since higher order polynomials are also possible.

Let us begin with the even parity representation $\Gamma_1^+$ $(A_{1g})$. The corresponding gap function is a scalar in the point group. In contrast with the spherically symmetric case, not only the $s$-state with $l = 0$, but functions such as $1 - 3\cos^2\theta = k_x^2 + k_y^2 - 2k_z^2$ with $l = 2$ behave as scalar. Hence we parameterize the general scalar function $g_0$ as

$$g_0(\hat{k}) = a + b(k_x^2 + k_y^2 - 2k_z^2) + \ldots, \tag{5.67}$$

where $a, b, \ldots$ are the real parameters. Because of the lower symmetry, components of different angular momenta may mix. As long as the $s$-component is dominant with $|a| \gg |b|$, $g_0(\hat{k})$ does not have nodes. In the opposite case of $|a| \ll |b|$, dominance of the $d$-component may lead to nodes in $g_0(\hat{k})$.

We move on to another even parity representation $\Gamma_3^+$ $(B_{1g})$ which is relevant to Cu oxide superconductors. The gap function is given by

$$D_1(\hat{k}) = (k_x^2 - k_y^2)g_0(\hat{k}) = \mathrm{Re}k_+^2 g_0(\hat{k}), \tag{5.68}$$

with $k_+ = k_x + ik_y$. Evidently $D_1(\hat{k})$ has nodes at $k_x = \pm k_y$. If we take $g_0 = a$, $D_1(\hat{k})$ reduces to the $d$-state. Hence the pair in Cu oxide is usually called the $d$-wave pair. Another $d$-wave pair is possible with the $\Gamma_4^+$ $(B_{2g})$ representation. The basis functions are given by

$$D_2(\hat{k}) = 2k_x k_y g_0(\hat{k}) = \mathrm{Im}k_+^2 g_0(\hat{k}), \tag{5.69}$$

which have nodes at $k_x = 0$ and $k_y = 0$. These nodes correspond to $\pi/4$ rotation of the nodes of $D_1(\hat{k})$. Table 5.1 includes other possible representations of even parity pairing.

In the case of odd parity pairing, we have already seen in Eq. (5.66) that the spin and orbital states are entangled in the order parameter. If the spin–orbit interaction is present, the Kramers pair consists of mixture of spin and orbital states, which are called pseudo-spins. Then we only have to change the basis set from the real spin to the pseudo-spin to describe the Kramers pair for a given crystal momentum. The simplest is the $\Gamma_1^-$ $(A_{1u})$ symmetry, which corresponds to the point-group version of the BW state. The $d$-vector is given for lower orders of $\boldsymbol{k}$ by

$$\begin{aligned} \boldsymbol{d}(\hat{k}) &= \hat{z}k_z g_0(\hat{k}) + (\hat{x}k_x + \hat{y}k_y)g_1(\hat{k}) + (\hat{x}k_x - \hat{y}k_y)(k_x^2 - k_y^2)g_2(\hat{k}) \\ &= \hat{z}k_z g_0(\hat{k}) + (\mathrm{Re}k_+\hat{r}_-)g_1(\hat{k}) + (\mathrm{Re}k_+^3\hat{r}_+)g_2(\hat{k}), \end{aligned} \tag{5.70}$$

where $\hat{x}$ is the unit vector along the $x$-axis, and $\hat{r}_\pm = \hat{x} \pm \mathrm{i}\hat{y}$. The functions $g_i(\hat{k})$ $(i = 0, 1, 2)$ all behave as scalar in the point group. An example has been given by Eq. (5.67). The condition of minimum free energy determines the form of $g_i(\hat{k})$. In the case of $g_i \neq 0$ in Eq. (5.70), the parts in front of $g_0$ and $g_1$ represent the $p$-wave, while the part in front of $g_2$ represents the $f$-wave $(l = 3)$. On the other hand, in the case of $g_1 = g_2 = 0$, a state with $S_z = 0$ is realized. This corresponds to the polar state in Eq. (5.66) with line nodes at $k_z = 0$.

Another one-dimensional representation has the symmetry $\Gamma_3^-$ $(B_{1u})$. The $d$-vector in the lowest order in $\boldsymbol{k}$ is given by

$$\boldsymbol{d}(\boldsymbol{k}) = (\hat{x}k_x - \hat{y}k_y)g_0(\hat{k}) = (\mathrm{Re}k_+\hat{r}_+)g_0(\hat{k}), \tag{5.71}$$

**Table 5.2** Irreducible representations and examples of basis functions of an odd parity pair in the point group $D_{4h}$

| Representation | Basis functions |
|---|---|
| $\Gamma_1^-$ ($A_{1u}$) | $\hat{z}k_z$; $\mathrm{Re}k_+\hat{r}_-$; $\mathrm{Re}k_+^3\hat{r}_+$ |
| $\Gamma_2^-$ ($A_{2u}$) | $\mathrm{Im}k_+\hat{r}_-$; $\mathrm{Im}k_+^3\hat{r}_+$; $k_z\mathrm{Im}k_+^4\hat{z}$ |
| $\Gamma_3^-$ ($B_{1u}$) | $\mathrm{Re}k_+\hat{r}_+$; $k_z\mathrm{Re}k_+^2\hat{z}$; $\mathrm{Re}k_+^3\hat{r}_-$ |
| $\Gamma_4^-$ ($B_{2u}$) | $\mathrm{Im}k_+\hat{r}_+$; $k_z\mathrm{Im}k_+^2\hat{z}$; $\mathrm{Im}k_+^3\hat{r}_-$ |
| $\Gamma_5^-$ ($E_u$) | $k_\pm^{n+1}\hat{z}$; $k_zk_\pm^n\hat{r}_\pm$; $k_zk_\pm^{n+2}\hat{r}_\mp$ $(n = 0, 2)$ |

Here $\hat{r}_\pm \equiv \hat{x} \pm i\hat{y}$, and, in the last row, the pair of functions with suffices $\pm$ or $\pm\mp$ make the basis functions of the 2D representation

with nodes at $k_x = k_y = 0$, which correspond to north and south poles on the Fermi surface. Table 5.2 shows all the irreducible representations for $D_{4h}$, and examples of basis functions. Similar analysis can be carried out for other point groups with cubic and hexagonal symmetries [19].

## 5.9 Pairing with Repulsive Interaction

In Cu oxides and Fe pnictides/chalcogenides, some compounds have high transition temperatures $T_c$ in spite of large Coulomb repulsion between electrons. The first material in the high-$T_c$ family was found in 1986; $Ba_2CuO_4$ doped with La with $T_c \sim 35$ K [5]. The highest $T_c$ so far reported is 164 K in $HgBa_2Ca_2Cu_3O_x$ $(x \sim 8)$ under pressure. On the other hand, the first Fe-based superconductor found in 2008 is LaFeAsO doped with F (fluorine) [6]. The highest $T_c$ in this family now reaches about 60 K [7]. The mechanism for the high $T_c$ in these materials has been studied intensively. The common feature is the antiferromagnetic order without doping, which remains to some extent of doping. With further increase of the carrier density, superconductivity appears. Thus it is natural to expect a strong interplay between superconductivity and antiferromagnetism. Another common feature is that both Cu and Fe superconductors have layer-type tetragonal structures, and electrons with quasi-2D character play a dominant role in the superconductivity. As the zero-th approximation, we may take the square lattice for investigating mechanism of superconductivity.

The tetragonal structure of the Cu oxide superconductors is shared with another class of superconductor $Sr_2RuO_4$, which has $T_c \sim 1.5$ K. The relevant 4$d$ electrons from Ru are quasi-two dimensional and close to magnetic orders. With Ca doping, for example, not only antiferromagnetism but also ferromagnetism tends to appear. The odd parity pairing has long been suspected on the basis of the NMR Knight shift [11], which does not seem to decrease below $T_c$.[2]

[2]However, recent accurate NMR experiment by A. Pustogow et al: Nature **574**, 72 (2019) has shown that the Knight shift *does* decrease below $T_c$ as in singlet superconductors.

Some materials with $f$ electrons become superconductor with heavy effective mass of electrons [20]. Among these materials $PuRhGa_5$ has the highest $T_c \sim 16\,K$ under pressure. The crystal structure consists of layers of $PuGa_2$ sandwiched by another layers $RhGa_3$. The quasi-2D $5f$ electrons in $PuGa_2$ may play a dominant role in the pairing. There are several Ce compounds with the same crystal structure such as $CeRhIn_5$ with $T_c \sim 2\,K$ under pressure of a few GP, and antiferromagnetic order with zero pressure [21].

In this section we discuss superconductivity by Coulomb repulsion by taking the Hubbard model. For simplicity the electron–phonon interaction is neglected. In terms of the perturbation theory in $U$, the effective attraction occurs in $O(U^2)$. The transition temperature is very small in the range of $U$ accessible by perturbation theory. On the other hand, actual materials mentioned at the beginning have large $U$ which is comparable to the conduction bandwidth or even more. Hence it is necessary to employ a non-perturbative method such as numerical simulation for realistic argument. Nevertheless, we rely on the weak-coupling theory since it provides the most transparent idea as to the origin of the attractive force.

We consider the situation where the number of electrons is slightly smaller than one per lattice site. If the number is just one, which corresponds to half-filling of the conduction band, antiferromagnetism is favored, and the system becomes an insulator. This is because the antiferromagnetism makes reduction of the Brillouin zone due to the doubled unit cell, and the energy gap is formed along the new boundary of the zone. The lower conduction band is then completely filled.

The effective interaction is derived in the general framework discussed in Chap. 1. As the perturbation $H'$ we take the repulsion term with $U$, which is written in the momentum space as

$$H' = \frac{U}{N} \sum_{\boldsymbol{k},\boldsymbol{p}} \sum_{\boldsymbol{q}} c_\uparrow^\dagger(\boldsymbol{k}+\boldsymbol{q}) c_\downarrow^\dagger(-\boldsymbol{k}) c_\downarrow(-\boldsymbol{p}) c_\uparrow(\boldsymbol{p}+\boldsymbol{q}). \tag{5.72}$$

This representation makes explicit the annihilation of the singlet pair by $c_\downarrow c_\uparrow$, and the creation by $c_\uparrow^\dagger c_\downarrow^\dagger$. In the following we only consider the part with the zero pair momentum $\boldsymbol{q} = 0$. In deriving the effective interaction between two electrons, we choose the model space as the BCS-like states given by Eq. (5.8) with arbitrary values of parameters $u_{\boldsymbol{k}}$, $v_{\boldsymbol{k}}$, and $\phi$. Then the second-order effective interaction

$$PH' \frac{1}{E_0 - H_0} QH'P \tag{5.73}$$

describes the interaction between a pair of electrons forming a spin singlet. Here the model space consists of singlet pairs of electrons excited from the ground state of many electrons. Consequently the projection operator $Q$ excludes the states where only singlet pairs are excited.

Evaluation of the effective interaction Eq. (5.73) is reduced to taking the average, or the vacuum expectation value, of the product of the creation and annihilation operators of electrons with the same momentum $\boldsymbol{k}$ and spin. Since such operator pairs occur only once for each spin, the average of the product of operators is equal to the product of the average of each spin component. This factorization property can be generalized to arbitrary number of operators. The details are discussed in Chap. 9 as Wick's theorem.

For the moment we restrict to terms of $O(U^2)$, and represent each average by a line where the arrow points to the annihilation operator. Thus each line represents either the occupation number $f(\epsilon_{\boldsymbol{k}}) = \langle c^\dagger_{\boldsymbol{k}\sigma} c_{\boldsymbol{k}\sigma}\rangle$ of the particle at $T = 0$, or that of the hole $1 - f(\epsilon_{\boldsymbol{k}}) = \langle c_{\boldsymbol{k}\sigma} c^\dagger_{\boldsymbol{k}\sigma}\rangle$, depending on the arrow direction relative to sequence of the process. Such pictorial representation of perturbation processes is called the Goldstone diagram [22], a simpler example of which has already appeared in Fig. 1.2. Since a given diagram faithfully follows the perturbation sequence, one can reconstruct the corresponding mathematical expression rather directly, as will be demonstrated in Eq. (5.76) below. There is another diagram method due to Feynman, which uses the single-particle Green function as a building block. The Feynman diagram deals with particles and holes symmetrically, and is valid also at finite temperature by use of the Matsubara frequency. The Matsubara scheme has been explained in Chap. 3. Since the Feynman method requires more elaborate preparation, its detailed exposition is deferred to Chap. 9. For representing an effective Hamiltonian with explicit use of projections operators $P$ and $Q$, the Goldstone diagram is simpler and more convenient.

For small $U$, the lowest-order term dominates the higher order ones for the isotropic part of the interaction. In other words, all effects that do not provide anisotropy of the effective interaction can be neglected. For example, we neglect the screening of $U$ by a particle–hole pair. We instead concentrate on the processes that give rise to anisotropic effective interaction. Since the anisotropy is not present in the original interaction $U$, the second-order term may dominate depending on the symmetry of the pair wave function.

In second-order scattering processes of a pair, we have either (a) parallel or (b) antiparallel lines between the action of two $U$'s. The case (a) represents the successive scattering of a singlet pair, which keeps the isotropy of the interaction. If the projection operator $P$ includes all energy range of the singlet pair, (a) is rejected as an effective interaction by $Q$. If one redefines $P$ so as to project the singlet pair onto the narrow energy range near the Fermi level, however, the case (a) represents renormalization of $U$. This process is illustrated in Fig. 5.6 in the solution of Problem 5.5. On the other hand, the case (b) which is illustrated in Fig. 5.4, gives an anisotropic effective interaction that has a strong dependence on the transferred momentum $\boldsymbol{p} - \boldsymbol{k}$.

Assuming electrons near the Fermi surface for both initial and final states, we put $\epsilon_{\boldsymbol{k}} = \epsilon_{\boldsymbol{p}} = 0$ in Fig. 5.4. Then the denominator in Eq. (5.73) is given by

$$E_0 - H_0 \to \epsilon_{-\boldsymbol{p}+\boldsymbol{q}} - \epsilon_{\boldsymbol{k}+\boldsymbol{q}}. \tag{5.74}$$

Furthermore the Pauli principle imposes the constraint

$$\left[1 - f(\epsilon_{\boldsymbol{k}+\boldsymbol{q}})\right] f(\epsilon_{\boldsymbol{p}-\boldsymbol{q}}) \tag{5.75}$$

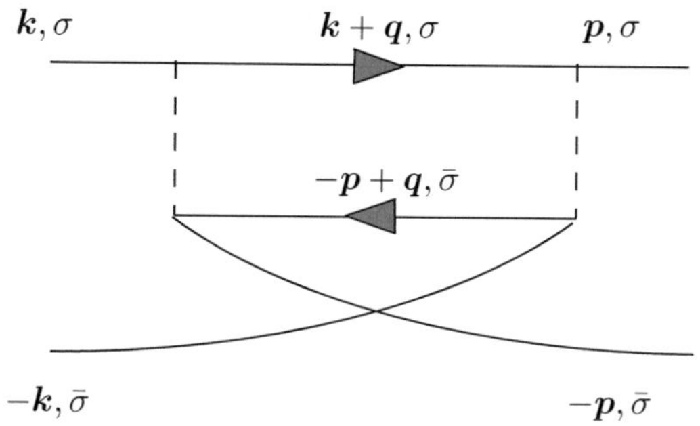


**Fig. 5.4** Illustration of the most relevant term in Eq. (5.73). The dashed lines represent the repulsive interaction $U$, and the solid lines with arrows specify either an electron state with positive energy, or a hole state otherwise. Provided the perturbation sequence goes from left to right in the diagram, the arrows indicate $\epsilon_{\boldsymbol{k}+\boldsymbol{q}} > 0$ and $\epsilon_{-\boldsymbol{p}+\boldsymbol{q}} < 0$

for the intermediate states with $f(\epsilon)$ the Fermi distribution function. To obtain the effective interaction $V(\boldsymbol{k}, \boldsymbol{p})$, we have to include another term where $(\boldsymbol{k}, \sigma)$ and $(\boldsymbol{p}, \sigma)$ are interchanged by $(-\boldsymbol{k}, \bar{\sigma})$ and $(-\boldsymbol{p}, \bar{\sigma})$ in Fig. 5.4. The latter term gives $\epsilon_{-\boldsymbol{p}+\boldsymbol{q}} - \epsilon_{\boldsymbol{k}+\boldsymbol{q}}$ for $E_0 - H_0$ and

$$\left[1 - f(\epsilon_{\boldsymbol{p}-\boldsymbol{q}})\right] f(\epsilon_{\boldsymbol{k}+\boldsymbol{q}})$$

for the constraint. Then the two processes combine to give

$$\frac{1}{N}\sum_{\boldsymbol{q}} \frac{f(\epsilon_{-\boldsymbol{p}+\boldsymbol{q}}) - f(\epsilon_{\boldsymbol{k}+\boldsymbol{q}})}{\epsilon_{-\boldsymbol{p}+\boldsymbol{q}} - \epsilon_{\boldsymbol{k}+\boldsymbol{q}}} \equiv -\chi_0(\boldsymbol{k}+\boldsymbol{p}), \tag{5.76}$$

with $\chi_0(\boldsymbol{k}+\boldsymbol{p})$ $(> 0)$ being the static polarization function. We now obtain the effective interaction, which is a generalized version of $V(\boldsymbol{k}-\boldsymbol{p})$ in the gap equation (5.27), as

$$V(\boldsymbol{k}, \boldsymbol{p}) = -U - U^2\chi_0(\boldsymbol{k}+\boldsymbol{p}), \tag{5.77}$$

up to the second order in $U$. The sign of $V(\boldsymbol{k}, \boldsymbol{p})$ is taken so that $V(\boldsymbol{k}-\boldsymbol{p}) > 0$ means the attraction as defined in Eq. (5.16). Hence, $\chi_0(\boldsymbol{k}+\boldsymbol{p})$ works as a repulsive force for the $s$-wave. However, this situation may change for anisotropic cases.

We take the 2D model in view of high-$T_c$ superconductors based on Cu and Fe. Although the sign of $V(\boldsymbol{k}, \boldsymbol{p})$ corresponds to repulsion for the $s$-wave pair, its dependence on relative angles between $\boldsymbol{k}$ and $\boldsymbol{p}$ favors an anisotropic electron pair. To be more specific, we take the nearest-neighbor hopping in the square lattice. Then the kinetic energy is given by

$$\epsilon_{\boldsymbol{k}} = -2t\left(\cos k_x + \cos k_y,\right) \tag{5.78}$$

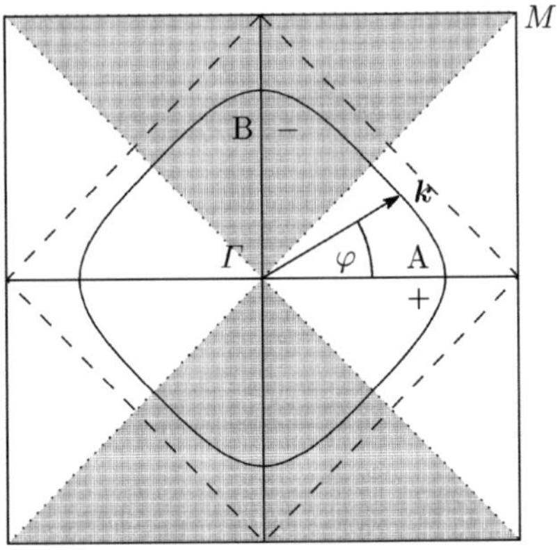


**Fig. 5.5** Illustration of the Fermi surface and sign of the stable gap function $\Delta(\boldsymbol{k})$. The region A around $\boldsymbol{k} = (k_x, 0)$ and the region B (shaded) around $\boldsymbol{p} = (0, p_y)$ have opposite signs so that $\Delta(\boldsymbol{p})\Delta(\boldsymbol{k}) < 0$. Then $\boldsymbol{k} + \boldsymbol{p}$ takes the value close to $\boldsymbol{Q} = (\pi, \pi)$ that corresponds to $M$ point in the Brillouin zone. The dashed lines show the Fermi surface for the half-filled case

where $-t$ is the transfer integral, and the lattice constant is taken to be unity. If the electron density is slightly smaller than one per site, the Fermi surface takes the shape as shown by the solid line in Fig. 5.5.

For the half-filled conduction band with the square Fermi surface, $\chi_0(\boldsymbol{q})$ is divergent logarithmically at $\boldsymbol{q} = (\pi, \pi) \equiv \boldsymbol{Q}$ and its equivalent points, called $M$, in the Brillouin zone. The divergence corresponds to the antiferromagnetism at $T = 0$. The divergence disappears with deviation of the density, but the large magnitude of $\chi_0(\boldsymbol{q})$ at $\boldsymbol{q} = \boldsymbol{Q}$ and its neighborhood remains. Then with $\boldsymbol{k} \sim (\pi, 0)$, $\boldsymbol{p} \sim (0, \pi)$, $\chi_0(\boldsymbol{k} + \boldsymbol{p}) \sim \chi_0(\boldsymbol{Q})$ in Eq. (5.77) becomes large. If the gap function has opposite signs in the regions A, B as indicated in Fig. 5.5, the large magnitude of $\chi_0(\boldsymbol{Q})$ contributes to formation of the anisotropic electron pair. The different regions of the Brillouin zone are analogous to different bands in the multiband model discussed in Sect. 5.4. By analogy with the pair transfer term that increases $T_c$, the transfer between the regions A, B contributes to the pairing. The symmetry of the pair corresponding to Fig. 5.5 is identified as $\Gamma_3^+$ $(= B_{1g})$. Namely, we obtain the singlet gap function

$$\Delta(\boldsymbol{k}) \propto k_x^2 - k_y^2, \tag{5.79}$$

which belongs to the $d$-wave. The node occurs at $k_x = \pm k_y$, which is indicated by dotted lines in Fig. 5.5. The singlet $d$-wave pairing captures a characteristic feature in Cu oxide superconductors from the weak-coupling side.

Let us proceed to solution of the gap equation with $V(\boldsymbol{k}, \boldsymbol{p})$. We assume that $T_c$ is much smaller than the Fermi energy, and use the value of $V(\boldsymbol{k}, \boldsymbol{p})$ at $T = 0$. As indicated in Fig. 5.5, it is convenient to take the argument $\varphi$ of $\boldsymbol{k}$ at the Fermi surface, and regard $\Delta(\boldsymbol{k})$ as a function of $\varphi$ [23]. To decompose into components of angular momentum, we introduce the angle-resolved density of states $\tilde{\rho}(\varphi)$ at the Fermi surface as

$$\rho(\mu) = \frac{1}{N}\sum_{\boldsymbol{k}} \delta(\mu - \epsilon_{\boldsymbol{k}}) \equiv \int \frac{\mathrm{d}\varphi}{2\pi}\tilde{\rho}(\varphi). \tag{5.80}$$

In terms of the angle-dependent Fermi momentum $k(\varphi)$ and the Fermi velocity $v(\varphi) = |\nabla_{\boldsymbol{k}}\epsilon_{\boldsymbol{k}}|$, we obtain $\tilde{\rho}(\varphi) = k(\varphi)/[2\pi v(\varphi)]$. To deal with the $\varphi$-dependence, we introduce the weight function $w(\varphi)$ such that $\tilde{\rho}(\varphi) \equiv \rho(\mu)w(\varphi)$. Then we use the modified Fourier components defined by

$$A_m(\varphi) = \exp(im\varphi)/\sqrt{w(\varphi)}, \tag{5.81}$$

which has the orthogonal property

$$\int \frac{\mathrm{d}\varphi}{2\pi} w(\varphi) A_l(\varphi)^* A_m(\varphi) = \delta_{lm}, \tag{5.82}$$

with the weight function $w(\varphi)$. Then setting $\Delta(\boldsymbol{k})|_{\mathrm{FS}} \equiv \tilde{\Delta}(\varphi)$ at the Fermi surface (FS), we make the Fourier-like decomposition

$$\tilde{\Delta}(\varphi) = \sum_{m=-\infty}^{\infty} \Delta_m A_m(\varphi), \tag{5.83}$$

$$\Delta_m = \int \frac{\mathrm{d}\varphi}{2\pi} w(\varphi) A_m(\varphi)^* \tilde{\Delta}(\varphi). \tag{5.84}$$

Similarly we introduce $V(\boldsymbol{k}, \boldsymbol{k}')|_{\mathrm{FS}} \equiv \tilde{V}(\varphi, \varphi')|_{\mathrm{FS}}$, and makes its Fourier decomposition

$$\rho(\mu)V(\varphi, \varphi') = \sum_{lm} g_{lm} A_l(\varphi) A_m(\varphi')^*, \tag{5.85}$$

$$g_{lm} = \int \frac{\mathrm{d}\varphi}{2\pi} w(\varphi) A_l(\varphi)^* \int \frac{\mathrm{d}\varphi'}{2\pi} w(\varphi') A_m(\varphi') \rho(\mu) V(\varphi, \varphi'). \tag{5.86}$$

The property $V(\varphi, \varphi') = V(\varphi', \varphi) = V(-\varphi, -\varphi')$, imposes the constraint on the coupling constant $g_{lm}$ as $g_{lm} = g_{ml} = g_{\bar{l}\bar{m}}$ with $\bar{l} = -l$. Furthermore the tetragonal symmetry requires $g_{lm}$ to vanish except for $l - m = 4n$ ($n =$ integer). This is because $V(\varphi, \varphi')$ is invariant against $\pi/2$ rotation in the $\boldsymbol{k}$-space: $\varphi, \varphi' \to \varphi + \pi/2, \varphi' + \pi/2$.

With use of components for angular momentum introduced above, the gap equation is rewritten as

$$\Delta_l = \ln \frac{D}{T_c} \sum_m g_{lm} \Delta_m, \tag{5.87}$$

where $D$ is of the order of bandwidth. The transition temperature $T_c$ is determined by the condition that the matrix composed of elements $\delta_{lm} - g_{lm}\ln(D/T_c)$ has zero determinant. The $s$-wave pair corresponds to $l = m = 0$, which is unfavorable because the bare term $-U$ in Eq. (5.77) makes $g_{00} < 0$. To the contrary, in $g_{lm}$

with $l, m \neq 0$, the isotropic bare term drops out through angular average. This is an example that anisotropic pairs are favored in the presence of short-range Coulomb repulsion.

As the simplest approximation to describe the anisotropic singlet pair, we keep only the lowest components $l, m = \pm 2$. Namely, we neglect possible mixture with $|l| = 6, 10, 14, \ldots$. Accordingly, we expand $\tilde{\rho}(\varphi)$ in terms of angular momentum, and keep only the isotropic ($l = 0$) component, which is nothing but the density of states $\rho(\mu)$. The gap equation (5.87) becomes diagonal for irreducible representations. The case $\Delta_{-2} = \Delta_2$ gives irreducible representations $B_{1g}$, while another one $\Delta_{-2} = -\Delta_2$ gives $B_{2g}$. In respective cases, we obtain

$$B_{1g}: \quad \tilde{\Delta}(\varphi) = 2\Delta_2 \cos 2\varphi \propto k_x^2 - k_y^2, \tag{5.88}$$

$$B_{2g}: \quad \tilde{\Delta}(\varphi) = 2i\Delta_2 \sin 2\varphi \propto 2k_x k_y, \tag{5.89}$$

in accordance with Table 5.1. The coupling constant for $B_{1g}$ is given by

$$g(B_{1g}) = \frac{1}{2}\sum_{\alpha\beta} g_{\alpha\beta} = 2\rho(\mu)\int \frac{\mathrm{d}\varphi}{2\pi}\cos 2\varphi \int \frac{\mathrm{d}\varphi'}{2\pi}\cos 2\varphi' V(\varphi, \varphi'), \tag{5.90}$$

with $\alpha, \beta = \pm 2$. The favorable region for integration is $\varphi + \varphi' \sim (n + 1/2)\pi$ with respect to $V(\varphi, \varphi')$, which corresponds to $\boldsymbol{k} + \boldsymbol{p} \sim (\pi, \pi)$ and equivalent points. The region matches with $\varphi \sim 0$ and $\varphi' \sim \pi/2$ where the cosine functions also have large weight with $\cos 2\varphi \sim 1$ and $\cos 2\varphi' \sim -1$, and the combination gives $g(B_{1g}) > 0$. The transition temperature is given by

$$T_c = D \exp[-1/g(B_{1g})]. \tag{5.91}$$

On the other hand, the $B_{2g}$-pairing has the coupling constant

$$g(B_{2g}) = 2\rho(\mu)\int \frac{\mathrm{d}\varphi}{2\pi}\sin 2\varphi \int \frac{\mathrm{d}\varphi'}{2\pi}\sin 2\varphi' V(\varphi, \varphi'). \tag{5.92}$$

The sine functions favor the region such as $\varphi \sim \pi/4$ and $\varphi' \sim 3\pi/4$, and this combination gives $\varphi + \varphi' \sim \pi$. However, this is not the best region for $V(\varphi, \varphi')$ as seen from Fig. 5.5. Hence $g(B_{2g})$ should be smaller than $g(B_{1g})$.

As the electron density becomes smaller, the Fermi surface tends to a circle as that in the free 2D space. In this case $\chi_0(\boldsymbol{q})$ is a constant for $|\boldsymbol{q}| < 2k_F$. Then various types of spin singlet and triplet compete for the stability, and it is not easy to determine the most stable pairing symmetry. On the other hand, the spherical Fermi surface is relevant to superfluid $^3$He. If we take the short-range repulsion between fermions, the $p$-wave pair is most favored since $\chi_0(\boldsymbol{q})$ becomes the largest at $\boldsymbol{q} = 0$. Let us consider the case $k_z = 0$ to simplify the situation. If the gap function changes sign by $\pi$ rotation around the $z$-axis, the effective interaction becomes attractive, and takes advantage of the peak of $\chi_0(\boldsymbol{k} + \boldsymbol{p})$ at $\boldsymbol{p} = -\boldsymbol{k}$. This is realized by a triplet with the $d$-vector

$$\boldsymbol{d}(\boldsymbol{k}) \propto \hat{z}(k_x \pm ik_y), \tag{5.93}$$

which corresponds to the ABM state in Eq. (5.66). It can be shown that BW state gives the same $T_c$ in the present approximation. If higher order effects of $U$ are taken into account. the component $\boldsymbol{q} = 0$ contributes even stronger in the presence of strong ferromagnetic fluctuations. In actual superfluid $^3$He, ferromagnetic fluctuations seem to contribute to stability of the $p$-wave pairing.

## Problems

**5.1** Derive the norm of the BCS state $\Psi_{\mathrm{BCS}}$ as given by Eq. (5.8).

**5.2** Diagonalize the $2 \times 2$ matrix $h_{\mathrm{BCS}}(\boldsymbol{k})$ in Eq. (5.22).

**5.3** Construct $\Psi_{\mathrm{BCS}}$ in terms of Bogoliubov quasi-particles.

**5.4** Show that the state with $\Delta_{\mathrm{A}}\Delta_{\mathrm{B}} \neq 0$ is already present at temperature higher than $T_{\mathrm{A}}$ and $T_{\mathrm{B}}$.

**5.5** * After learning the scaling method to be explained in Chap. 6, derive the effective repulsion $U^*$ given by Eq. (5.47).

**5.6*** Derive the eigenvalues of $\hat{h}(\boldsymbol{k})$ given by Eq. (5.64).

## Solutions to Problems

### Problem 5.1

The pair creation operator $c_\uparrow^\dagger(\boldsymbol{k})c_\downarrow^\dagger(-\boldsymbol{k})$ commutes with another with momentum different from $\boldsymbol{k}$. Then the expectation value can be taken for each $\boldsymbol{k}$, and we obtain

$$\begin{aligned}\langle\Psi_{\mathrm{BCS}}|\Psi_{\mathrm{BCS}}\rangle &= \prod_{\boldsymbol{k}}\langle 0|u_{\boldsymbol{k}}^2 + v_{\boldsymbol{k}}^2 c_\downarrow(-\boldsymbol{k})c_\uparrow(\boldsymbol{k})c_\uparrow^\dagger(\boldsymbol{k})c_\downarrow^\dagger(-\boldsymbol{k})|0\rangle \\ &= \prod_{\boldsymbol{k}}\left(u_{\boldsymbol{k}}^2 + v_{\boldsymbol{k}}^2\right).\end{aligned} \tag{5.94}$$

Namely, the normalization condition is $u_{\boldsymbol{k}}^2 + v_{\boldsymbol{k}}^2 = 1$. Note that the norm is independent of the phase $\phi$. This independence holds also for matrix element of gauge-invariant operators such as Hamiltonian. Hence the energy of the BCS state is independent of $\phi$.

**Problem 5.2**

We first consider the case where $\Delta(\boldsymbol{k})$ is real. The inversion symmetry requires $\epsilon_{-\boldsymbol{k}} = \epsilon_{\boldsymbol{k}}$. In the unitary transformation defined by Eq. (5.23), we make the parameterization

$$u_{\boldsymbol{k}} = \cos\theta, \quad v_{\boldsymbol{k}} = \sin\theta, \tag{5.95}$$

in the normalized BCS state. Then the diagonalization of the matrix is equivalent to the coordinate rotation to make hyperbola specified by

$$f(x, y) = \epsilon(x^2 - y^2) + 2\Delta xy = C \tag{5.96}$$

to the standard form without the cross term $xy$. Here $C$ is a real constant, and $\epsilon$ and $\Delta$ are in the simplified notation without $\boldsymbol{k}$. The rotation angle $\theta$ should be chosen as $\tan 2\theta = \Delta/\epsilon$, or equivalently

$$\sin^{-2} 2\theta = 1 + \cot^2 2\theta = \frac{\epsilon^2 + \Delta^2}{\Delta^2}. \tag{5.97}$$

With Eqs. (5.95) and (5.97), we obtain

$$2u_{\boldsymbol{k}}v_{\boldsymbol{k}} = \sin 2\theta = |\Delta(\boldsymbol{k})|/E_+(\boldsymbol{k}), \tag{5.98}$$

where

$$E_\pm(\boldsymbol{k}) = \pm\sqrt{\epsilon_{\boldsymbol{k}}^2 + |\Delta(\boldsymbol{k})|^2} \tag{5.99}$$

is the eigenvalues of $h_{\mathrm{BCS}}$.

If $\Delta(\boldsymbol{k})$ is a complex number with argument $\phi$, the unitary transformation is chosen as

$$\boldsymbol{\psi}(\boldsymbol{k}) = \begin{pmatrix} u_{\boldsymbol{k}} & e^{i\phi}v_{\boldsymbol{k}} \\ -e^{-i\phi}v_{\boldsymbol{k}} & u_{\boldsymbol{k}} \end{pmatrix} \tilde{\boldsymbol{\psi}}(\boldsymbol{k}). \tag{5.100}$$

By using the same $\theta$ as given by Eq. (5.95) we obtain

$$\boldsymbol{\psi}^\dagger(\boldsymbol{k}) \begin{pmatrix} \epsilon_{\boldsymbol{k}} & \Delta(\boldsymbol{k}) \\ \Delta(\boldsymbol{k})^* & -\epsilon_{-\boldsymbol{k}} \end{pmatrix} \boldsymbol{\psi}(\boldsymbol{k}) = \tilde{\boldsymbol{\psi}}^\dagger(\boldsymbol{k}) \begin{pmatrix} E_+(\boldsymbol{k}) & 0 \\ 0 & E_-(\boldsymbol{k}) \end{pmatrix} \tilde{\boldsymbol{\psi}}(\boldsymbol{k}). \tag{5.101}$$

Here the two-component field $\tilde{\psi}^{\dagger}(\boldsymbol{k}) = (\alpha^{\dagger}(\boldsymbol{k}), \beta^{\dagger}(\boldsymbol{k}))$ corresponds to creation operators of quasi-particles. From Eq. (5.23), we obtain the inverse transformation

$$\alpha_{\boldsymbol{k}} = u_{\boldsymbol{k}} c_{\uparrow}(\boldsymbol{k}) - e^{i\phi} v_{\boldsymbol{k}} c_{\downarrow}(-\boldsymbol{k})^{\dagger}, \tag{5.102}$$

$$\beta_{\boldsymbol{k}} = e^{-i\phi} v_{\boldsymbol{k}} c_{\uparrow}(\boldsymbol{k}) + u_{\boldsymbol{k}} c_{\downarrow}(-\boldsymbol{k})^{\dagger}, \tag{5.103}$$

which is often called the Bogoliubov transformation.

**Problem 5.3**

Because of the property $E_{-}(\boldsymbol{k}) = -E_{+}(\boldsymbol{k}) < 0$, the ground state is characterized by $\langle \beta_{\boldsymbol{k}}^{\dagger} \beta_{\boldsymbol{k}} \rangle = 1$ and $\langle \alpha_{\boldsymbol{k}}^{\dagger} \alpha_{\boldsymbol{k}} \rangle = 0$. Hence the ground state should be constructed from the vacuum $|0\rangle$ by

$$\prod_{\boldsymbol{k}} \beta_{\boldsymbol{k}}^{\dagger} \alpha_{\boldsymbol{k}} |0\rangle = \prod_{\boldsymbol{k}} \left(-e^{i\phi}\right) v_{\boldsymbol{k}} \left[u_{\boldsymbol{k}} + e^{i\phi} v_{\boldsymbol{k}} c_{\uparrow}^{\dagger}(\boldsymbol{k}) c_{\downarrow}^{\dagger}(-\boldsymbol{k})\right] |0\rangle, \tag{5.104}$$

which is indeed proportional to $\Psi_{\mathrm{BCS}}$ given by Eq. (5.8).

**Problem 5.4**

Nonzero order parameter requires the zero determinant of the matrix in Eq. (5.41). This condition is equivalent to

$$1 - g_{\mathrm{A}} L_{\mathrm{A}} = \frac{J_{\mathrm{H}}^2 \rho_{\mathrm{A}} \rho_{\mathrm{B}} L_{\mathrm{A}} L_{\mathrm{B}}}{1 - g_{\mathrm{B}} L_{\mathrm{B}}} \quad (> 0). \tag{5.105}$$

With decreasing temperature, $L_{\mathrm{A}}(T)$ increases, and the LHS becomes zero at $T = T_{\mathrm{A}}$ with $g_{\mathrm{A}} > 0$. Hence there is a temperature $T_c$ above $T_{\mathrm{A}}$ where Eq. (5.105) is satisfied. Note that the result $T_c > T_{\mathrm{A}}$ does not depend on the signs of $J_{\mathrm{H}}$ and $g_{\mathrm{B}}$. Namely, the pair transfer always increases $T_c$, even if the band $B$ does not favor superconductivity. This is analogous to lowering of the ground state energy in the second-order perturbation theory.

**Problem 5.5***

Although the topic is included in superconductivity, it is appropriate to challenge the problem after learning the scaling theory to be discussed in Sect. 6.2. Following the standard procedure to derive the effective Hamiltonian in the scaling theory, we change the bandwidth by the infinitesimal amount $\delta D < 0$ on both high and low energy ends. Correspondingly, in Fig. 5.6 which represents $\Pi(\boldsymbol{p})$, both of two electrons in the intermediate state have energies in the narrow range of $[D+\delta D, D]$ or $[-D, D - \delta D]$. Then the change of the effective interaction is given by

$$\delta U = \frac{\delta D}{D} U^2 \rho_{\mathrm{B}}. \tag{5.106}$$

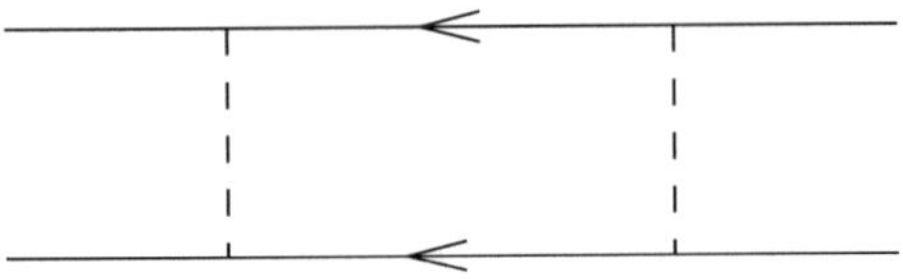

**Fig. 5.6** Correction to the effective interaction which corresponds to renormalization of the Coulomb repulsion. The energy of electrons with arrows are both near the upper or lower band edges

corresponding to Eq. (6.16). By integrating both sides of Eq. (5.106) and putting $D_{\rm eff} = \omega_{\rm D}$, we obtain Eq. (5.47).

**Problem 5.6***
It is convenient to deal first with $\hat{h}(\boldsymbol{k})^2$, which is easily obtained as

$$\hat{h}(\boldsymbol{k})^2 = \begin{pmatrix} \epsilon_{\boldsymbol{k}}^2 + \bar{\Delta}\bar{\Delta}^\dagger & 0 \\ 0 & \epsilon_{\boldsymbol{k}}^2 + \bar{\Delta}^\dagger\bar{\Delta} \end{pmatrix}. \tag{5.107}$$

With use of quantities defined by Eq. (5.61), we obtain

$$\bar{\Delta}\bar{\Delta}^\dagger = |D|^2 + |\boldsymbol{d}|^2 + \boldsymbol{w}\cdot\boldsymbol{\sigma}, \quad \bar{\Delta}^\dagger\bar{\Delta} = |D|^2 + |\boldsymbol{d}|^2 - \boldsymbol{w}\cdot\boldsymbol{\sigma},$$

where $\boldsymbol{w} = \mathrm{i}\boldsymbol{d} \times \boldsymbol{d}^* = \boldsymbol{w}^*$ is a real vector. Diagonalization of $2 \times 2$ matrices in Eq. (5.107) can easily be performed. From the eigenvalues for $\hat{h}(\boldsymbol{k})^2$, we take the square root to obtain the eigenvalues of $\hat{h}(\boldsymbol{k})$ as

$$E_{\tau\sigma}(\boldsymbol{k}) = \tau\sqrt{\epsilon(\boldsymbol{k})^2 + |D(\hat{k})|^2 + |\boldsymbol{d}(\hat{k})|^2 + \sigma|\boldsymbol{w}(\hat{k})|}, \tag{5.108}$$

with $\tau, \sigma = \pm 1$. This result suggests that $\boldsymbol{w}$ behaves as a kind of magnetic field.

## References


1. Landau, L.D., Lifshitz, E.M.: Statistical Physics, 3rd edn, Part 2, p. 85. Elsevier, Oxford (2005)
2. Bogoliubov, N.: J. Phys. USSR **11**, 23 (1947)
3. Bardeen, J., Cooper, L., Schrieffer, J.: Phys. Rev. **108**, 1175 (1957)
4. Kondo, J.: Prog. Theor. Phys. **29**, 1 (1963)
5. Bednorz, J.G., Müller, K.A.: Z. Phys. B **64**, 189 (1986)
6. Kamihara, Y., Watanabe, T., Hirano, M., Hosono, H.: J. Am. Chem. Soc. **130**, 3296 (2008)
7. Ishida, K., Nakai, Y., Hosono, H.: J. Phys. Soc. Jpn. **78**, 62001 (2009)
8. Nagamatsu, J., Nakagawa, N., Muranaka, T., Zenitani, Y., Akimitsu, J.: Nature **410**, 63 (2001)
9. Vollhardt, D., Wölfle, P.: The Superfluid Phases of Helium 3. Dover Books on Physics. Dover, New York (2013)
10. Gladstone, G., Jensen, M.A., Schrieffer, J.R.: In: Superconductivity, Parks, R.D. (ed.), p. 665. Dekker, New York, (1969)

11. Mackenzie, A.P., Maeno, Y.: Rev. Mod. Phys. **75**, 657 (2003)
12. Ashcroft, N.W.: Phys. Rev. Lett. **92**, 187002 (2004)
13. Peng, F., Sun, Y., Pickard, C.J., Needs, R.J., Wu, Q., Yanming, M.: Phys. Rev. Lett. **119**, 107001 (2017)
14. Drozdov, A.P., Eremets, M.I., Troyan, I.A., Ksenofontov, V., Shylin, S.I.: Nature (London) **525**, 73 (2015)
15. Somayazulu, M., Ahart, M., Mishra, A.K., Geballe, Z.M., Baldini, M., Meng, Y., Struzhkin, V.V., Hemley, R.J.: Phys. Rev. Lett. **122**, 027001 (2019)
16. Drozdov, A.P., Kong, P.P., Minkov, V.S., Besedin, S.P., Kuzovnikov, M.A., Mozaffari, S., Balicas, L., Balakirev, F.F., Graf, D.E., Prakapenka, V.B., Greenberg, E., Knyazev, D.A., Tkacz, M., Eremets, M.I.: Nature **569**, 528 (2019)
17. Schilling, A., Cantoni, M., Guo, J.D., Ott, H.R.: Nature **363**, 56 (1993)
18. Bauer, E., Sigrist, M. (ed.): Non-Centrosymmetric Superconductors: Introduction and Overview, Lecture Notes in Physics, vol. 847. Springer, Berlin (2012)
19. Sigrist, M., Ueda, K.: Rev. Mod. Phys. **63**, 239 (1991)
20. Steglich, F., Aarts, J., Bredl, C.D., Lieke, W., Meschede, D., Franz, W., Schafer, H.: Phys. Rev. Lett. **43**, 1892 (1979)
21. Pfleiderer, C.: Rev. Mod. Phys. **81**, 1551 (2009)
22. Lindgren, I., Morrison, J.: Atomic Many-Body Theory. Springer, Berlin (1986)
23. Hlubina, R.: Phys. Rev. B**59**, 9600 (1999)

# Chapter 6
# Kondo Effect

**Abstract** This chapter discusses Kondo effect for a magnetic impurity in metals. The concept of renormalization is crucial in understanding Kondo effect. Simply speaking, the exchange interaction between the impurity and conduction electrons becomes endlessly large in the effective Hamiltonian, even though its magnitude is originally tiny. As a result, the impurity spin is completely screened in the ground state, resulting in a local Fermi liquid. However, if the impurity has also an orbital degrees of freedom, another ground state may appear which is *not* a Fermi liquid but with a residual entropy. We begin with description of the model and proceed to actual procedure of renormalization. The effective Hamiltonian formalism presented in Chap. 1 is used extensively. Also discussed is how to understand the ground state, which is either Fermi or non-Fermi liquid.

## 6.1 Hybridization and Exchange Interactions

The most popular model to describe a magnetic impurity in metals is called the Anderson model [1]. The simplest version takes a non-degenerate orbital for local electrons at the origin, together with a non-degenerate conduction band. The annihilation operators of these electrons are written as $f_\sigma$ with spin $\sigma$ and $c_{\boldsymbol{k}\sigma}$ with momentum $\boldsymbol{k}$, respectively. The local and conduction electrons are called $f$- and $c$-electrons, respectively, in the following. The Anderson model is given by

$$H_{\rm A} = \sum_{k\sigma} \epsilon_{\boldsymbol{k}} c^\dagger_{\boldsymbol{k}\sigma} c_{\boldsymbol{k}\sigma} + \epsilon_f \sum_\sigma n_{f\sigma} + \frac{1}{2} U \sum_{\sigma \neq \sigma'} n_{f\sigma} n_{f\sigma'} + H_{\rm hyb}, \tag{6.1}$$

$$H_{\rm hyb} = \frac{1}{\sqrt{N}} \sum_{k\sigma} V (c^\dagger_{\boldsymbol{k}\sigma} f_\sigma + f^\dagger_\sigma c_{\boldsymbol{k}\sigma}), \tag{6.2}$$

where $U$ represents the Coulomb repulsion between local electrons with the number operator $n_{f\sigma} = f^\dagger_\sigma f_\sigma$. The two kinds of electrons mix (or hybridize) with strength $V$. The factor $1/\sqrt{N}$, with $N$ being the number of lattice points in the system, enters in superposing the Bloch state to build the Wannier state localized at the origin. The



Y. Kuramoto, *Quantum Many-Body Physics*, Lecture Notes in Physics 934,
https://doi.org/10.1007/978-4-431-55393-9_6

actual hybridization of a local electron mostly occurs with a surrounding ligand orbital, which accompanies $\boldsymbol{k}$-dependence in hybridization. Equation (6.2) neglects this aspect for simplicity. The Anderson model has rich content encompassing the atomic (localized electron) limit with $U/V \gg 1$ (or $UD/V^2 \gg 1$ for the large bandwidth $D$), and the single-particle resonance with $U = 0$.

Let us consider the situation where the $f$-electron level $\epsilon_f$ is much below the Fermi level, while $\epsilon_f + U$, which is the energy for the second electron to occupy, is much above because of the large $U$. Then the average occupation number of $f$-electrons is almost one, and the fluctuation from the average is negligible. In this situation it is reasonable to take the model space where the occupancy of the local level is fixed to one. In the model space the spin $\boldsymbol{S}$ is the only degrees of freedom for $f$-electrons. We derive the effective Hamiltonian explicitly for this model space.

The hybridization $H_{\text{hyb}}$ connects the model space to the complementary space projected by $Q$, which has either zero or double occupancy of $f$-electrons. Then the effective interaction in the lowest order is given by

$$H_{\text{int}} = P H_{\text{hyb}}(E_{\text{g}} - H_c - H_f)^{-1} Q H_{\text{hyb}} P, \tag{6.3}$$

where $E_{\text{g}}$ is the zeroth-order energy for the ground state, and $H_c$, $H_f$ represent $c$- and $f$-electron parts without hybridization, respectively. Figure 6.1 illustrates the perturbation processes in the effective interaction. Since the double occupancy is possible only for singlet of local spins, hybridization to the doubly occupied state works only for singlet pair of $f$- and $c$-electrons. Hence the processes (a), (b) shown in Fig. 6.1 accompany the singlet projection operator

$$P_{\text{s}} = -\boldsymbol{S} \cdot \boldsymbol{s}_c + \frac{1}{4} n_c, \tag{6.4}$$

where $\boldsymbol{s}_c$ and $n_c$ are, respectively, spin and number operators of $c$-electrons at the origin. In the process (a), the $c$-electron spin remains the same through

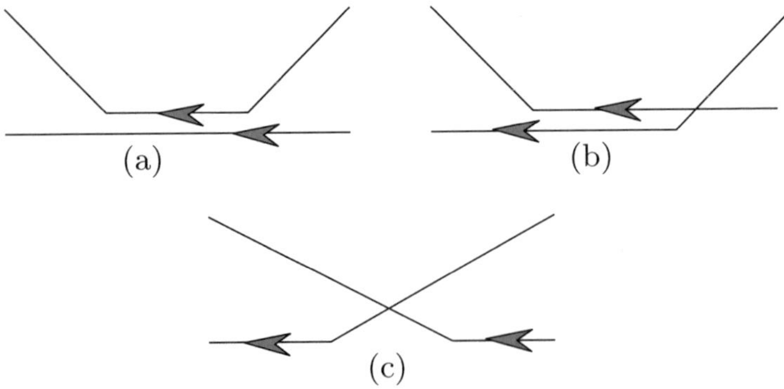


**Fig. 6.1** Second-order processes of hybridization. The horizontal lines with arrow represent occupied $f$-states with any spin, and the slant lines represent $c$-electron states

hybridization, while in (b), a spin exchange occurs with $f$ states. On the other hand, the process (c) in Fig. 6.1 involves the vacant $f$ state, and accompanies the spin exchange which allows triplet states of $f$- and $c$-electrons as well. The latter kind of exchange is called permutation, and is described by the operator

$$\mathcal{P}_{\text{spin}} = 2\boldsymbol{S} \cdot \boldsymbol{s}_c + \frac{1}{2} n_c, \tag{6.5}$$

which gives $\pm 1$ when applied to triplet and singlet states, respectively. In this way we rearrange the processes (a), (b), and (c) in Fig. 6.1 into the exchange interaction including $\boldsymbol{S} \cdot \boldsymbol{s}_c$, and the potential scattering including $n_c$. The effective Hamiltonian thus takes the form

$$H_{\text{eff}} = H_c + \frac{1}{N} \sum_{k\sigma} \sum_{k'\sigma'} \left( \frac{1}{2} J_{kk'} \boldsymbol{S} \cdot \boldsymbol{\sigma}_{\sigma'\sigma} c^{\dagger}_{k'\sigma'} c_{k\sigma} + K_{kk'} \delta_{\sigma\sigma'} c^{\dagger}_{k'\sigma'} c_{k\sigma} \right), \tag{6.6}$$

where the exchange interaction is given by

$$J_{kk'} = 2V^2 \left[ \frac{1}{\epsilon_{k'} - \epsilon_f} - \frac{1}{\epsilon_k - \epsilon_f - U} \right], \tag{6.7}$$

and the potential scattering by

$$K_{kk'} = \frac{1}{2} V^2 \left[ \frac{1}{\epsilon_{k'} - \epsilon_f} + \frac{1}{\epsilon_k - \epsilon_f - U} \right]. \tag{6.8}$$

For simplicity, we consider the situation where the energy $\epsilon_k$ of conduction electrons is negligible as compared with $\epsilon_f$ and $\epsilon_f + U$. In the special case with the condition $\epsilon_f + U = |\epsilon_f|$, which is called the symmetric case, the potential scattering $K_{kk'}$ vanishes. The exchange interaction in this case is given by

$$J = 4V^2/|\epsilon_f| = 8V^2/U. \tag{6.9}$$

In summary, by neglecting charge fluctuations of $f$-electrons in the Anderson model, we have obtained the effective model

$$H_{\text{K}} = H_c + J\boldsymbol{S} \cdot \boldsymbol{s}_c = H_c + H_{\text{ex}}, \tag{6.10}$$

which is called the Kondo model. In the momentum representation, the spin operator $\boldsymbol{s}_c$ is given by

$$\boldsymbol{s}_c = \frac{1}{2N} \sum_{kk'} \sum_{\sigma\sigma'} c^{\dagger}_{k\sigma} \boldsymbol{\sigma}_{\sigma\sigma'} c_{k'\sigma'}. \tag{6.11}$$

The Kondo model has even simpler appearance than the Anderson model. However, the model has the astonishing property that perturbation theory in $J$ breaks down at zero temperature, no matter how small $J$ is. At high temperatures, on the other hand, the perturbation theory works. Here the meaning of "high temperatures" is highly nontrivial, as explained later. We note that the expansion parameters $U$ in the Anderson model, and $J$ in the Kondo model assume different states as the starting point. These different states can be connected continuously by the renormalization, which is discussed in the next section.

## 6.2 Renormalization in Kondo Model

Renormalization of Kondo model has been studied by a variety of methods. We apply the effective Hamiltonian formalism, which is equivalent to the method often called the scaling. Namely, the effective bandwidth in the model space is reduced by infinitesimal amount $|\delta D|$ in each step of renormalization. Then the renormalization takes the form of a differential equation for the effective exchange interaction $J$, and the integration gives a finite change of $J$. The process how the effective interaction changes is called the renormalization (or scaling) flow. An example will be illustrated later in Fig. 6.3.

We assume the simplest form of the conduction band with constant density of states (per spin) $\rho_c = (2D)^{-1}$ in the interval $[-D, D]$, and zero otherwise. The Fermi level is located at the center of the band: $\mu = 0$. With this setting we consider a new model space in the Kondo model where the $c$-electron states near the upper and lower band edges are excluded by an infinitesimal amount. Namely, we choose the projection operator $Q$ that includes only such $c$-electron states with energy $[D + \delta D, D]$ or $[-D, -D - \delta D]$ $(\delta D < 0)$. The model space projected by $P = 1 - Q$ excludes those states near the band edges. In the lowest order with respect to $J$, the effective Hamiltonian in the new model space is given by

$$H_{\text{eff}} = P(H_c + H_{\text{ex}})P + PH_{\text{ex}}(E_{\text{g}} - H_c)^{-1}QH_{\text{ex}}P, \tag{6.12}$$

where $E_{\text{g}}$ is the ground state energy of $H_c$.

Let us explicitly derive the change of $H_{\text{eff}}$ by the infinitesimal change $\delta D$. Figure 6.2 illustrates the second-order terms in Eq. (6.12) which gives $\delta J$. The $c$-electron has spin $\sigma$ for the incident state, $\xi$ for intermediate states, and $\sigma'$ for the scattered state. Let us focus on the component $JS^\beta s_c^\beta$ in the first scattering by $QH_{\text{ex}}P$, and another one $JS^\alpha s_c^\alpha$ in the second scattering by $PH_{\text{ex}}$. In Fig. 6.2a, which is called the direct scattering diagram, the matrix elements of $\boldsymbol{s}_c$ are given by

$$\langle\sigma'|s_c^\alpha|\xi\rangle\langle\xi|s_c^\beta|\sigma\rangle. \tag{6.13}$$

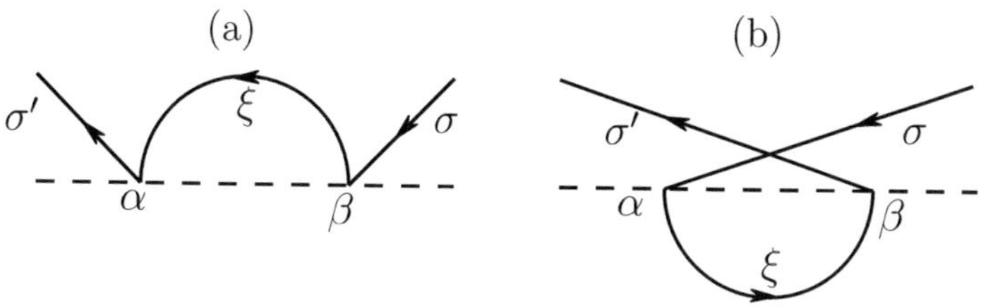


**Fig. 6.2** Second-order correction to the effective interaction. The solid lines represent $c$-electron states, and the dashed lines impurity spin states

On the other hand, the crossed diagram shown in Fig. 6.2b has matrix elements

$$\langle \sigma' | s_c^{\beta} | \xi \rangle \langle \xi | s_c^{\alpha} | \sigma \rangle. \tag{6.14}$$

Summation over $\xi$ can be taken independent of the energy denominator. Taking care of the minus sign in the process (b) that involves commutation of fermion operators, we combine (a) and (b) to obtain the second-order correction to the effective interaction as

$$\frac{J^2}{-D} \sum_{\alpha\beta} S^{\alpha} S^{\beta} \left[ s_c^{\alpha}, s_c^{\beta} \right] |\delta D| \rho_c = -\frac{\delta D}{D} J^2 \rho_c \mathbf{S} \cdot \mathbf{s}_c. \tag{6.15}$$

Some details how to use the spin commutation relation are the subject of Problem 6.1. We emphasize that the correction has precisely the same form as the original interaction $H_{\text{ex}}$ except for the coupling strength. Hence the renormalization appears as the change $\delta J$ of the exchange interaction given by

$$\delta J = -\frac{\delta D}{D} J^2 \rho_c, \tag{6.16}$$

which is called the scaling equation. It is also called the renormalization group equation since each step of scaling is regarded as an element of a group. However, the renormalization group has no inverse element, and is not an ordinary group in the mathematical sense. We remark that the spin commutation rule plays the essential role in the renormalization. If the perturbation $H_{\text{ex}}$ in Eq. (6.12) is replaced by a potential scattering $V$, the second-order processes (a) and (b) in Fig. 6.2 cancel each other, and renormalization of $V$ does not occur.

We can repeat the renormalization procedure as long as the effective bandwidth $2D_{\text{eff}}$ remains much larger than the characteristic energy scale, which is to be determined later. Under this restriction, integration of the scaling equation leads to

$$J_{\text{eff}} = \frac{J}{1 - J\rho_c \ln(D/D_{\text{eff}})}, \tag{6.17}$$

where the boundary condition is $J_{\text{eff}} = J$ at $D_{\text{eff}} = D$. Most importantly, $J_{\text{eff}}$ increases with the decrease of $D_{\text{eff}}$. The divergence at $D_{\text{eff}} = D\exp[-1/(J\rho_c)]$ should be taken with care since the result is outside the valid range of second-order renormalization. The Kondo temperature $T_{\text{K}}$ is defined by this energy in units of temperature. With the Boltzmann constant $k_{\text{B}}$ set to unity, $T_{\text{K}}$ is given explicitly by

$$T_{\text{K}} = D\exp\left(-\frac{1}{J\rho_c}\right) \to D\exp\left(-\frac{U}{8V^2\rho_c}\right), \tag{6.18}$$

where the rightmost expression corresponds to parameters in the symmetric Anderson model. Note that the scale $T_{\text{K}}$ cannot be obtained by perturbation theory in $J$, since it is not analytic around $J = 0$. On the other hand, $T_{\text{K}}$ is analytic around $U = 0$ according to Eq. (6.18). Although the result has been obtained with the assumption of large $U$, the analyticity remains valid also down to small $U$.

The divergence of $J_{\text{eff}}$ at $D_{\text{eff}} = T_{\text{K}}$ in Eq. (6.17) should not be taken literally. It should rather be interpreted that the perturbation theory breaks down below the energy of the order of $T_{\text{K}}$. In this sense, the characteristic energy scale of the system is given by $T_{\text{K}}$. As we shall discuss later, the Kondo temperature $T_{\text{K}}$ characterizes physical quantities such as resistivity, susceptibility, and specific heat.

At finite temperature $T$, $c$-electrons are thermally excited in the range of $\pm T$ around the Fermi level. As long as $D_{\text{eff}} \gg T$, there is no effect of temperature in the renormalization. However, $D_{\text{eff}}$ cannot be made smaller than $T$ because of thermal excitations. Then the effective exchange at $T$ is given by setting $D_{\text{eff}} = T$ as

$$J_{\text{eff}}(T) = \frac{J}{1 - J\rho_c \ln(D/T)}. \tag{6.19}$$

In deriving physical quantities such as resistivity and susceptibility at finite $T$, one can use the straightforward perturbation theory in $J$ to arbitrary order. The equivalent result is reproduced simply by replacing the bare $J$ by $J_{\text{eff}}(T)$, as long as the most divergent logarithmic terms are concerned. This situation is best illustrated in the electric resistivity $\rho(T)$ at $T$ sufficiently larger than $T_{\text{K}}$. In the simplest transport theory, the conductivity $\sigma = 1/\rho$ is determined in terms of the relaxation time $\tau$ as follows:

$$\sigma = \frac{ne^2\tau}{m^*}, \tag{6.20}$$

where $n$ is the density of conduction electrons with the effective mass $m^*$. In the lowest order for scattering by $J$, which is called the Born approximation, we obtain

$$\frac{1}{\tau} = 2\pi c_{\text{imp}} v_{\text{cell}} J^2 \rho_c \sum_{\alpha} \left\langle S_\alpha^2 s_\alpha^2 \right\rangle = \frac{3\pi}{8} c_{\text{imp}} v_{\text{cell}} J^2 \rho_c, \tag{6.21}$$

where $c_{\text{imp}}$ is the density of magnetic impurities and $v_{\text{cell}}$ is the volume of unit cell. Derivation of Eq. (6.21) is the subject of Problem 6.2. The Born approximation does not take into account higher order effect of $J$, which affects the low-energy physics via intermediate states with high excitation energies. The effects of these intermediate states are taken into account by using $J_{\text{eff}}(T)$ instead of $J$. The replacement gives

$$\rho(T) = \frac{\rho_0}{[1 - J\rho_c \ln(D/T)]^2} = \frac{\rho_0}{[J\rho_c \ln(T/T_{\text{K}})]^2}, \tag{6.22}$$

where $\rho_0$ represents the resistivity in the Born approximation. Equation (6.22) reproduces the original result of Kondo who derived the logarithmic term in the lowest order, $O(J^3)$ [2], and also more elaborate calculation which sums up the most divergent perturbation series [3]. The result is, however, justified only for $T \gg T_{\text{K}}$, but breaks down as the temperature approaches $T_{\text{K}}$.

The actual $\rho(T)$ has an upper bound, in contrast to the divergent behavior at $T \to T_{\text{K}}$ suggested by Eq. (6.22). For more precise treatment than the Born approximation, we deal with the transition matrix (called the $t$-matrix) of conduction electrons. According to the scattering theory [4], the $S$- and $t$-matrix elements are related by

$$S_{ab} = \delta_{ab} - 2\pi \mathrm{i}\delta(E_a - E_b)t_{ab}, \tag{6.23}$$

where $E_a$ is the kinetic energy of the one-particle state $a$. Since the $S$-matrix is unitary, $S^\dagger S = 1$, the $t$-matrix is constrained to be

$$\operatorname{Im} t_{aa} = -\pi \sum_b \delta(E_a - E_b)|t_{ab}|^2, \tag{6.24}$$

which is called the optical theorem. The LHS describes the forward scattering, while the RHS gives the total cross section of scattering. The latter is proportional to $1/\tau$. In the angular momentum representation, the phase shift $\delta$ for the $s$-wave scattering characterizes the $t$-matrix. Then the dimensionless quantity $(J\rho_c)^2$ in the Born approximation is replaced by $\sin^2\delta$ in Eq. (6.21) for $1/\tau$. The upper bound corresponds to $\delta = \pi/2$, which is called the unitarity limit. Hence the divergence of $\rho(T)$ is a result of unjustified approximation. Another consequence of the unitarity constraint will be discussed later in Sect. 10.6.

## 6.3 Anisotropic Kondo Model

The exchange interaction in the Kondo model is positive ($J > 0$), as long as it is regarded as the effective Hamiltonian of the Anderson model. In actual cases, other origins of $J$ are present such as the Coulomb repulsion between local and conduction electrons. The latter is ferromagnetic in sign ($J < 0$). Since the result given by Eq. (6.17) is valid for both signs of $J$, the negative renormalized exchange becomes smaller in magnitude than the bare one. This means that the impurity spin tends to be decoupled from conduction electrons at low temperature. In other words, the case $J < 0$ is connected continuously to the special point $J = 0$, which leads to the Curie law $\chi = C/T$ for the magnetic susceptibility at low temperature.

A natural question then arises as to what happens if the exchange is anisotropic, including the case of positive and negative components. The resultant anisotropic Kondo model is useful to visualize the renormalization flows how the localized limit obeying the Curie law changes to the Pauli law similar to itinerant electron systems. The simplest model including the anisotropy is given by

$$H_{\rm ex} = J_\perp(S_x s_x + S_y s_y) + J_z S_z s_z, \tag{6.25}$$

with axial symmetry. The scaling equation is modified from Eq. (6.16) as

$$\delta J_z = -\frac{\delta D}{D} J_\perp^2 \rho_c, \quad \delta J_\perp = -\frac{\delta D}{D} J_z J_\perp \rho_c. \tag{6.26}$$

Combining the two equations above we obtain another differential equation

$$\mathrm{d}J_z/\mathrm{d}J_\perp = J_\perp/J_z, \tag{6.27}$$

which is integrated as

$$J_z^2 - J_\perp^2 = C, \tag{6.28}$$

with $C$ being a real number fixed by the bare exchange.

Figure 6.3 illustrates the scaling flows with various values of the bare exchange interaction. The arrows indicate scaling flows as $D_{\rm eff}$ decreases. These scaling flows do not depend on the sign of $J_\perp$. The end point of each arrow is called the fixed point of renormalization. It can be seen that the fixed point is either $J_\perp = 0$ or $|J_\perp| = \infty$. The former ($J_\perp = 0$) represents the absence of spin flip that leads to the

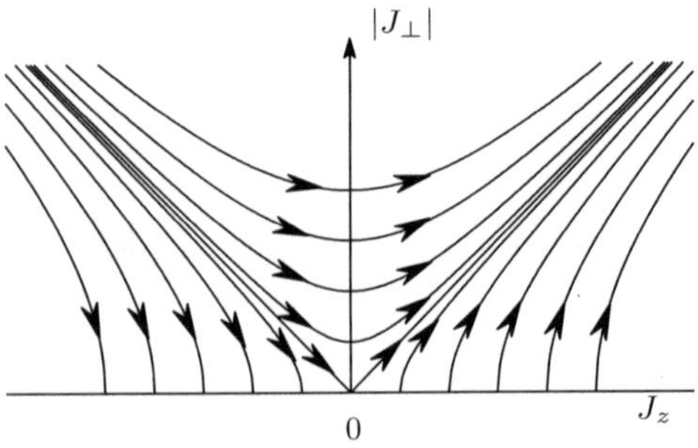


**Fig. 6.3** Renormalization flows of anisotropic exchange. The same flow occurs for both signs of $J_\perp$. The arrows represent the direction of renormalization as the effective bandwidth is reduced from the bare ones

Curie law. The other fixed point $|J_\perp| = \infty$ represents the complete screening as in the isotropic case.

The Kondo temperature $T_K$ in the anisotropic Kondo model is defined as the energy where the effective exchange diverge according to Eq. (6.26). Since the integration constant remains finite, the divergence occurs simultaneously in $J_z$ and $J_\perp$. With some algebra which is the subject of Problem 6.3 [5], we obtain the following result for $J_z > J_\perp > 0$:

$$T_K = D \exp\left(-\frac{1}{J_{z\perp}\rho_c} \tanh^{-1} \frac{J_{z\perp}}{J_z}\right), \tag{6.29}$$

with $J_{z\perp} \equiv \sqrt{J_z^2 - J_\perp^2}$, and $\tanh^{-1}$ being the inverse hyperbolic tangent function. In the isotropic limit $J_{z\perp} \to 0$, we recover Eq. (6.18). In the Ising limit $J_\perp = 0$ or $J_{z\perp} = J_z$, on the other hand, we obtain $T_K \to 0$. This is physically obvious because Kondo effect is absent without spin flip. For the opposite anisotropy $J_\perp > J_z > 0$, the result is given by

$$T_K = D \exp\left(-\frac{1}{J_{\perp z}\rho_c} \tan^{-1} \frac{J_{\perp z}}{J_z}\right), \tag{6.30}$$

with $J_{\perp z} \equiv \sqrt{J_\perp^2 - J_z^2}$. We again recover the isotropic limit with $J_{\perp z} \to 0$. On the other hand, $T_K$ remains finite in the limit $J_z = 0$, since the exponent does not diverge in the limit. This corresponds to the scaling behavior around $J_z \sim 0$ in Fig. 6.3.

## 6.4 Ground State of Kondo Systems

Because of the mapping discussed in Sect. 6.1, it is reasonable to refer to both Kondo and Anderson models as Kondo systems. The fixed point of the Kondo model with $J \to \infty$ is understood intuitively in terms of the Anderson model. For the symmetric case $2\epsilon_f + U = 0$, the state with completely screened local spin is connected to the trivial limit $U \to 0$ in the Anderson model, as suggested by the relation $J \sim V^2/U$. It is known [6] that the ground state of the Anderson model has the spin singlet irrespective of the magnitude of $U$ $(> 0)$. Namely, the singlet ground state is connected continuously both to the $U = 0$ state with finite occupation of $f$ state and to the vacant $f$ state. It is obvious that the vacant state has the spin singlet even with $U \neq 0$.

In the following, we shall discuss the other end of extreme case with $U \to \infty$. We generalize the spin degeneracy to an arbitrary integer $n$, which is called the SU($n$) Anderson model. The generalization actually has a relation to realistic rare-earth magnetic impurities. Because of the large spin–orbit coupling, the $4f^1$ configuration of $Ce^{3+}$, for example, has the lowest level with the total angular momentum $J = L - S = 5/2$, where $L = 3$ and $S = 1/2$ is the orbital and spin angular momenta, respectively. Hence we have $n = 2J + 1 = 6$ in this case. In $Yb^{3+}$ with $4f^{13}$ configuration, the lowest level has $J = 7/2$ and $n = 8$. Such large values of $n$ make it useful to start from the result obtained in the limit of $n \to \infty$ [7].

We modify the original Anderson model $H_A$ in such a way that the spin index $\sigma$ runs from 1 to $n$, and takes the limit $U \to \infty$. In the resultant SU($n$) Anderson model

$$H_{SU(n)} = H_c + H_f + H_{hyb}, \tag{6.31}$$

the characteristic temperature, which corresponds to $T_K$, depends on the hybridization intensity $W_0(\epsilon)$ as defined by

$$W_0(\epsilon) = \frac{1}{N}\sum_{\boldsymbol{k}} |V|^2 \delta(\epsilon - \epsilon_{\boldsymbol{k}}). \tag{6.32}$$

We make the simplest model where $W_0(\epsilon) = W_0$ is a constant for $-D < \epsilon < D$, and 0 otherwise. In order to organize the perturbation series we take $nW_0$ as the unit of energy.

According to the Brillouin–Wigner perturbation theory, as discussed in Chap. 1, the singlet energy $E_0$ is given in the lowest order by

$$\begin{aligned} E_0 &= \langle 0|H_{hyb}(E_0 - H_c - H_f)^{-1} H_{hyb}|0\rangle \\ &= nW_0 \int_{-D}^{0} \frac{d\epsilon}{E_0 + \epsilon - \epsilon_f}, \end{aligned} \tag{6.33}$$

where the origin of the energy is taken to be the ground state of conduction electrons without $f$ states. The energy shift from the unperturbed state is called the effective potential, or the self-energy, which is illustrated in the left panel of Fig. 6.4. In the extreme case of $nW_0 \ll |\epsilon_f| \sim D$, we can derive $E_0$ analytically. The result is given by

$$E_0 - \epsilon_f \equiv -T_0 \simeq -D \exp\left(\frac{\epsilon_f}{nW_0}\right), \tag{6.34}$$

**Fig. 6.4** Self-energies in the lowest order of hybridization. The left panel shows the singlet case with vacant $f$ state (wavy line) as the model space, while the right panel shows the $n$-fold degenerate case with a singly occupied $f$ state (dashed line) as the model space

the derivation of which is the subject of Problem 6.4 [8]. In the case of $n = 2$ and $U = \infty$, the energy $T_0$ is the same as $T_K$ given by Eq. (6.18) with $J\rho_c = 2W_0/|\epsilon_f|$ according to Eq. (6.7). Note that $T_0$ here is defined at zero temperature without any reference to divergence.

On the other hand, any of the $n$-fold degenerate states has the self-energy shown in the right panel of Fig. 6.4, with the magnitude of $O(W_0)$. Note that the self-energy is of $O(1/n)$ in units of $nW_0$, since the hybridization works only for the same spin index of $f$- and $c$-electrons. This is to be contrasted with the singlet case where the hybridization involves $n$ different species, making the self-energy of $O(1)$ with the same unit. This difference between the two self-energies causes the reversal of renormalized energy levels from the bare levels, even though the local level $\epsilon_f$ is deep inside the Fermi level.

In order to compare the renormalized energies of singlet and multiplet states, we need actually another dimensionless parameter. Provided $|\epsilon_f|$ and $D$ are of the same order, the relevant parameter is the ratio $r \equiv |\epsilon_f|/(nW_0)$ $(> 1)$. Then we require the condition $r \ll \ln n$, which leads to the inequality

$$T_0 = D\exp(-r) \gg D/n > W_0. \tag{6.35}$$

In the large $n$ limit, the multiplet energy $E_1 \sim \epsilon_f - O(W_0)$ is higher than the singlet energy $E_0 = \epsilon_f - T_0$. Moreover, higher order terms for the singlet self-energy are at most of $O(1/n)$, and can be neglected in the large $n$ limit. It is remarkable that such simple calculation has identified the characteristic energy scale $T_0$.

We proceed to the case where the average occupation number $n_f$ of $f$-electrons is not an integer. Such situation in actual rare-earth systems including Ce or Yb is called the mixed valence or valence fluctuation. In the present framework, $n_f$ is derived as

$$n_f = \frac{\partial E_0}{\partial \epsilon_f} = nW_0 \int_{-D}^{0} d\epsilon \frac{1-n_f}{(E_0+\epsilon-\epsilon_f)^2} = \frac{nW_0}{T_0}(1-n_f), \tag{6.36}$$

which means

$$n_f = \left(1+\frac{T_0}{nW_0}\right)^{-1}. \tag{6.37}$$

Hence $T_0$ becomes larger as $n_f$ becomes smaller. The case $n_f \to 1$ is often called the Kondo limit.

The magnetic susceptibility $\chi$ at zero temperature can be derived by the second derivative of the groundstate energy $E_0(H)$ with respect to the magnetic field $H$. We associate the total angular momentum $J = (n-1)/2$ with the degeneracy $n$, and $J_z$ with each of the $n$ degenerate components. Since $E_0(H)$ is given by

$$E_0 = W_0 \sum_{J_z} \int_{-D}^{0} \frac{d\epsilon}{E_0+\epsilon-\epsilon_f - g_J\mu_B J_z H}, \tag{6.38}$$

the second derivative leads to

$$\chi = C_J n_f / T_0, \tag{6.39}$$

where $C_J = (g_J \mu_B)^2 J(J+1)/3$ is the Curie constant. Derivation of Eq. (6.39) is the subject of Problem 6.5. Note that $T_0$ enters in $\chi$ in place of temperature $T$ for the Curie law.

## 6.5 Local Fermi Liquid

The Kondo systems lack the translational symmetry, but still the idea of the Fermi liquid theory is applicable, as developed by Nozières [9]. In the singlet ground state of a Kondo system, conduction electrons feel the spherically symmetric potential at the impurity site, which is taken as the origin. Hence it is convenient to take the spherical wave as the eigenbasis for conduction states. In addition, mutual interaction works only locally around the origin. Hence the system is properly called the local Fermi liquid. We apply the framework discussed in Chap. 4 to the local Fermi liquid.

Since the conduction electrons feel mutual interactions only for the $s$-wave part, the quasi-particles are specified by the radial momentum $p$ ($> 0$) and spin $\sigma$. Correspondingly the Landau interactions take the form $f(p\sigma, k\sigma')$, which carry all the information about the $f$-electron degrees of freedom. At the ground state, the distribution function $n_{p\sigma}$ of quasi-particles is one for $p < p_F$ with $p_F$ being the radial Fermi momentum, and 0 for $p > p_F$. In terms of the deviation $\delta n_{p\sigma}$, the low-lying excited states are characterized by the energy

$$E = E_{\mathrm{g}} + \sum_{p\sigma} \epsilon_{p\sigma} \delta n_{p\sigma} + \frac{1}{2} \sum_{p\sigma} \sum_{k\sigma'} f(p\sigma, k\sigma') \delta n_{p\sigma} \delta n_{k\sigma'} + O\left(\delta n_{p\sigma}^3\right), \tag{6.40}$$

where $E_{\mathrm{g}}$ is the groundstate energy. As in ordinary Fermi liquids, it is sufficient to consider up to the second order in $\delta n_{p\sigma}$ in order to describe the specific heat and susceptibility at low temperatures.

For low-energy excitations, $\delta n_{p\sigma}$ has a sharp peak at the Fermi level. In $f(p\sigma, k\sigma')$, therefore, $p$ and $k$ can be replaced by $p_F$ since the change of $f(p\sigma, k\sigma')$ is negligible in the small variation $\Delta p$ with $\Delta p \ll p_F$. By further restriction from the spherical symmetry, the independent parameters for $f(p\sigma, k\sigma')$ are reduced to the following two:

$$F = \frac{1}{2} [f(p_F\sigma, p_F\sigma) + f(p_F\sigma, p_F\bar{\sigma})] \rho^*,$$

$$Z = \frac{1}{2} [f(p_F\sigma, p_F\sigma) - f(p_F\sigma, p_F\bar{\sigma})] \rho^*, \tag{6.41}$$

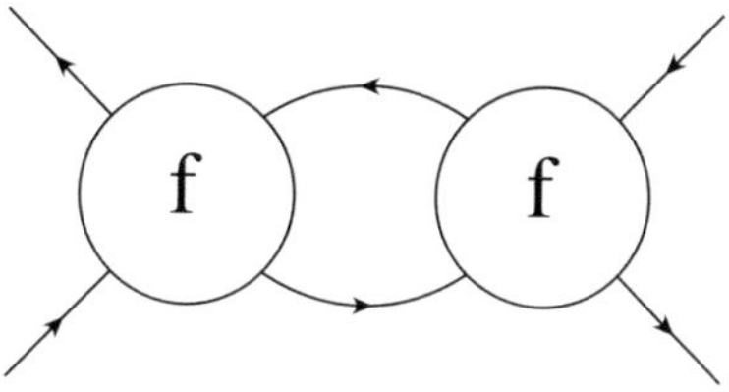


**Fig. 6.5** Scattering process of quasi-particles in the second order of the effective interaction $f(p\sigma, p'\sigma')$. The central part with two arrowed lines corresponds to $\Pi_k(q,\omega)$

with $\bar{\sigma} = -\sigma$ and where $\rho^*$ is the density of states (sum of spin components) at the Fermi level. Let us derive the order of magnitude of the Landau parameters in local Fermi liquid. The double momentum summation in Eq. (6.40) amounts to the factor $O(N^2)$. The correction to $E_g$ by the single impurity should be of $O(1)$, implying that the Landau interactions $f(p\sigma, k\sigma')$ have the order of magnitude $f \sim O(1/N^2)$. Consequently the dimensionless quantities $F$ and $Z$ are both of $O(1/N)$ with $\rho^* \sim O(N)$.

If we regard Eq. (6.40) as the effective Hamiltonian of a hypothetical one-dimensional system, $f(p\sigma, k\sigma')$ corresponds to the effective interaction. The momenta $p, k$ of two quasi-particles remain the same after the interaction event, which is the forward scattering. More generally, we consider successive scattering processes with momentum and energy transfers $q$ and $\omega$, respectively, as shown in Fig. 6.5. Here the intermediate state with a particle–hole pair excitation is described by

$$\Pi_k(q,\omega) = \frac{qv}{\omega - qv}\delta(\epsilon_k), \tag{6.42}$$

where $v = \partial\epsilon_k/\partial k \sim v_F$ with $v_F$ being the Fermi velocity. This quantity is the one-dimensional version of Eq. (4.20), and depends on the dimensionless parameter $|v_F q/\omega|$. The case $v_F|q| \gg |\omega|$ is called the $q$-limit, with $\Pi_k(q,0) = -\delta(\epsilon_k)$. The opposite case $v_F|q| \ll |\omega|$ is called the $\omega$-limit with $\Pi_k(0,\omega) = 0$.

The Landau interaction parameter $f(p\sigma, p'\sigma')$ corresponds to the $\omega$-limit of the general scattering processes. The scattering amplitude in the opposite limit ($q$-limit) is written as $a(p\sigma, p'\sigma')$, which is related to $f(p\sigma, p'\sigma')$ by

$$a(p\sigma, p'\sigma') = f(p\sigma, p'\sigma') - \sum_{k\tau} f(p\sigma, k\tau)\delta(\epsilon_k)a(k\tau, p'\sigma'), \tag{6.43}$$

where $\Pi_k(q,\omega)$ is replaced by its $q$-limit. In Fig. 6.5, the leftmost arrowed lines have momentum $p$ for both particle and hole, and the rightmost arrowed lines have $p'$. By analogy with Eq. (6.41), we combine the spin components of $a(p\sigma, p'\sigma')$ to make the spin symmetric part $A$, and the antisymmetric part $B$. Then we obtain from Eq. (6.43)

$$A = F/(1+F), \quad B = Z/(1+Z). \tag{6.44}$$

Furthermore a restriction on $A$ and $B$ arises from the Pauli principle. In order to see the consequence, we consider the most general scattering amplitude:

$$\langle p_1\sigma_1, p_2\sigma_2|a|p_3\sigma_3, p_4\sigma_4\rangle,$$

where $p_3\sigma_3, p_4\sigma_4$ are the incoming quantum numbers, while $p_1\sigma_1, p_2\sigma_2$ are the outgoing ones. Then antisymmetry associated with exchange of fermions imposes the constraint:

$$\langle p_1\sigma_1, p_2\sigma_2|a|p_3\sigma_3, p_4\sigma_4\rangle = -\langle p_1\sigma_1, p_2\sigma_2|a|p_4\sigma_4, p_3\sigma_3\rangle. \tag{6.45}$$

In the special case of the forward scattering, we set $p_i = p$ for all $i$ and obtain $a(p\sigma, p\sigma) = 0$, which is equivalent to

$$A + B = 0. \tag{6.46}$$

In the translationally invariant Fermi liquid, the relation analogous to Eq. (6.46) is called the forward-scattering sum rule [10]. With the constraint Eq. (6.46), only a single parameter remains to characterize the local Fermi liquid. We shall give the most standard parameterization in the end of this section.

Let us derive specific heat and magnetic susceptibility by the local Fermi liquid theory. With thermal excitation of quasi-particles, $\delta n_p$ is an odd function of $p - p_F$. Then in Eq. (6.40), the term linear in $\delta n_p$ is dominant, since the linear term contributes to $O(T^2)$, while the second-order term contributes to $O(T^4)$. Thus the interaction between quasi-particles can be neglected in the low temperature limit, as in the case of the standard Fermi liquid. The impurity contribution to the specific heat is extracted from the change in density of states $\rho^*$. We introduce the parameter $\alpha = O(1/N)$ by

$$\rho^* = N\rho_c(1+\alpha). \tag{6.47}$$

Then the impurity specific heat $C$ is of $O(1)$ and is given by

$$C = \frac{1}{3}\pi^2 N\rho_c \alpha T \equiv \gamma T. \tag{6.48}$$

We proceed to derive impurity contribution $\chi_s$ to the total spin magnetic susceptibility $\chi_s^{\mathrm{total}}$. The latter is given by

$$\chi_s^{\mathrm{total}} = \frac{\rho^*}{4(1+Z)} = \frac{N}{4}\rho_c(1+\alpha-Z) + O\left(\frac{1}{N}\right), \tag{6.49}$$

where we have used the property $Z = O(1/N)$. The impurity contribution, which corresponds to the $O(1)$ correction, is extracted as

$$\chi_{\rm s} = \frac{1}{4} N\rho_c(\alpha - Z). \tag{6.50}$$

Similarly, the impurity contribution $\chi_{\rm c}$ to the charge susceptibility is derived as

$$\chi_{\rm c} = N\rho_c(\alpha - F). \tag{6.51}$$

These results together with the constraint $A + B = F + Z + O(1/N^2) = 0$ lead to the significant relation

$$4\chi_{\rm s} + \chi_{\rm c} = 2N\rho_c\alpha = 6\gamma/\pi^2. \tag{6.52}$$

By using the corresponding quantities of free quasi-particles with the suffix 0, we obtain the alternative expression

$$\chi_{\rm s}/\chi_{\rm s0} + \chi_{\rm c}/\chi_{\rm c0} = 2. \tag{6.53}$$

The dimensionless quantity $R \equiv \chi_{\rm s}/\chi_{\rm s0}$ is called the Wilson ratio. From Eq. (6.53), we obtain the constraint $0 < R < 2$. In the Kondo limit of the Anderson model, the charge fluctuation of $f$-electrons is negligible with $\chi_{\rm c} \to 0$. In this limit we obtain $R \to 2$. In the opposite limit of free quasi-particles, we obtain $R = 1$. In more general cases including attractive interaction, the charge fluctuation may dominate over the spin fluctuation. Then we have $R < 1$. Hence $R$ is a measure of correlation of local electrons, and the only parameter to characterize the local Fermi liquid.

## 6.6 Mean Field Theory for Kondo Systems

We have seen that the singlet ground state of Kondo systems is effectively described by the Anderson model. The strength of effective hybridization, however, is renormalized from the bare value. It is possible to carry out the renormalization by a kind of mean field theory [11]. The $n$-fold degenerate Anderson model with $U \to \infty$ is most suitable for this purpose, since the characteristic energy $T_0$ is correctly reproduced for large $n$. We introduce a fictitious bosonic creation operator $b^\dagger$ that creates the physical vacuum $|0\rangle$ with no $f$ electrons. In addition, fictitious fermionic operators $f_\sigma^\dagger$ create the singly occupied state with spin $\sigma$. Namely, we define

$$|0\rangle\langle 0| = b^\dagger b, \quad |\sigma\rangle\langle\sigma| = f_\sigma^\dagger f_\sigma, \quad |\sigma\rangle\langle 0| = f_\sigma^\dagger b. \tag{6.54}$$

Note that $b^\dagger$ operates on the fictitious "vacuum" that has no impurity states. Any of physical $f$-electron states corresponds to either $|0\rangle$ or $|\sigma\rangle$, since plural occupation of $f$ electrons is prohibited by the condition $U \to \infty$. Hence the operator constraint $\sum_\sigma f_\sigma^\dagger f_\sigma + b^\dagger b = 1$ takes account of the infinite repulsion. With this constraint, the operators $f_\sigma$ and $f_\sigma^\dagger$ no longer satisfy the fermionic commutation rule. On the other hand, it is also possible to regard the constraint as a selection rule for the states in the Fock space. Then operators $b, f$ obey the bosonic and fermionic commutation rules. In the latter approach, the entities represented by $b, f$ are called auxiliary (or slave) particles.

In the mean field theory with auxiliary particles, the original constraint is replaced by a looser one:

$$\sum_\sigma \langle f_\sigma^\dagger f_\sigma \rangle + \langle b^\dagger b \rangle = 1, \tag{6.55}$$

where the constraint holds only as average. Furthermore, we make the approximation

$$\langle b^\dagger b \rangle \to |\langle b \rangle|^2. \tag{6.56}$$

Here the average $\langle b \rangle \equiv r$ should vanish in the exact theory with inclusion of the phase fluctuation of $b$. However, the mean field theory makes formal analogy to the Bose condensation represented by Eq. (5.6). It is obvious that the approximation is justified only in the case of large number of Bose condensed particles, which is not the case here. We discuss meaning of the mean field theory in the end of this section.

In the mean field theory, the SU($n$) Anderson model with $U \to \infty$ is simulated by another Anderson model $H_{\rm MF}$ without interaction:

$$H_{\rm MF} = \sum_{k\sigma} \left[ \epsilon_k c_{k\sigma}^\dagger c_{k\sigma} + \frac{1}{\sqrt{N}} V r \left( c_{k\sigma}^\dagger f_\sigma + f_\sigma^\dagger c_{k\sigma} \right) \right] + \epsilon_f \sum_\sigma f_\sigma^\dagger f_\sigma + \lambda (n_f + r^2 - 1), \tag{6.57}$$

where $\lambda$ is the Lagrange multiplier to impose the constraint Eq. (6.55), and $V$ and $r$ are taken to be real. Let us consider the general feature of optimization for the Hamiltonian without two-body interactions. The argument is valid for arbitrary temperature including $T = 0$. The statistical operator $\rho_{\rm MF} \equiv \exp(-\beta H_{\rm MF})$ determines the corresponding thermodynamic potential as $\Omega_{\rm MF} = -T \ln {\rm Tr}\, \rho_{\rm MF}$. We rewrite $H_{\rm MF}$ formally as

$$H_{\rm MF} = \sum_{ij} \epsilon_{ij} d_i^\dagger d_j + C, \tag{6.58}$$

where the operator $d$ denotes either $c$ or $f$ electron, and $C$ is a constant. Then the thermodynamic potential $\Omega_{\rm MF}$ is written as

$$\Omega_{\rm MF} = -T \ln {\rm Tr}\left[1 + \exp(-\beta\hat{\epsilon})\right] + C, \tag{6.59}$$

where $\hat{\epsilon}$ is the matrix composed of $\epsilon_{ij}$. Variation of $\Omega_{\rm MF}$ against change of $H_{\rm MF}$ is given by

$$\delta\Omega_{\rm MF} = \sum_{ij}\langle d_i^\dagger d_j\rangle\delta\epsilon_{ij} + \delta C, \tag{6.60}$$

which can be checked most easily by using the eigenbasis of $H_{\rm MF}$. Note that $\delta\Omega_{\rm MF}$ does not involve variation $\delta\langle d_i^\dagger d_j\rangle$, even though the average depends on $H_{\rm MF}$. This observation simplifies enormously the optimization procedure. For example, variation of $r$ to minimize $\Omega_{\rm MF}$ leads to the relation

$$\frac{1}{\sqrt{N}}\sum_{k\sigma} V\langle f_\sigma^\dagger c_{k\sigma} + c_{k\sigma}^\dagger f_\sigma\rangle + 2\lambda r = 0. \tag{6.61}$$

There is a nontrivial solution $r \neq 0$ below a characteristic temperature $T_B$ ($\sim T_{\rm K}$), which gives a lower value for $\Omega_{\rm MF}$ than the trivial solution. We shall give the nontrivial solution soon at $T = 0$. However, only the trivial solution $r = 0$ is possible for $T > T_B$. The second-order phase transition at $T = T_B$ is the artifact of the mean field theory. Therefore we confine the following discussion to the case $T = 0$, where the mean field theory is qualitatively correct for describing the fixed point of the Anderson model.

We use the Green function to discuss both dynamics and statistical average. The $f$ electron Green function $G_f^*(z)$ associated with $H_{\rm MF}$ is derived as in Eq. (3.123), and given for $z$ in the upper half plane with $|z| \ll D$ by

$$G_f^*(z) = [z - \tilde{\epsilon}_f + {\rm i}\tilde{\Delta}]^{-1}, \tag{6.62}$$

where renormalized quantities are defined by $\tilde{\epsilon}_f = \epsilon_f + \lambda$, and $\tilde{\Delta}/\Delta = r^2$. The asterisk (*) suggests that we are dealing with quasi-particles. The density of states $\rho_f^*(\epsilon)$ per spin is given by

$$\rho_f^*(\epsilon) = -\frac{1}{\pi}{\rm Im}\, G_f^*(\epsilon + {\rm i}0_+) = \frac{\tilde{\Delta}}{\pi}\frac{1}{(\epsilon - \tilde{\epsilon}_f)^2 + \tilde{\Delta}^2}. \tag{6.63}$$

The occupation number $n_f$ is derived by

$$n_f = n\int_{-\infty}^{0} {\rm d}\epsilon\rho_f^*(\epsilon) = \frac{n}{\pi}\arctan\left(\frac{\tilde{\Delta}}{\tilde{\epsilon}_f}\right) = 1 - r^2, \tag{6.64}$$

where the last equality comes from the constraint given by Eq. (6.55). Hence with given $n_f$ (or $\epsilon_f$) and $V$, Eq. (6.64) determines $r$ and $\lambda = \tilde{\epsilon}_f - \epsilon_f$. In the case of large $n$ with $n_f \sim 1$, we should have $\tilde{\Delta}/\tilde{\epsilon}_f \sim O(1/n)$, and $\tilde{\epsilon}_f > 0$. This feature is in contrast to the original Kondo model with $n = 2$ where the resonance is at the Fermi level: $\tilde{\epsilon}_f = 0$.

We next derive the average $\langle f_\sigma^\dagger c_{k\sigma}\rangle$ from the Green function defined by $G_{cf}(\boldsymbol{k}, z) = \langle\{c_{\boldsymbol{k}\sigma}^\dagger, f_\sigma\}\rangle(z)$. Using the relation Eqs. (3.102) and (3.104) we obtain at $T = 0$

$$\langle f_\sigma^\dagger c_{\boldsymbol{k}\sigma}\rangle = G_{cf}(\boldsymbol{k}, \tau = -0) = \int_{-\infty}^{0} d\epsilon\ \rho_{cf}(\boldsymbol{k}, \epsilon), \tag{6.65}$$

where $\rho_{cf}(\boldsymbol{k}, \epsilon) = (-1/\pi)\,\mathrm{Im}\, G_{cf}(\boldsymbol{k}, \epsilon)$. On the other hand, $G_{cf}(\boldsymbol{k}, z)$ is decomposed as

$$G_{cf}(\boldsymbol{k}, z) = G_{fc}(\boldsymbol{k}, z) = Vr(z - \epsilon_{\boldsymbol{k}})^{-1} G_f^*(z). \tag{6.66}$$

Provided the bandwidth $2D$ is much larger than $(\tilde{\epsilon}_f^2 + \tilde{\Delta}^2)^{1/2}$, we can use the approximation

$$\rho_{cf}(\boldsymbol{k}, \epsilon) \sim Vr\delta(\epsilon - \epsilon_{\boldsymbol{k}})\,\mathrm{Re}\, G_f^*(\epsilon), \tag{6.67}$$

with relative error of order $(\tilde{\epsilon}_f^2 + \tilde{\Delta}^2)^{1/2}/D$ after summation over $\boldsymbol{k}$. Then we obtain the compact result

$$\frac{1}{\sqrt{N}}\sum_{\boldsymbol{k}\sigma} V\langle f_\sigma^\dagger c_{\boldsymbol{k}\sigma}\rangle = nW_0 r \ln\left(\frac{\sqrt{\tilde{\epsilon}_f^2 + \tilde{\Delta}^2}}{D}\right). \tag{6.68}$$

Together with Eq. (6.61), the energy scale of the impurity is derived as

$$\sqrt{\tilde{\epsilon}_f^2 + \tilde{\Delta}^2} = D\exp\left(-\frac{\lambda}{nW_0}\right) \sim D\exp\left(\frac{\epsilon_f}{nW_0}\right), \tag{6.69}$$

where we have used $|\tilde{\epsilon}_f| \ll |\epsilon_f|$. The rightmost quantity reproduces $T_0$ defined by Eq. (6.34). If we apply the result to $n = 2$ with $n_f = 1$, the leftmost quantity becomes $\tilde{\Delta}$ with $\tilde{\epsilon}_f = 0$. Thus the energy scale is correctly derived in the mean field theory including the case of $n = 2$.

The mean field theory can easily derive static quantities. For example, the linear specific heat due to the impurity is characterized by

$$\gamma = \frac{1}{3}\pi^2 \rho_f^*(0), \tag{6.70}$$

where $\rho_f^*(0)$ is given by

$$\rho_f^*(0) = \frac{n\tilde{\Delta}}{\pi(\tilde{\epsilon}_f^2 + \tilde{\Delta}^2)} = \frac{n}{\pi\tilde{\Delta}} \sin^2\left(\frac{\pi n_f}{n}\right). \tag{6.71}$$

On the other hand, the magnetic susceptibility is given by

$$\chi = C_J \rho_f^*(0) \to C_J n_f^2 / T_0 \quad (n \to \infty), \tag{6.72}$$

where $C_J$ is the Curie constant that has appeared in Eq. (6.39). This result recovers the correct one Eq. (6.39) in the case of $n_f = 1$, but deviates to smaller value for $n_f < 1$.

Regarding $G_f^*(z)$ as describing quasi-particles, we discuss its relation to the exact Green function $G_f(z)$. The latter is represented by

$$G_f(z) = [z - \epsilon_f + \mathrm{i}\Delta - \Sigma_f(z)]^{-1}, \tag{6.73}$$

in terms of the self-energy $\Sigma_f(z)$, and with $\Delta = \pi W_0$. The contribution of quasi-particles is extracted by expansion of $\Sigma_f(z)$ around $z = 0$ as

$$\Sigma_f(z) = \Sigma_f(0) + z \left.\frac{\partial \Sigma_f(z)}{\partial z}\right|_{z=0} + O(z^2), \tag{6.74}$$

where the derivative $\Sigma_f'(0)$ is real in the Fermi liquid. Then we obtain

$$G_f(z) = a_f G_f^*(z), \tag{6.75}$$

where $a_f = [1 - \Sigma_f'(0)]^{-1}$ is called the renormalization factor. Thus the mean field quantities in $G_f^*(z)$ are given as

$$\lambda = \Sigma_f(0), \quad r^2 = a_f = 1 - n_f. \tag{6.76}$$

As $n_f$ becomes closer to unity, the effective hybridization and $a_f$ become smaller. In this way the fixed point of Kondo systems is correctly described by the mean field theory. It is the simplest framework to carry out the renormalization at $T = 0$, and is accurate provided the degeneracy $n$ is large.

Finally we discuss the meaning of the mean field theory, especially how to regard the unjustifiable approximation Eq. (6.56). The key is the variational principle which does not rely on the analogy with the Bose condensation. The exact thermodynamic potential $\Omega$ of the original SU($n$) Anderson model satisfies the following inequality:

$$\Omega \leq \Omega_{\mathrm{tr}} + \langle H_{\mathrm{SU}(n)} - H_{\mathrm{tr}} \rangle_{\mathrm{tr}}, \tag{6.77}$$

where $H_{tr}$ is an arbitrary trial Hamiltonian which determines $\Omega_{tr}$ and the statistical average $\langle \cdots \rangle_{tr}$. Equation (6.77) is often called the Feynman (or Peierls–Bogoliubov) inequality, the proof of which is the subject of Problem 6.6. In the case of $H_{tr}$ taken as the same as $H_{MF}$, the quantity $r$ now is a variational parameter, which needs no reference to Bose condensation. The parameters $\lambda$ and $r$ should be determined so that the RHS of Eq. (6.77) takes the minimum. Then the solution agrees with the one given by the mean field theory.

## 6.7 Dynamical Susceptibility of Kondo Impurity

The local Fermi liquid theory provides a useful relation for the dynamical susceptibility. For example, the imaginary part $\mathrm{Im}\chi(\omega)$ that describes the magnetic relaxation process is related to the static magnetic susceptibility $\chi$ as

$$\lim_{\omega\to 0} \mathrm{Im}\frac{\chi(\omega)}{\omega\chi^2} = \frac{\pi}{nC_J}, \tag{6.78}$$

with $C_J$ the Curie constant. This relation is called the Korringa–Shiba relation, and is widely used in analysis of the NMR experiment. It is not easy to derive the result fully microscopically [12]. Here we take a phenomenological approach that includes the $f$-electron states as local quasi-particles. This inclusion is along the line of the mean field theory discussed in the previous section, but deviates from the original local Fermi liquid theory in Sect. 6.5 where the quasi-particles consist of conduction electrons only.

Let $\chi_1(\omega)$ be a hypothetical dynamical susceptibility for which the interactions between local quasi-particles are neglected. In terms of the local density of states $\rho_f^*(\epsilon_1)$ due to $f$-electrons, $\chi_1(\omega)$ is given by

$$\chi_1(\omega) = nC_J \int \mathrm{d}\epsilon_1 \int \mathrm{d}\epsilon_2 \rho_f^*(\epsilon_1)\rho_f^*(\epsilon_2)\frac{f(\epsilon_1) - f(\epsilon_2)}{\omega - \epsilon_1 + \epsilon_2 + \mathrm{i}0_+}. \tag{6.79}$$

The dynamical susceptibility $\chi(\omega)$, which incorporates interaction between quasi-particles, is related to $\chi_1(\omega)$ as

$$nC_J\chi(\omega)^{-1} = nC_J\chi_1(\omega)^{-1} - U_{\mathrm{eff}}, \tag{6.80}$$

where $U_{\mathrm{eff}}$ is the effective repulsion between the local quasi-particles, which is related to $-Z_0$ in the local Fermi liquid theory. In the present scheme, however, we have $U_{\mathrm{eff}}\rho_f^*(0) \sim O(1)$. In the case of Lorentzian density of states given by Eq. (6.63), we obtain

$$\chi_1(0) = nC_J\rho_f^*(0), \quad \mathrm{Im}\chi_1(\omega) \to nC_J\pi\omega\rho_f^*(0)^2, \tag{6.81}$$

the derivation of which is the subject of Problem 6.7. In the limit $\omega \to 0$, we then obtain

$$\mathrm{Im}\chi_1(\omega)^{-1} \to -\mathrm{Im}\chi_1(\omega)/\chi_1(0)^2 = -\pi\omega/(nC_J), \tag{6.82}$$

which amounts to $\mathrm{Im}\chi(\omega)^{-1}$ since $U_{\mathrm{eff}}$ is real in Eq. (6.80). In this way we obtain Eq. (6.78). This derivation relies on the fact that the magnetic relaxation is solely due to independent quasi-particles in the low-energy limit. Hence, analogous relation holds for charge and orbital susceptibilities as well.

We remark that the finite bandwidth causes deviation from the Korringa–Shiba relation. Namely, the cutoffs at $\pm D$ modify $\rho_f^*(\epsilon)$ out of the Lorentzian form at high energies, which brings about correction to $\chi_1(0)$ in Eq. (6.81). Similar correction arises also in the fully microscopic theory [13]. Therefore one should be careful in applying the Korringa–Shiba relation to systems with large hybridization, where possible error is of order $(V/D)^2$ in terms of the Anderson model.

## 6.8 Multi-Channel Kondo Model

The singlet formation in the Kondo model is made possible only if the spin degrees of freedom share the same number for both conduction and local electrons. Namely, if the localized electrons have the spin $S$, the screening by the single conduction band leads to the ground state with spin $S - 1/2$. Hence the finite spin remains except for the special case of $S = 1/2$. More generally, we consider the situation [14] where the conduction electrons have the additional degeneracy $n$ for all momentum and spin, which is referred to as channels.

The orbital degrees of freedom indexed by $l$ $(= 1, 2, \ldots, n)$ is assigned as the origin of such degeneracy. The $n$-channel Kondo model in the simplest form is thus defined by

$$H_{n\mathrm{K}} = \sum_{\boldsymbol{k}\sigma}\sum_{l=1}^{n} \epsilon_{\boldsymbol{k}} c_{\boldsymbol{k}l\sigma}^\dagger c_{\boldsymbol{k}l\sigma} + J\boldsymbol{S}\cdot\sum_{l=1}^{n}\boldsymbol{s}_l, \tag{6.83}$$

where $\boldsymbol{S}$ denotes the localized spin with magnitude $S$, and $\boldsymbol{s}_l$ is given by

$$\boldsymbol{s}_l = \frac{1}{2N}\sum_{\boldsymbol{k}\boldsymbol{k}'}\sum_{\alpha\beta} c_{\boldsymbol{k}l\alpha}^\dagger \boldsymbol{\sigma}_{\alpha\beta} c_{\boldsymbol{k}'l\beta}, \tag{6.84}$$

with $N$ being the number of $\boldsymbol{k}$-states and $\boldsymbol{\sigma}$ the Pauli matrix. The energy $\epsilon_{\boldsymbol{k}}$ is common to all spin $\sigma$ and orbitals of conduction electrons. Notice that the simplest model does not reflect the realistic situation that the degeneracy occurs only for the particular crystal momentum such as $\boldsymbol{k} = 0$ in the Brillouin zone.

The ground states depend on relative magnitudes of localized spins and the sum of itinerant ones, and are classified as follows:

1. singlet in the case of $S = n/2$,
2. residual localized spin if $S > n/2$,
3. overscreening of the localized spin in the case of $S < n/2$.

In the $n$-channel Kondo model, the case (3) is most interesting since it realizes an exotic ground state. Let us first follow the argument [14] from the strong-coupling limit $J/D \gg 1$, where $D$ is the bandwidth of conduction electrons. Suppose that the impurity spin has $S_z = S$ initially. Then conduction electrons with spin down are attracted to the impurity site because of the energy gain of $O(J)$. If all channels of them gather, the screening cloud has too much negative spin, resulting in the total spin with magnitude $n/2 - S$. This state then creates a new antiferromagnetic interaction of $O(D^2/J)$ with nearby conduction electrons with spin up. These electrons also gather at the impurity and partially cancel the negative total spin. As a result, the strong-coupling fixed point becomes unstable. Since the weak-coupling fixed point is also unstable, a nontrivial fixed point should appear somewhere between $J = 0$ and $J = \infty$. This intuitive argument has been confirmed by the exact solution [6, 15], which uses the technique called the Bethe ansatz [16].

As the simplest quantitative route to confirm the nontrivial fixed point, we extend the weak-coupling scaling theory to the third order. Such perturbative renormalization is justified if the orbital degeneracy is large: $n \gg 1$, since the resultant fixed point then comes inside the valid range of the weak-coupling theory: $J\rho \ll 1$ [14]. In the Rayleigh–Schrödinger perturbation theory, we deal with the third-order effective Hamiltonian:

$$\langle a|H_3|b\rangle = \left\langle a|V\frac{1}{\epsilon_b - H_0}QV\frac{1}{\epsilon_b - H_0}QV|b\right\rangle - \sum_c \left\langle a\left|V\frac{1}{\epsilon_b - H_0}\frac{1}{\epsilon_c - H_0}QV\right|c\right\rangle\langle c|V|b\rangle, \tag{6.85}$$

which has been given by Eq. (1.14). The Hamiltonian $H_0$ corresponds to the kinetic energy of conduction electrons, and $V$ to the exchange interaction. Here the states $a, b, c$ belong to the model space. As in the lowest-order scaling, the projection operator $Q$ requires one of the intermediate conduction-electron states to have energies near the band edges.

We concentrate on such contributions that become dominant for large $n$. The first line in the RHS of Eq. (6.85) has a form familiar in the Brillouin–Wigner perturbation theory. Fig. 6.6a shows one of the corresponding Goldstone diagrams with a loop of conduction-electron lines. Such a loop acquires the factor $n$ by summing over degenerate orbitals, and the corresponding diagram becomes dominant over non-loop diagrams. With the one-particle energies assigned in Fig. 6.6a, the energy

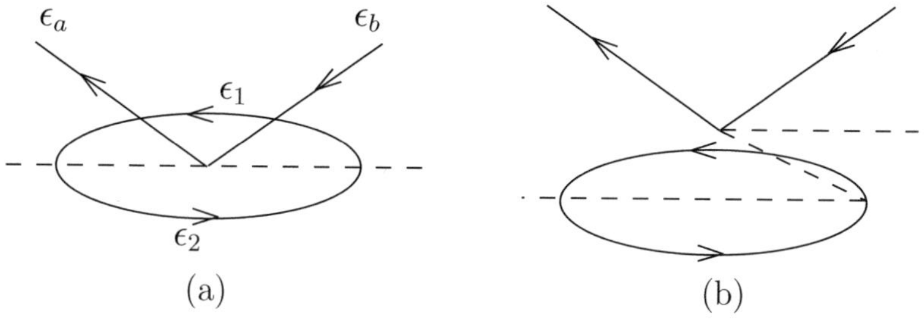


**Fig. 6.6** Exchange scattering processes in the third order. See text for the meaning of the folded dashed line in (**b**)

denominator on the right has the value

$$\epsilon_b - H_0 \to \epsilon_b - (\epsilon_b + \epsilon_1 - \epsilon_2) = \epsilon_2 - \epsilon_1, \tag{6.86}$$

while the other energy denominator on the left has the value

$$\epsilon_b - H_0 \to \epsilon_b - (\epsilon_a + \epsilon_1 - \epsilon_2). \tag{6.87}$$

On the other hand, the term in the second line of Eq. (6.85) also has a loop contribution. The energy denominators on the right has the value

$$\epsilon_c - H_0 \to \epsilon_c - (\epsilon_c + \epsilon_1 - \epsilon_2) = \epsilon_2 - \epsilon_1, \tag{6.88}$$

which is the same as given by Eq. (6.86). As a convenient device, one can represent the term as in Fig. 6.6b with the folded dashed line. Then by reading off the one-particle energies at the vertical section including the folded line, the same result as in Eq. (6.88) follows by regarding $\epsilon_b$ as the initial energy. This situation commonly occurs for such terms in Rayleigh–Schrödinger perturbation series that contain model space as intermediate states [17]. In Fig. 6.6b, the energy denominator on the left has the value

$$\epsilon_b - H_0 \to \epsilon_b - (\epsilon_c + \epsilon_1 - \epsilon_2), \tag{6.89}$$

which again becomes the same as that in Eq. (6.87) since both $\epsilon_c$ and $\epsilon_a$ represent the same energy of the scattered electron.

If the perturbation $V$ were the potential scattering, the product of matrix elements in Fig. 6.6a, b would be the same. Then the two contributions cancel each other since the overall signs are opposite according to Eq. (6.85) while the magnitudes determined by energy denominators are identical. We have encountered an analogous cancellation in the second-order potential scatterings described in Fig. 6.2a, b. In the case of exchange interaction, products of matrix elements of $V$ are different between (a) and (b) in Fig. 6.6. Let us derive the products of spin operators focusing on the $l = 1$ component in $\boldsymbol{S} \cdot \boldsymbol{s}_l$ and neglect the index $l$. In terms of the spin components $\alpha, \beta, \gamma$ to be summed over, the spin product in (a) and (b) takes the respective form:

$$\text{(a)} \quad S^\alpha S^\beta S^\gamma \mathrm{Tr}(s^\alpha s^\gamma) s^\beta, \tag{6.90}$$

$$\text{(b)} \quad S^\alpha S^\gamma S^\beta \mathrm{Tr}(s^\alpha s^\gamma) s^\beta, \tag{6.91}$$

where the trace is over the spin states of conduction electrons, resulting in $\mathrm{Tr}(s^\alpha s^\gamma) = \delta_{\alpha\gamma}/2$. Then the spin product in (a) minus that in (b) becomes

$$\frac{1}{2} \sum_{\alpha\beta} S^\alpha \left[ S^\beta, S^\alpha \right] s^\beta = -\frac{1}{2} \boldsymbol{S} \cdot \boldsymbol{s}, \tag{6.92}$$

as a result of the commutation rule: $[S^\alpha, S^\beta] = \mathrm{i}\epsilon_{\alpha\beta\gamma} S^\gamma$ with $\epsilon_{\alpha\beta\gamma}$ being the completely antisymmetric unit tensor. Thus the third-order effective Hamiltonian keeps the same exchange form as in the original Hamiltonian.

In the weak-coupling scaling, the energies of incoming and outgoing conduction electrons are much smaller than $D$ and are neglected. In both Fig. 6.6a, b, the energy denominators are obtained by putting either $\epsilon_1 = D$ or $\epsilon_2 = -D$. Then we combine the particle and hole contributions at the band edges as

$$\int_0^D \mathrm{d}\epsilon_1 \int_{-D}^0 \mathrm{d}\epsilon_2 \frac{1}{(\epsilon_1 - \epsilon_2)^2} \to \delta D \left[ \int_{-D}^0 \frac{\mathrm{d}\epsilon_2}{(D - \epsilon_2)^2} + \int_0^D \frac{\mathrm{d}\epsilon_1}{(\epsilon_1 + D)^2} \right] = \frac{\delta D}{D}. \tag{6.93}$$

We thus obtain the third-order renormalization $\delta J^{(3)}$ as

$$\delta J^{(3)} = \frac{\delta D}{2D} n J^3 \rho_c^2, \tag{6.94}$$

where the factor $n$ comes from summation over the orbital index $l$.

We shall express the result in a more general context of renormalization. Let us introduce the scaling variable $\ell = \ln D$, and the dimensionless interaction $g(D) \equiv J(D)\rho_c$ with a given cutoff energy $D$. Together with the second-order scaling obtained previously, Eq. (6.94) is put into the form:

$$\frac{\partial g}{\partial \ell} \equiv \beta(g) = -g^2 + \frac{n}{2} g^3, \tag{6.95}$$

where $\beta(g)$ is called the beta-function, which does not depend explicitly on $\ell$. The zero of $\beta(g)$ gives a fixed point of renormalization. A trivial fixed point of the model is given by $g = 0$ which is unstable, as we have seen in the original Kondo model. Namely, with either sign of $g$ ($\sim 0$), decreasing $\ell$ drives $g$ to the positive direction. In addition to the trivial one, a new fixed point at $g = g_c$ appears due to the $O(g^3)$

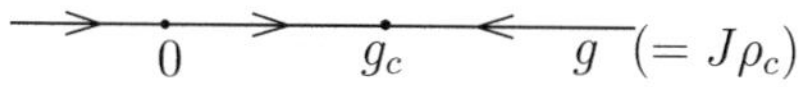


**Fig. 6.7** The scaling flow shown by arrows in the orbitally degenerate Kondo model. The fixed point at $g_c$ is stable against renormalization, while another fixed point at $g = 0$, which also appears in the original Kondo model, is unstable

term. We obtain

$$g_c = 2/n, \tag{6.96}$$

which is much smaller than unity with large $n$. Thus emergence of the new fixed point is genuine since the magnitude $g_c$ is within the reach of the perturbative renormalization. Namely, higher order terms of $O(g^4)$, which are neglected in Eq. (6.95), only modify the magnitude of $g_c$ by a relative amount of $1/n$. This is because $O(g^4)$ terms are smaller by the factor $g_c$ than the terms kept in Eq. (6.95).

Linearization of the scaling equation around $g_c = 2/n$ leads to

$$\frac{\partial g}{\partial \ell} \equiv \beta(g) \sim \frac{2}{n}(g - g_c), \tag{6.97}$$

which shows that the scaling flow changes sign around the fixed point. The combination of signs corresponds to the stable fixed point. Figure 6.7 illustrates the stable and unstable fixed points.

One may naturally ask how the nontrivial fixed point at $g = g_c$ behaves with decreasing degeneracy $n$. This question is related to another one; whether the strong-coupling fixed point at $g = \infty$ remains or disappears with increasing $n$ from unity. It is impossible to answer these questions within the weak-coupling renormalization theory, since one has to deal with $g \sim O(1)$ or even larger. As explained at the beginning of this section, however, the intuitive approach from the strong-coupling limit has given the correct solution. Namely, the nontrivial fixed point remains down to $n = 2$, although the value of $g_c$ is no longer given by Eq. (6.96). On the other hand, the Fermi liquid fixed point at $g = \infty$ is unstable for all $n \geq 2$.

## 6.9 Realization of Multi-Channel Kondo Systems

We discuss whether the multi-channel Kondo effect can be observed experimentally. The most notable feature is that the ground state is not a Fermi liquid, but with finite entropy. In reality, the remaining entropy should be removed by some interactions neglected in the theoretical model. For example, the Kondo impurities, however dilute, will eventually interact with one another to make an ordered state. If the ordering temperature is sufficiently low, there should be a temperature region where the non-Fermi liquid behavior is dominant. The temperature dependence of the resistivity $\rho(T)$ distinguishes the difference between ground states. Namely, while the Fermi liquid has the $T^2$ law: $\Delta\rho(T) \equiv \rho(T) - \rho(0) \propto T^2$, the two-channel Kondo system has the peculiar behavior [18, 19]

$$\Delta\rho(T) \propto \pm T^{1/2}, \tag{6.98}$$

where the sign depends on the value of the bare coupling constant; the plus sign applies to the case $g > g_c$ in Fig. 6.7. Otherwise the minus sign applies. Most experimental work concentrates on $\Delta\rho(T)$. However, due care is necessary about the role of disorder and other effects which may give rise to apparently similar behavior.

We discuss in the following some exemplary systems which may show the multi-channel Kondo effect.

### 6.9.1 Orbital as Source of Channels

In the original argument of Nozières and Blandin [14], a possible candidate is an impurity having two ($S = 1$) or three ($S = 3/2$) local electrons in the $d$ shell. In the spherical symmetry, the orbital degeneracy amounts to $n = 5$, which satisfies the overscreening condition $n > 2S$. Actually, the crystalline anisotropy in solids makes the effective degeneracy smaller. In the cubic symmetry, for example, we have $n = 2$ for the $E_g$ ($\Gamma_3$) state, while $n = 3$ for the $T_{2g}$ ($\Gamma_5$) state. Here the spin–orbit coupling is neglected. The local ground state with two electrons in the $E_g$ state has $S = 1$ according to the Hund's rule. The hybridization between local and itinerant electrons occurs most easily for states with the same symmetry. Hence in the case of $S = 1$, the conduction electrons with the $E_g$ symmetry are dominant for the spin exchange, resulting in the singlet ground state because of $n = 2S = 2$. Similarly, exchange interaction in the case with $S = 3/2$ is dominated by conduction electrons with the $T_g$ symmetry, leading again to the singlet with $n = 2S = 3$. Hence, Nozières and Blandin [14] were not optimistic about realization of the multi-channel Kondo effect in crystalline solids.

### 6.9.2 Spin as Source of Channels

In Chap. 1 we have introduced the concept of the pseudo-spin for orbital degrees of freedom. The concept remains valid for the plural number of localized electrons and with full account of the spin–orbit interaction. For example, in the case of $f^2$-configuration as in $Pr^{3+}$, the Hund's rule gives the ground state with $S = 1, L = 5, J = 4$ for spin, orbital, and total angular momenta. The cubic crystalline anisotropy splits the $J = 4$ multiplets into several levels each of which is at most threefold degenerate. The degeneracy is described in terms of the pseudo-spin 1/2 for the doublet case, and the pseudo-spin 1 for the triplet case. To be specific, let us consider the case of $\Gamma_3$ doublet. The charge distribution of a member state is different from that of another member. This contrasts with the spin doublet where the time reversal changes a member state to another with the same charge density. Thus the $\Gamma_3$ doublet is an example of the non-Kramers doublet.

Among various interactions between the $\Gamma_3$ doublet and conduction electrons, most relevant here is the hybridization via intermediate configurations $f^1$ and $f^3$. Since both intermediate states have level structures with possible degeneracy, the matrix element of hybridization is much more complicated than that in the Anderson model. Provided any orbital degeneracy is present in the hybridizing conduction band, the spin degrees of freedom in conduction electrons should cause the overscreening. With the cubic symmetry in the crystal, an orbital degeneracy is present in the $\Gamma_8$ states in the conduction band. Here, the parity must be odd ($\Gamma_8^-$) in order to mix with $f$ states.

On the basis of the situation described above, Cox [20] proposed that a non-Kramers doublet may realize the two-channel Kondo effect if the relevant degeneracy is present in the conduction band. For a candidate, Cox referred to $UBe_{13}$, which is a cubic system with possible $5f^2$ configuration for $U$. The Kondo effect with spin as the source of the two channels was originally called the "quadrupolar Kondo effect." However, the non-Kramers doublet may involve other multipoles such as hexadecapoles under the point group. Hence it seems appropriate to use the more general term "orbital Kondo effect," which includes also nonmagnetic triplet such as $\Gamma_5$ in the cubic symmetry.

There are a lot of experimental reports on observation of non-Fermi liquid behaviors in $d$- and $f$-electron systems, some of which have been ascribed to the orbital Kondo effect. However, it is difficult to exclude other possibilities responsible for apparent behaviors of the non-Fermi liquid. One of the main obstacles is the difficulty to separate from the spin Kondo effect involving the higher local states above the non-Kramers doublet, and the interplay of the ordinary Kondo effect and disorder effect.

### 6.9.3 Nano-Scale Reservoir as Source of Channels

So far the most convincing result for the two-channel Kondo effect has been reported in an artificial semiconductor nanostructure on GaAs/GaAlAs heterostructure that is widely used as field effect transistors (FET) [21]. The two-dimensional layer has a structure called the quantum dot where a few electrons are trapped by the potential lower than the surrounding. Because the Coulomb interaction fixes the number of local electrons in the quantum dot, a local spin is realized for an odd number of electrons. The quantum dot is weakly coupled to the lead on the left by tunneling. The dot is also weakly coupled to the right that forms a nanoscale reservoir of electrons. The Fermi level of the right reservoir can be controlled by the gate voltage relative to the quantum dot.

Through hybridizations on both sides, the local spin in the dot has the exchange interaction $J_l$ with electrons in the left leads, while $J_r$ with electrons in the right reservoir. As an advantage of the artificial structure, one can control the exchange interactions $J_r$ by applying a bias voltage on the nanoscale reservoir. This is because the exchange interaction changes in analogy to Eq. (6.7), where energies are measured from the Fermi level.

In general, either left or right channel dominates over the other depending on the magnitudes of $J_l\rho_l$ and $J_r\rho_r$ with $\rho_l$ and $\rho_r$ being the density of states of both sides. Thus the ordinary Kondo effect should take place in general. With appropriate tuning of the bias voltage, however, one may fortunately achieve the condition $J_l\rho_l = J_r\rho_r$ for the renormalized exchanges. Then the two-channel Kondo effect may be realized. Instead of the orbital degrees of freedom, the nanoscale reservoir on the right provides the additional channel. The difference between the ordinary and two-channel Kondo effects appears in the scaling behaviors of the conductance $g(V_b, T)$ as a function of the bias voltage $V_b$ and temperature $T$. More details of experiment and theoretical background are referred to the original paper [21] and references therein.

## Problems

**6.1** Derive the second-order effective interaction Eq. (6.15) by using the commutation relation of spins.

**6.2** Derive Eq. (6.21) obtained by the Born approximation for scattering.

**6.3*** Derive the Kondo temperature $T_K$ for the anisotropic Kondo model.

**6.4** Derive the singlet groundstate energy $E_0$ as given by Eq. (6.34).

**6.5** Derive the zero-temperature susceptibility given by Eq. (6.39).

**6.6*** Derive the Feynman inequality given by Eq. (6.77).

**6.7** Derive the results for the susceptibility given by Eq. (6.81).

## Solutions to Problems

**Problem 6.1**

In terms of the completely antisymmetric unit tensor $\epsilon_{\alpha\beta\gamma}$, we represent the spin commutation relation as

$$\left[s_c^\alpha, s_c^\beta\right] = \mathrm{i}\sum_\gamma \epsilon_{\alpha\beta\gamma} s_c^\gamma .$$

Substitution of the result into the LHS of Eq. (6.15), and the use of

$$\sum_{\alpha\beta} \epsilon_{\alpha\beta\gamma} S^{\alpha} S^{\beta} = \mathrm{i} S^{\gamma}$$

lead to the RHS. Hence we confirm that the second-order correction is proportional to $\boldsymbol{S} \cdot \boldsymbol{s}_c$. This means that none of new type's interaction, such as the potential scattering, is generated by renormalization except for the change of effective $J$.

**Problem 6.2**

A magnetic impurity located at $\boldsymbol{R}$ causes scattering of conduction electrons which is described by the Hamiltonian $H_{\mathrm{ex}}(\boldsymbol{R})$. Contributions of each impurity sum up to the scattering probability of a conduction electron. In the Born approximation, the transition probability, which corresponds to the inverse of the lifetime $\tau$, is given by the Golden rule as

$$\frac{1}{\tau} = 2\pi \sum_{\boldsymbol{R}} \sum_{f} |\langle f | H_{\mathrm{ex}}(\boldsymbol{R}) | i \rangle|^2 \, \delta(E_i - E_f), \tag{6.99}$$

where $i, f$ refer to initial and final states of the scattering. The energy $E_i = \epsilon_{\boldsymbol{k}}$ of the conduction electron is near the Fermi level, and summation over final states with $E_f = E_i$ is performed in terms of the density of states $\rho_c$ per unit volume. Assuming that impurities are located randomly in the system, we make the following replacement:

$$\sum_{\boldsymbol{R}} \sum_{f} \rightarrow c_{\mathrm{imp}} V_{\mathrm{cr}} N \rho_c \int \mathrm{d}E_f, \tag{6.100}$$

where $V_{\mathrm{cr}}$ is the volume of the whole system. According to Eq. (6.11), $H_{\mathrm{ex}}$ has the factor $1/N$ with use of the plane waves. Hence the factor $(V_{\mathrm{cr}} N/N^2 = v_{\mathrm{cell}}$ appears. Furthermore, only the same spin components contribute to $|\langle f | H_{\mathrm{ex}}(\boldsymbol{R}) | i \rangle|^2$, which leads to the numerical factor

$$\sum_{\alpha} \left\langle S_{\alpha}^2 s_{\alpha}^2 \right\rangle = 3 \cdot \frac{1}{4} \cdot \frac{1}{4} = \frac{3}{16}. \tag{6.101}$$

In this way we obtain Eq. (6.21).

**Problem 6.3***

We first consider the technically easier case $J_{\perp} > J_z > 0$, and introduce the parameters $J_{\perp z}, \alpha$ so that $J_z = J_{\perp z} \sinh \alpha$ and $J_{\perp} = J_{\perp z} \cosh \alpha$. Then using $\delta J_{\perp} = J_{\perp z} \sinh \alpha \, \delta\alpha$, we integrate the scaling equation (6.26) to obtain

$$\int_{\alpha_0}^{\infty} \frac{\mathrm{d}\alpha}{\cosh \alpha} = -J_{\perp z} \rho_c \int_{D_0}^{T_{\mathrm{K}}} \frac{\mathrm{d}D}{D}, \tag{6.102}$$

where the suffix 0 is attached to indicate the bare values in the integration range. As the renormalization of $D$ proceeds down to $T_K$, we should have $\alpha \to \infty$. The integral can be performed explicitly by introducing $u = \exp\alpha$ as follows:

$$\int \frac{\mathrm{d}\alpha}{\cosh\alpha} = 2\int \frac{\mathrm{d}u}{u^2+1} = \frac{1}{\mathrm{i}}\ln\left(\frac{u-\mathrm{i}}{u+\mathrm{i}}\right) = \frac{1}{2\mathrm{i}}\ln\left(\frac{\sinh\alpha-\mathrm{i}}{\sinh\alpha+\mathrm{i}}\right), \tag{6.103}$$

where we have taken the square inside the first logarithm to obtain the rightmost expression. The logarithm there is pure imaginary since the argument has the unit modulus with the phase given by $-2\tan^{-1}(1/\sinh\alpha)$. Then we obtain to Eq. (6.30) after exponentiating $\ln(D/T_K)$ in Eq. (6.102).

In the case of $J_z > J_\perp > 0$, we put $J_z = J_{z\perp}\cosh\alpha$ and $J_\perp = J_{z\perp}\sinh\alpha$. Then we obtain a variant of Eq. (6.102) with cosh $\to$ sinh in the LHS, and $J_{\perp z} \to J_{z\perp}$ in the RHS. With use of the variable $u = \exp\alpha$, the integral can be performed similarly. Using the identity

$$\ln\left(\frac{u-1}{u+1}\right) = -\tanh^{-1}\left(\frac{1}{\cosh\alpha}\right), \tag{6.104}$$

which corresponds to analytic continuation of the arctangent function, we obtain the result in Eq. (6.29) [22].

**Problem 6.4**
We carry out the integration in Eq. (6.34) to obtain

$$\frac{E_0}{nW_0} = \ln\frac{E_0-\epsilon_f}{E_0-\epsilon_f-D} \sim \ln\frac{E_0-\epsilon_f}{-D}, \tag{6.105}$$

where we have anticipated the result $|E_0-\epsilon_f| \sim T_0 \ll D$. By exponentiating both sides, we obtain

$$E_0-\epsilon_f = -D\exp\left(\frac{E_0}{nW_0}\right) \sim -D\exp\left(\frac{\epsilon_f}{nW_0}\right). \tag{6.106}$$

In the final result we have approximated the exponent using $T_0/nW_0 \ll 1$.

**Problem 6.5**
Putting $h \equiv g_J\mu_B H$ we take the derivative of both sides of Eq. (6.38) as

$$\frac{\partial E_0}{\partial h} = W_0\sum_{J_z}\int_{-D}^{0}\frac{\mathrm{d}\epsilon}{(E_0+\epsilon-\epsilon_f-J_z h)^2}\left(J_z-\frac{\partial E_0}{\partial h}\right). \tag{6.107}$$

Taking the derivative once more, and setting $h = 0$ we obtain

$$\frac{\partial^2 E_0}{\partial h^2} = W_0 \sum_{J_z} \int_{-D}^{0} d\epsilon \left[ \frac{2J_z^2}{(E_0 + \epsilon - \epsilon_f)^3} - \frac{1}{(E_0 + \epsilon - \epsilon_f)^2} \frac{\partial^2 E_0}{\partial h^2} \right]. \tag{6.108}$$

The integration is easily carried out. The summation over $J_z$ for the first term with $J_z^2$ in the RHS leads to $C_J$. Namely, using Eq. (6.37) we obtain

$$\chi = -\frac{\partial^2 E_0}{\partial H^2} = n W_0 \left( \frac{C_J}{T_0^2} - \frac{1}{T_0} \chi \right) = \frac{n_f}{1 - n_f} \left( \frac{C_J}{T_0} - \chi \right). \tag{6.109}$$

Rearranging the result in terms of $\chi$, we obtain Eq. (6.39).

**Problem 6.6***

A function $f(x)$ with positive second derivative, i.e., $f''(x) > 0$, is called convex (downward). The exponential function belongs to this category. Let $M$ be a Hermitian matrix with eigenvalues $M_k$ and corresponding normalized eigenvectors $|k\rangle$. As a preliminary, we shall prove the inequality:

$$f(M)_{nn} \geq f(M_{nn}), \tag{6.110}$$

where $M_{nn}$ denotes a diagonal element of $M$ with any normalized vector $|n\rangle$. Similarly, $f(M)_{nn}$ denotes the diagonal element of $f(M)$, which is a matrix defined by the Taylor expansion:

$$f(x) = f_0 + f_1 x + f_2 x^2 + \cdots,$$

with $x = M$. We can regard the diagonal element as an average over eigenstates with the weight factor $w_k = |\langle n|k\rangle|^2$. Note that $\sum_k w_k = 1$ by normalization. We may then write Eq. (6.110) as

$$\langle e^M \rangle \geq e^{\langle M \rangle}, \tag{6.111}$$

by choosing $f(x) = e^x$. For the proof of Eq. (6.110), we start with the inequality

$$f(M_k) \geq f(M_{nn}) + (M_k - M_{nn}) f'(M_{nn}), \tag{6.112}$$

which follows from the convex property $f''(x) > 0$ for any $x$. Multiplying both sides of Eq. (6.112) by $w_k$ and summing over $k$, we obtain for the LHS:

$$\sum_k \langle n|k\rangle \langle k|f(M)|k\rangle \langle k|n\rangle = f(M)_{nn}. \tag{6.113}$$

In the RHS, we note the relation

$$\sum_k \langle n|k\rangle M_k \langle k|n\rangle = \langle n|M|n\rangle \equiv M_{nn}. \tag{6.114}$$

Hence the coefficient of $f'(M_{nn})$ in Eq. (6.112) vanishes by the $k$-average. We thus obtain Eq. (6.110).

In the classical case where the Hamiltonians $H$ and $H_{\rm MF}$ commute, we can take the common eigenbasis $|k\rangle$ of these Hamiltonians, and $|n\rangle$ is chosen so that $w_k$ becomes the Boltzmann factor. By setting $M \to -\beta(H - H_{\rm MF})$, we obtain from Eq. (6.111)

$$\left\langle e^{-\beta(H-H_{\rm MF})}\right\rangle = \left\langle e^{-\beta H} e^{\beta H_{\rm MF}}\right\rangle \geq \exp\left(-\beta\langle H - H_{\rm MF}\rangle\right). \tag{6.115}$$

The average over the mean field Boltzmann weight gives the LHS as

$$\frac{1}{Z_{\rm MF}}{\rm Tr}\left(e^{-\beta H_{\rm MF}} e^{-\beta(H-H_{\rm MF})}\right) = \frac{Z}{Z_{\rm MF}}, \tag{6.116}$$

where $Z = {\rm Tr}\exp(-\beta H)$ is the partition function. Putting this result into Eq. (6.115) and taking the logarithm of both sides, we obtain Eq. (6.77).

In the quantum case with non-commuting $H$ and $H_{\rm MF}$, the proof is more complicated. Regarding $V \equiv H - H_{\rm MF}$ as perturbation, we take the expansion as in Eq. (3.5), but now to infinite order to obtain

$$\exp(-\beta H) = \exp(-\beta H_{\rm MF})\mathcal{U}(\beta), \tag{6.117}$$

$$\mathcal{U}(\beta) = T_\tau \exp\left[-\int_0^\beta {\rm d}\tau\, V(\tau)\right], \tag{6.118}$$

with $V(\tau) = \exp(\tau H_{\rm MF})V\exp(-\tau H_{\rm MF})$ and where the time-ordering operator $T_\tau$ arranges the operators appearing in the Taylor expansion in such a way that an operator with larger $\tau$ always sits left of another with smaller $\tau$. Taking the trace of both sides of Eq. (6.117), we obtain

$$Z = Z_{\rm MF}\langle \mathcal{U}(\beta)\rangle \tag{6.119}$$

where $\langle\cdots\rangle$ is the statistical average with $H_{\rm MF}$. In terms of the cumulant average $\langle\cdots\rangle_c$ defined in Chap. 3, alternative representation is given by

$$\begin{aligned} \ln\langle \mathcal{U}(\beta)\rangle &= \left\langle T_\tau \exp\left[-\int_0^\beta {\rm d}\tau\, V(\tau)\right]\right\rangle_c - 1 \\ &= -\beta\langle V\rangle + \frac{\beta}{2}\int_0^\beta {\rm d}\tau\,\langle V(\tau)V\rangle_c + O(V^3), \end{aligned} \tag{6.120}$$

where $\langle V(\tau)\rangle \equiv \langle \exp(\tau H_{\rm MF}) V \exp(-\tau H_{\rm MF})\rangle$ does not depend on $\tau$ in the average with respect to $H_{\rm MF}$. The $O(V^2)$ term is non-negative. This can be recognized if one regards the quantity

$$\int_0^\beta \mathrm{d}\tau \langle V(\tau)V\rangle_c = \int_0^\beta \mathrm{d}\tau \langle [V(\tau) - \langle V\rangle][V - \langle V\rangle]\rangle, \tag{6.121}$$

as a susceptibility or a relaxation function with $t = 0$, which is represented by the spectral intensity as in Eq. (3.54). By way of Eq. (3.32) we can confirm its non-negativity. Using the relation $\ln\langle \mathcal{U}(\beta)\rangle = \ln Z - \ln Z_{\rm MF}$, we obtain Eq. (6.77), provided the $O(V^3)$ terms can be neglected.

Actually the inequality holds even with higher order terms [23]. For the proof we scale $V \to \alpha V$ and construct the corresponding quantity $\langle \mathcal{U}_\alpha(\beta)\rangle$. It is sufficient to prove that the function $g(\alpha) = \ln\langle \mathcal{U}_\alpha(\beta)\rangle$ is convex downward for $0 \le \alpha \le 1$, which is equivalent to $g''(\alpha) > 0$ in the same range. What we have shown above is the convexity at $\alpha = 0$ with $g''(0) > 0$. Regarding $H_{\rm MF} + \alpha V$ as the new mean field Hamiltonian, we introduce the new scaled perturbation $\gamma V$. Then we repeat the same argument to take the second derivative in $\gamma$ which is set to zero in the end, and arrives at $g''(\alpha) > 0$. Hence Eq. (6.77) is valid without assumption of small $V$.

**Problem 6.7**

We make use of the identity

$$T\sum_n \frac{1}{(\mathrm{i}\epsilon_n - \epsilon_1)(\mathrm{i}\epsilon_n - \epsilon_2)}$$
$$= \frac{T}{\epsilon_2 - \epsilon_1}\sum_n \left(\frac{1}{\mathrm{i}\epsilon_n - \epsilon_1} - \frac{1}{\mathrm{i}\epsilon_n - \epsilon_2}\right) = \frac{f(\epsilon_1) - f(\epsilon_2)}{\epsilon_2 - \epsilon_1} \tag{6.122}$$

where $f(\epsilon)$ is the Fermi distribution function. The identity follows from application of Eq. (3.104). No difference arises whether $f(\epsilon)$ or $f(\epsilon) - 1$ is used in Eq. (3.104), since the summation over Matsubara frequencies $\epsilon_n$ is absolutely convergent here. Using this result in Eq. (6.79) we obtain

$$\chi_1(0) = -T\sum_n \left(G_f^*(\mathrm{i}\epsilon_n)\right)^2, \tag{6.123}$$

where the Green function is given by

$$G_f^*(\mathrm{i}\epsilon_n) = \int \mathrm{d}\epsilon \frac{\rho_f^*(\epsilon)}{\mathrm{i}\epsilon_n - \epsilon} = \frac{1}{\mathrm{i}\epsilon_n - \tilde{\epsilon}_f + \mathrm{i}\tilde{\Delta}\mathrm{sgn}\epsilon_n}, \tag{6.124}$$

which has been used in Eq. (6.62) for $\epsilon_n > 0$. In this Lorentzian form it is possible to represent $\chi_1(0)$ as

$$\chi_1(0) = -\frac{\partial}{\partial\tilde{\epsilon}_f} T \sum_n G_f^*(\mathrm{i}\epsilon_n) = -\frac{\partial n_f}{\partial\tilde{\epsilon}_f}, \tag{6.125}$$

Using Eq. (6.64), we obtain the first result in Eq. (6.81) at $T = 0$. The second result follows straightforwardly from Eq. (6.79) by noting $\mathrm{Im}(\omega - \epsilon + \mathrm{i}0_+)^{-1} = -\pi\delta(\omega - \epsilon)$.

## References

1. Anderson, P.W.: Phys. Rev. **124**, 41 (1961).
2. Kondo, J.: Prog. Theor. Phys. **32**, 37 (1964)
3. Abrikosov, A.A.: Physics **2**, 5 (1965)
4. Sakurai, J.J.: Modern Quantum Mechanics, revised edition. Addison Wesley, Reading (1993)
5. Shiba, H.: Prog. Theor. Phys. **43**, 601 (1970)
6. Tsvelick, A.M., Wiegmann, P.B.: Z. Phys. B **54**, 201 (1984)
7. Bickers, N.E., Rev. Mod. Phys. **59**, 845 (1987)
8. Rasul, J.W., Hewson, A.C.: J. Phys. **C17**, 2555, 3332 (1984)
9. Nozières, P.: J. Low Temp. Phys. **17**, 31 (1974)
10. Pines, D., Nozières, P.: The Theory of Quantum Liquids: Normal Fermi Liquids. Perseus Books Group, New York (1994)
11. Read, N., Newns, D.M.: J. Phys. C **16**, 3273 (1983)
12. Shiba, H. Prog. Theor. Phys. **54**, 967 (1975)
13. Otsuki, J., Kusunose, H., Werner, P., Kuramoto, Y.: J. Phys. Soc. Jpn. **76** (2007) 114707
14. Nozières, P., Blandin, A.: J. Phys. **41**, 193 (1980)
15. Andrei, N., Destri, C.: Phys. Rev. Lett. **52**, 364 (1984)
16. Bethe, H.: Z. Phys. **71**, 205 (1931)
17. Lindgren, I., Morrison, J.: Atomic Many-Body Theory. Springer, Berlin (1986)
18. Affleck, I.: Acta Phys. Polon. **26**, 1826 (1995)
19. Cox, D.L., Zawadowski, A.: Adv. Phys. **47**, 599 (1998)
20. Cox, D.L.: Phys. Rev. Lett. **59**, 1240 (1987)
21. Potok, R.M., Rau, I.G., Shtrikman, H., Oreg, Y., Goldhaber-Gordon, D.: Nature **446**, 167 (2007)
22. Shiba, H.: Prog. Theor. Phys. **43**, 601 (1970)
23. Feynman, R.P.: Statistical Mechanics. W.A. Benjamin, Reading (1972)

# Chapter 7
# One-Dimensional Fermions and Bosonization

**Abstract** This chapter deals with interaction effects in one-dimensional (1D) fermions. Because of the restricted spatial motion, interaction effects are stronger than in higher dimensional systems. Detailed treatment is provided for a powerful method called bosonization, which regards density fluctuation of 1D fermions as a bosonic object. In terms of bosonization, it is possible to discuss such interesting states as non-Fermi liquid with power-law decay of correlation functions, and separation of spin and charge degrees of freedom, both caused by mutual interactions. A state with an energy gap only for the charge sector is called the Mott insulating state, while the energy gap only for the spin sector corresponds to superconductivity. Finally we revisit the Kondo state from the 1D point of view.

## 7.1 Quantum Theory of String Oscillation

In order to deal with interacting quantum particles, we start from quantization of the string oscillation, which is regarded as collection of harmonic oscillators. The Hamiltonian is given by

$$H_{\mathrm{B}} = \sum_{q \neq 0} \omega_q \left( n_q + \frac{1}{2} \right), \quad n_q = b_q^\dagger b_q, \tag{7.1}$$

where $b_q^\dagger$ and $b_q$ are creation and annihilation operators of a boson with momentum $q$ and energy $\omega_q$. We shall show how the Hamiltonian $H_{\mathrm{B}}$ with $\omega_q \propto |q|$ is derived from a harmonically oscillating string.

We take a string with length $L$ and mass density $\rho$. For simplicity we consider only the longitudinal displacement along the string, and neglect the transverse displacements from the straight line. With tension $T_{\mathrm{s}}$ of the string, the classical energy density $\mathcal{H}$ is given by

$$\mathcal{H} = \frac{1}{2} T_{\mathrm{s}} (\nabla \Phi)^2 + \frac{1}{2} \rho \dot{\Phi}^2, \tag{7.2}$$



Y. Kuramoto, *Quantum Many-Body Physics*, Lecture Notes in Physics 934,
https://doi.org/10.1007/978-4-431-55393-9_7

where $\Phi$ is the displacement at space-time point $(x, t)$, $\nabla = \partial/\partial x$ and $\dot{\Phi} = \partial\Phi/\partial t$. The classical equation of motion gives a solution

$$\Phi(x,t) \propto \exp[\mathrm{i}q(x - vt)] \tag{7.3}$$

with $v = \sqrt{T_s/\rho}$. To simplify the notation we choose $\rho T_s = 1$ with appropriate unit for the tension $T_s$. Then the Hamiltonian can be written as

$$H_{\text{string}} \equiv \int_0^L \mathrm{d}x\, \mathcal{H} = \frac{v}{2}\int_0^L \mathrm{d}x \left[(\nabla\Phi)^2 + \Pi(x)^2\right], \tag{7.4}$$

where $\Pi(x,t) \equiv \rho\dot{\Phi}(x,t)$ is the momentum density.

The quantization is achieved by imposition of the canonical commutation rule:

$$[\Phi(x), \Pi(y)] = \mathrm{i}\delta(x - y). \tag{7.5}$$

Here we have put $t = 0$, which is always the case unless stated otherwise. With the periodic boundary condition, we make the Fourier decomposition in terms of bosonic creation and annihilation operators as

$$\Phi(x) = \sum_{q\neq 0} \frac{-\mathrm{i}}{\sqrt{2|q|L}}\left(b_q - b_{-q}^\dagger\right)\exp(\mathrm{i}qx), \tag{7.6}$$

$$\Pi(x) = -\sum_{q\neq 0} \sqrt{\frac{|q|}{2L}}\left(b_q + b_{-q}^\dagger\right)\exp(\mathrm{i}qx). \tag{7.7}$$

By changing the phase of boson operators such as $b_q \to \mathrm{i}b_q$, and $b_q^\dagger \to -\mathrm{i}b_q^\dagger$, we obtain different but equivalent expression. The present choice of the phase is most convenient to make connection to bosonization of fermion systems. It can be checked by direct substitution that the commutation rule $[b_q, b_p^\dagger] = \delta_{pq}$ is equivalent to Eq. (7.5). After quantization, $H_{\text{string}}$ becomes equivalent to $H_{\mathrm{B}}$ in Eq. (7.1) with $\omega_q = v|q|$. The demonstration is the subject of Problem 7.1.

For later convenience, we introduce the scaled displacement field $\theta(x)$ and its conjugate $\Pi_\theta(x)$ as

$$\theta(x) = \sqrt{4\pi}\,\Phi(x), \quad \Pi_\theta(x) = (4\pi)^{-1/2}\,\Pi(x), \tag{7.8}$$

which satisfy the same canonical commutation rule. In terms of the new fields we can write

$$H_{\text{string}} = \frac{v}{2}\int_0^L \mathrm{d}x \left[\frac{1}{4\pi}\left(\frac{\partial\theta}{\partial x}\right)^2 + 4\pi\,\Pi_\theta^2\right]. \tag{7.9}$$

Let us proceed to some useful relations involving $b^\dagger$ and/or $b$ in the exponential function. First, with arbitrary complex number $\alpha$ we obtain

$$[b, \exp(\alpha b^\dagger)] = \alpha \exp(\alpha b^\dagger), \tag{7.10}$$

which can be confirmed by Taylor expansion of the exponential and using $[b, (b^\dagger)^n] = n(b^\dagger)^{n-1}$ repeatedly. As a consequence, we obtain

$$b \exp(\alpha b^\dagger)|0\rangle \equiv b|\alpha\rangle = [b, \exp(\alpha b^\dagger)]|0\rangle = \alpha|\alpha\rangle \tag{7.11}$$

with $|0\rangle$ being the vacuum annihilated by $b$, and $|\alpha\rangle \equiv \exp(\alpha b^\dagger)|0\rangle$. Since $|\alpha\rangle$ is a coherent state defined by Eq. (5.3), we conclude that *the coherent state is an eigenstate of* $b$.

The next useful relation concerns factorization such as

$$\begin{aligned} \exp\left(\alpha b^\dagger + \gamma b\right) &= \exp(\alpha b^\dagger) \exp\left(\gamma b + \frac{1}{2}\alpha\gamma\right) \\ &= \exp(\gamma b) \exp\left(\alpha b^\dagger - \frac{1}{2}\alpha\gamma\right), \end{aligned} \tag{7.12}$$

which results from $[b, b^\dagger] = 1$. Equation (7.12) is often called the BCH (Baker–Campbell–Hausdorff) formula. Derivation of the formula is the subject of Problem 7.2. Concerning the statistical average $\langle \cdots \rangle$, we obtain the relation

$$\langle \exp\left(\alpha b^\dagger + \gamma b\right)\rangle = \exp\left[\alpha\gamma\left(\langle b^\dagger b\rangle + \frac{1}{2}\right)\right]. \tag{7.13}$$

Problem 7.3 deals with the derivation using a little trick.

If a pair of operators $A, B$ are both linear combination of $b$ and $b^\dagger$, the commutator $[A, B]$ becomes an ordinary number, usually called a c-number. Then combination of Eqs. (7.12) and (7.13) leads to valuable formulae:

$$\langle e^A\rangle = \exp\frac{1}{2}\langle A^2\rangle, \quad \langle e^A e^B\rangle = \exp\langle\frac{1}{2}\left(A^2 + B^2\right) + AB\rangle, \tag{7.14}$$

which play a central role in deriving correlation functions in Sect. 7.5.

## 7.2 Bosonization of Free Fermi Gas

The idea to deal with collective motion of one-dimensional fermions in terms of bosonic degrees of freedom has a long history. The pioneering work of Tomonaga [1] has already established the framework how to include the interaction effect with small momentum transfer. We first take the spinless free Fermi gas in order to explain the Tomonaga theory. The reason for solving the trivial problem by an elaborate framework is because the same method is useful in much more complicated systems. The Hamiltonian is simply given by

$$H_0 = \sum_k \epsilon_k c_k^\dagger c_k, \tag{7.15}$$

where we take, following Tomonaga, the free fermion spectrum $\epsilon_k = (k^2 - k_F^2)/(2m)$ with appropriate units. On the other hand, in the tight-binding approximation for the lattice case, the spectrum is given by $\epsilon_k = -2t \cos k$ with the lattice constant set to unity. The wave number $k$ is restricted inside the boundary $\pm\pi$ of the Brillouin zone. In the low-energy excitations, only those $k$ near the Fermi wave number $\pm k_F$ are important. Then we expand the spectrum around $\epsilon_k = 0$ as

$$\begin{aligned} \epsilon_k &\sim (k - k_F)v_F, \quad (k \sim k_F) \\ \epsilon_k &\sim -(k + k_F)v_F, \ (k \sim -k_F) \end{aligned} \tag{7.16}$$

which is illustrated in Fig. 7.1a, where the lines tangent to the parabola at $k = \pm k_F$ indicate the linear spectrum.

The key quantity in the Tomonaga's method is the density fluctuation operator as defined by

$$\rho_\pm(q) = \sum_k c_{\pm,k}^\dagger c_{\pm,k+q}, \tag{7.17}$$

where $\pm$ indicates the sign of the momentum near $\mu$. To indicate the momentum branches, the indices R and L instead of $\pm$ are also used as in $c_{\mathrm{R}k}, c_{\mathrm{L}k}$.

The density operators defined in the real space commute with each other. However, those defined for separate momentum branches satisfy the following

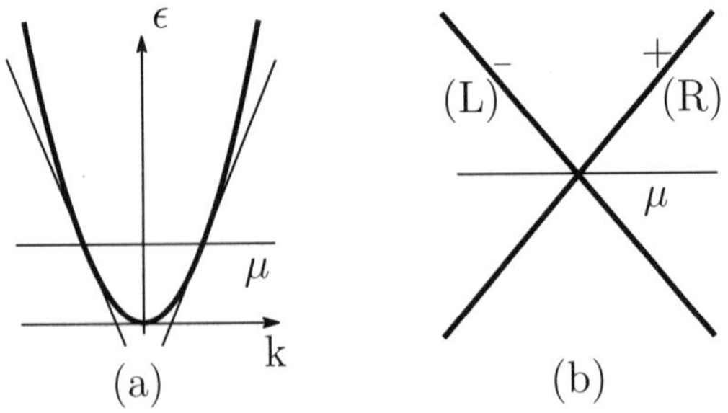


**Fig. 7.1** Spectrum of one-dimensional fermions: (**a**) Free fermions with the chemical potential $\mu$ set to zero, (**b**) Spectrum in the Luttinger model, which corresponds to extrapolation of linear approximation near the Fermi momentum in (**a**) to $\epsilon \to \pm\infty$ together with shift of momenta by $\pm k_F$ in left and right branches, respectively

commutation relations:

$$\left[\rho_\alpha(q), \rho_{\alpha'}(-q')\right] = \alpha\delta_{\alpha\alpha'}\delta_{qq'}N_q, \tag{7.18}$$

with $\alpha, \alpha' = \pm$ and $N_q \equiv qL/(2\pi)$. In order to identify the origin of non-commutativity, we remark that both R and L branches have the lower limit of energy corresponding to the bottom of the band. Equivalently, there is a cutoff $k_0 \sim 0$ in the momentum, which gives the lower limit for the the R branch and the upper limit for the L branch. Then the commutation relation for the R branch with $q > 0$ is derived as

$$[\rho_{\rm R}(q), \rho_{\rm R}(-q)] = \sum_{k>k_0} \left(\hat{n}_k - \hat{n}_{k+q}\right) = N_q, \tag{7.19}$$

which is independent of the cutoff $k_0$ and the Fermi momentum $k_F$. Similar calculation for the L branch brings the minus sign in the RHS.

Without introducing $k_0$, the summation over $k$ in Eq. (7.19) becomes ambiguous. Namely, one has to take the difference of two divergent terms in the thermodynamic limit. In fact, Luttinger took the RHS of Eq. (7.19) as zero by considering an infinite system with the linear spectrum [2], as illustrated in Fig. 7.1b. The R and L branches become independent of each other for the linear spectrum without cutoff. However, in order to have the well-defined ground state of fermions in the thermodynamic limit, the difference of infinite quantities requires a particular prescription. The choice of Luttinger did not match the physical requirement for the ground state. In Tomonaga's treatment, the linear spectrum is only an approximation of the parabolic spectrum as illustrated in Fig. 7.1a. Tomonaga arrived at the proper result (7.19) by assuming that the interaction does not affect the fermionic states near the band bottom. This paradoxical consequence is an example of the quantum anomaly, which may occur in field theory with infinite degrees of freedom [3]. Proper treatment in the thermodynamic limit has been examined in detail by Mattis and Lieb [4]. As long as lattice models are used, such delicate issue does not appear in condensed matter physics. In the continuum approximation, however, issues analogous to the quantum anomaly may appear.

Simple calculation using the kinetic energy (7.15) leads to the commutation relation:

$$[H_0, \rho_\alpha(q)] = v_{\rm F}\alpha q\rho_\alpha(q). \tag{7.20}$$

The same commutation relation ensues from another Hamiltonian

$$H_{\rm B} = \frac{2\pi v_{\rm F}}{L} \sum_{q>0,\alpha=\pm} :\rho_\alpha(q)\rho_\alpha(-q): \tag{7.21}$$

in place of $H_0$, which can be shown with use of Eq. (7.19). Here the colons (:...:) is called the N-product, or normal ordering, inside which any operator annihilating the ground state should be moved to the right of other operators. In the present case, the normal ordering results in

$$: \rho_\alpha(q)\rho_\alpha(-q) : \equiv \begin{cases} \rho_{\mathrm{R}}(-q)\rho_{\mathrm{R}}(q), & (\alpha = \mathrm{R}) \\ \rho_{\mathrm{L}}(q)\rho_{\mathrm{L}}(-q), & (\alpha = \mathrm{L}). \end{cases} \tag{7.22}$$

With this convention, $H_0$ and $H_{\mathrm{B}}$ give identical dynamics for the density fluctuations, although they may give different results for other quantities. Since the interaction to be discussed later is given in terms of density fluctuations, $H_{\mathrm{B}}$ is justified for studying interaction effects. The overall use of $H_{\mathrm{B}}$ is the crucial feature in the method of Tomonaga.

The density operators are related to the Bose operators $b_q, b_q^\dagger$. With $q > 0$, the correspondence reads

$$\rho_{\mathrm{R}}(q) = \sqrt{N_q}\, b_q, \quad \rho_{\mathrm{R}}(-q) = \sqrt{N_q}\, b_q^\dagger,$$
$$\rho_{\mathrm{L}}(q) = \sqrt{N_q}\, b_{-q}^\dagger, \quad \rho_{\mathrm{L}}(-q) = \sqrt{N_q}\, b_{-q}. \tag{7.23}$$

We further introduce the Hermitian phase operator $\theta_{\mathrm{R,L}}(x)$ by

$$\theta_{\mathrm{R}}(x) = \frac{1}{L}\sum_{q\neq 0} \rho_{\mathrm{R}}(q)\frac{e^{\mathrm{i}qx}}{\mathrm{i}q} = \sum_{q>0} \frac{-\mathrm{i}}{\sqrt{N_q}}\left(b_q e^{\mathrm{i}qx} - b_q^\dagger e^{-\mathrm{i}qx}\right) \tag{7.24}$$

$$\theta_{\mathrm{L}}(x) = \frac{1}{L}\sum_{q\neq 0} \rho_{\mathrm{L}}(q)\frac{e^{\mathrm{i}qx}}{\mathrm{i}q} = \sum_{q>0} \frac{\mathrm{i}}{\sqrt{N_q}}\left(b_{-q} e^{-\mathrm{i}qx} - b_{-q}^\dagger e^{\mathrm{i}qx}\right). \tag{7.25}$$

Using Eq. (7.23) we obtain the relations

$$\nabla\theta_\alpha(x) = 2\pi\rho_\alpha(x), \tag{7.26}$$
$$[\theta_\alpha(x), \rho_\beta(y)] = -\mathrm{i}\alpha\delta_{\alpha\beta}\delta(x-y), \tag{7.27}$$

with $\alpha = \pm 1$ in the RHS of Eq. (7.27) depending on R or L branch. Here we have defined the density operator $\rho_\alpha(x)$ in the real space by

$$\rho_\alpha(x) = \frac{1}{L}\sum_{q\neq 0} \rho_\alpha(q)\exp(\mathrm{i}qx). \tag{7.28}$$

Equation (7.27) shows that $\theta_\alpha(x)$ and $\rho_\beta(y)$ are canonically conjugate variables. The equivalent relation is given by the $y$-integral of Eq. (7.27) as

$$[\theta_\alpha(x), \theta_\beta(y)] = \alpha\pi\mathrm{i}\delta_{\alpha\beta}\mathrm{sgn}(x-y), \tag{7.29}$$

Note that $\rho_\alpha(x)$ and $\theta_\alpha(x)$ do not have the homogeneous component because $q=0$ is excluded in the definition of the phase operators.

By combining the right and left branches of the density fluctuations, we introduce the number and current operators by

$$N(q) = \rho_R(q) + \rho_L(q), \tag{7.30}$$

$$J(q) = \rho_R(q) - \rho_L(q). \tag{7.31}$$

In the real space we define accordingly

$$\rho_N(x) = \rho_R(x) + \rho_L(x), \tag{7.32}$$

$$\rho_J(x) = \rho_R(x) - \rho_L(x). \tag{7.33}$$

In defining the corresponding phase operators, it is convenient to include the homogeneous ($q=0$) component $N, \mathcal{J}$ and the global phase $\bar{\theta}_N, \bar{\theta}_J$ as

$$\theta_N(x) = \theta_R(x) + \theta_L(x) + \bar{\theta}_N + 2\pi x\mathcal{J}/L, \tag{7.34}$$

$$\theta_J(x) = \theta_R(x) - \theta_L(x) + \bar{\theta}_J + 2\pi xN/L. \tag{7.35}$$

The components $N = N_R + N_L$ and $\mathcal{J} = N_R - N_L$ correspond, respectively, to the total number and current of fermions. Each of them is not a dynamical variable but a topological one called the "winding number" [5]. The states with different winding numbers do not mix by any bosonic excitation. Although the local fermionic number increases with increase of $\theta_\alpha(x)$ by $2\pi$, such increase is always compensated for by the negative change of $\theta_\alpha(y)$ in other place. On the other hand, each component

$$\bar{\theta}_\alpha \equiv \frac{1}{2}\left(\bar{\theta}_N + \alpha\bar{\theta}_J\right) \tag{7.36}$$

can change the topological number $N_\alpha$. This is equivalent to the commutation rule:

$$[\bar{\theta}_\alpha, N_\beta] = -\mathrm{i}\delta_{\alpha\beta}, \quad [\bar{\theta}_\alpha, \bar{\theta}_\beta] = 0, \tag{7.37}$$

by analogy with Eq. (7.27). The commutation rule leads to $[\bar{\theta}_N, \mathcal{J}] = [\bar{\theta}_J, N] = 0$. The variables $\bar{\theta}_\alpha$ are useful in constructing fermion operators in terms of bosonic variables to be explained in the next section. We summarize the important relations involving number and current variables.

$$\nabla\theta_N(x) = 2\pi[\rho_N(x) + \bar{\rho}_N], \quad \nabla\theta_J(x) = 2\pi[\rho_J(x) + \bar{\rho}_J], \tag{7.38}$$

$$[\theta_N(x), \rho_J(y)] = [\theta_J(x), \rho_N(y)] = -2i\delta(x-y), \tag{7.39}$$

$$[\theta_N(x), \rho_N(y)] = [\theta_J(x), \rho_J(y)] = 0, \tag{7.40}$$

with $\bar{\rho}_N \equiv N/L$ and $\bar{\rho}_J \equiv \mathcal{J}/L$. Note the factor 2 in Eq. (7.38) which comes from our definition of $\theta_N, \theta_J$. We may alternatively use $\check{\theta}_a \equiv \theta_a/\sqrt{2}$ with $a = N, J$ for keeping the canonical commutation rule.

Thus the Hamiltonian $H_\mathrm{B}$ defined by Eq. (7.21) is also represented by

$$H_\mathrm{B} = \frac{\pi v_F}{L} \sum_{q>0} : N(q)N(-q) + J(q)J(-q) : \tag{7.41}$$

$$= \frac{v_F}{16\pi} \int_0^L \mathrm{d}x : (\nabla\theta_N)^2 + (\nabla\theta_J)^2 : . \tag{7.42}$$

Problem 7.4 is concerned with the derivation and identification of canonical variables.

## 7.3 Bosonic Representation of Fermions

We have seen that density excitations in free fermions with the linear spectrum are equivalently described by free bosons with the spectrum $\omega_q = v_F|q|$, provided the ground state of fermions is well-defined. However, the fermionic creation and annihilation operators cannot be described by bosonic operators alone, since the complete set of bosonic operators belong to the Hilbert space with a definite number of fermions. Together with an extra operator that connects the ground states with $N$ and $N \pm 1$ fermions, however, any fermionic operator can be constructed in terms of bosonic operators. This situation was ambiguous in the early stage of bosonization, and some confusion still remains in the literature. We take an approach [5] starting from finite size of the system, which is capable of dealing with delicate issues in the thermodynamic limit.

For R and L branches, we introduce the corresponding field operators by

$$\psi_\mathrm{R}(x) = \frac{1}{\sqrt{L}} \sum_k c_{\mathrm{R}k} \exp[\mathrm{i}(k + k_F)x], \tag{7.43}$$

$$\psi_\mathrm{L}(x) = \frac{1}{\sqrt{L}} \sum_k c_{\mathrm{L}k} \exp[\mathrm{i}(k - k_F)x], \tag{7.44}$$

where dominant contribution in dynamics comes from $k \sim 0$. According to Eq. (7.23), the density operators for the R branch are described by bosons with $q > 0$ only. We obtain

$$[b_q, \psi_\mathrm{R}(x)] = -\frac{1}{\sqrt{N_q}} e^{-\mathrm{i}qx} \psi_\mathrm{R}(x), \tag{7.45}$$

$$[b_q^\dagger, \psi_\mathrm{R}(x)] = -\frac{1}{\sqrt{N_q}} e^{\mathrm{i}qx} \psi_\mathrm{R}(x), \tag{7.46}$$

with the commutation relation $[c_{\mathrm{R}k}, \rho_{\mathrm{R}}(q)] = c_{\mathrm{R},k+q}$. Comparison with Eq. (7.10) reveals that $\psi_{\mathrm{R}}(x)$ behaves as exponentials of $b_q$ and $b_q^\dagger$ with $q > 0$. This observation leads to the representation

$$\psi_{\mathrm{R}}(x)e^{-\mathrm{i}k_F x} = F_{\mathrm{R}} \exp\left(-\sum_{q>0} \frac{1}{\sqrt{N_q}} b_q^\dagger e^{-\mathrm{i}qx}\right) \exp\left(\sum_{q>0} \frac{1}{\sqrt{N_q}} b_q e^{\mathrm{i}qx}\right), \tag{7.47}$$

where the unitary operator $F_{\mathrm{R}}$ is called the Klein factor, which decreases $N_{\mathrm{R}}$ by one but does nothing to the bosonic state. Hence $F_{\mathrm{R}}$ commutes with $b_q$ and $b_q^\dagger$, and has the commutation rule:

$$[F_{\mathrm{R}}, N_{\mathrm{R}}] = F_{\mathrm{R}}, \quad [F_{\mathrm{R}}^\dagger, N_{\mathrm{R}}] = -F_{\mathrm{R}}^\dagger. \tag{7.48}$$

The Klein factors $F_\alpha$ ($\alpha = \mathrm{R, L}$) can be constructed in terms of the homogeneous ($q = 0$) components of bosonic operators, which describe topological numbers $N_\alpha$ and the global phase $\bar{\theta}_\alpha$. With use of Eq. (7.37), we find that Eq. (7.48) is reproduced by the form:

$$F_\alpha = C_\alpha \exp(\mathrm{i}\bar{\theta}_\alpha) \tag{7.49}$$

with a unitary operator $C_\alpha^\dagger = C_\alpha^{-1}$, which is specified soon.

With use of the normal ordering, we obtain the concise form of Eq. (7.47) as

$$\psi_{\mathrm{R}}(x)e^{-\mathrm{i}k_F x} = \frac{1}{\sqrt{L}} F_{\mathrm{R}} : \exp[\mathrm{i}\theta_{\mathrm{R}}(x)] := \frac{1}{\sqrt{L}} C_{\mathrm{R}} : \exp[\mathrm{i}\theta_{\mathrm{R}}(x) + \mathrm{i}\bar{\theta}_{\mathrm{R}}] : . \tag{7.50}$$

We can also use the alternative form

$$\psi_{\mathrm{R}}(x) = \frac{1}{\sqrt{2\pi\eta}} F_{\mathrm{R}} \exp[\mathrm{i}\theta_{\mathrm{R}}(x) + \mathrm{i}k_F x], \tag{7.51}$$

where $\eta$ is positive infinitesimal. The equivalence between Eqs. (7.51) and (7.50) can be seen from the BCH formula (7.12). Namely, in evaluating $\sum_{q>0}[b_q, b_q^\dagger]/N_q$, which appears in the commutator in Eq. (7.10), we replace the summation over $q$ by the integral $[2\pi/L, \infty]$. In order to suppress the logarithmic divergence at $q \to \infty$, we introduce the convergence factor $\exp(-q\eta)$ and obtain

$$\int_{2\pi/L}^{\infty} \frac{\mathrm{d}q}{2q} \exp(-q\eta) = -\frac{1}{2} \ln(2\pi\eta/L). \tag{7.52}$$

After exponentiating the result we recover Eq. (7.51). This calculation shows that the infinitesimal quantity $\eta$ originates from a regularization that accompanies calculation of the N-product. Although Eq. (7.51) has an uncomfortable form with the infinitesimal denominator, its compactness is convenient for practical calculation. Due care should be taken for the meaning of $\eta$. The same line of argument is applicable also to the L branch of fermions which are represented by bosons with $q < 0$. Including the Klein factor we obtain

$$\psi_{\rm L}(x) = \frac{1}{\sqrt{2\pi\eta}} F_{\rm L} \exp[-{\rm i}\theta_{\rm L}(x) - {\rm i}k_F x]. \tag{7.53}$$

Any choice of $C_\alpha$ guarantees the anticommutation rule for the same branch $\alpha$: $\{\psi_\alpha(x), \psi_\alpha(y)\} = 0$ with $x \neq y$. This is confirmed by the BCH formula Eq. (7.12):

$$\exp[{\rm i}\theta_\alpha(x)]\exp[{\rm i}\theta_\alpha(y)] = \exp[{\rm i}\theta_\alpha(y)]\exp[{\rm i}\theta_\alpha(x)]\exp[\alpha\pi {\rm i}\,{\rm sgn}(x-y)], \tag{7.54}$$

where the last factor $(= -1)$ in the RHS results from $[\theta_\alpha(x), \theta_\alpha(y)]$ corresponding to Eq. (7.29). Since $F_\alpha$ commutes with $\theta_\alpha(x)$, the anticommutation rule $\{\psi_\alpha(x), \psi_\alpha(y)\} = 0$ is confirmed. If $C_\alpha$ is chosen to be Hermitian: $C_\alpha^\dagger = C_\alpha$, similar calculation confirms the anticommuation of $\psi_\alpha^\dagger(x)$ and $\psi_\alpha(y)$ for $x \neq y$. The general case including $x = y$ requires a way to deal with singular behavior, but straightforward calculation reproduces the proper relation including the $\delta(x-y)$ term [5].

In order to reproduce the anticommutation rule: $\{\psi_{\rm R}(x), \psi_{\rm L}(y)\} = 0 = \{\psi_{\rm R}(x)^\dagger, \psi_{\rm L}(y)\}$, on the other hand, it is nontrivial how to choose $C_\alpha$ properly. We try a choice

$$C_\alpha = \exp[{\rm i}\pi N(1-\alpha)/2], \tag{7.55}$$

which means $C_{\rm R} = 1$ and $C_{\rm L} = (-1)^N$. We obtain

$$\begin{aligned} F_{\rm R} F_{\rm L} &= e^{{\rm i}\bar\theta_{\rm R}}(-1)^N e^{{\rm i}-\bar\theta_{\rm L}} = (-1)^{N-1} e^{{\rm i}\bar\theta_{\rm R}} e^{{\rm i}-\bar\theta_{\rm L}} \\ &= (-1)^{N-1} e^{{\rm i}-\bar\theta_{\rm L}} e^{{\rm i}\bar\theta_{\rm R}} = -F_{\rm L} F_{\rm R}. \end{aligned} \tag{7.56}$$

Thus we confirm the anticommutation property of $\psi_{\rm R}$ and $\psi_{\rm L}$; the inhomogeneous part $\exp[{\rm i}\theta_\alpha(x)]$ commutes for different branches R and L. The anticommutation between $\psi_{\rm R}^\dagger$ and $\psi_{\rm L}$ also follows since $C_\alpha$ in Eq. (7.55) is Hermitian. Note that even if $N$ takes even integer in the ground state, $(-1)^N$ cannot be regarded as unity in the Fock space where $\psi_\alpha$ decreases $N$ by one. We further remark that the choice of $C_\alpha$ is not unique. For example, another choice: $C_{\rm R} = 1$ and $C_{\rm L} = (-1)^{N_{\rm R}}$ works equally well. The anticommutation property of different Klein factors is summarized as

$$\{F_i^\dagger, F_j\} = 2\delta_{ij}, \quad (1-\delta_{ij})\{F_i, F_j\} = 0, \tag{7.57}$$

together with $F_i^\dagger F_i = 1$.

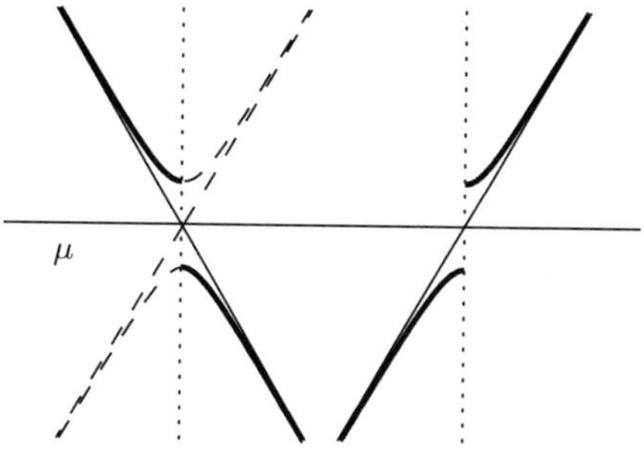


**Fig. 7.2** The gapped spectrum due to the hybridization of original fermions. The dotted lines show the momenta $\pm k_F$. The dashed line illustrates the R branch shifted by $-2k_F$

With these preliminaries we now deal with hybridization of R and L branches such as

$$H_{\text{hyb}} = m \int_0^L \mathrm{d}x \exp(2\mathrm{i}k_F x) : \psi_{\mathrm{R}}^{\dagger}(x)\psi_{\mathrm{L}}(x) : +\text{h.c.}, \tag{7.58}$$

where the normal ordering means subtraction of possible singular terms. The meaning is in common with that of the N-product in the bosonic case since both focus in the change from the ground state. With inclusion of kinetic energy and two-body interactions, the model with hybridization of R, L branches is called the massive Thirring model. Motivation of the notation $m$ as hybridization is its relation to the effective mass in the relativistic spectrum with the energy gap $2m$. If two-body interactions are absent, it is possible to take exact account of the hybridization effect. The resultant spectrum is illustrated in Fig. 7.2. After bosonization, the hybridization is written as

$$e^{2\mathrm{i}k_F x} : \psi_{\mathrm{R}}^{\dagger}(x)\psi_{\mathrm{L}}(x) : +\text{h.c.} = \pm\frac{2}{L} : \cos\left[\theta_N(x)\right] := \pm\frac{1}{\pi\eta}\cos\left[\theta_N(x)\right], \tag{7.59}$$

where $\theta_N(x)$ has been defined by Eq. (7.34). Here the product of Klein factors remains only as the phase factor $\pm 1$. If this factor is $-1$, we can absorb the minus sign by redefining $C_{\mathrm{L}}$ as $C_{\mathrm{L}} = (-1)^{N+1}$. Problem 7.5 concerns with the demonstration in more detail.

Thus the massive Thirring model without interaction is equivalent to the following model:

$$H_{\text{sG}} = H_{\mathrm{B}} + \frac{2m}{\pi L}\int_0^L \mathrm{d}x : \cos[a_N\theta_J(x)] : \tag{7.60}$$

with $a_N = 1$. We rewrite the bosonized kinetic energy $H_{\mathrm{B}}$ in the form

$$H_{\mathrm{B}} = \frac{v_F}{2}\int_0^L \mathrm{d}x \left[\frac{1}{4\pi}\left(\nabla\check{\theta}_N\right)^2 + 4\pi\check{\Pi}_N(x)^2\right], \tag{7.61}$$

where $\breve{\theta}_N(x) = \theta_N(x)/\sqrt{2}$ and $\breve{\Pi}_N(x) = -(4\pi)^{-1}\nabla\theta_J(x)/\sqrt{2}$. The canonical field variables for coordinate corresponds to $\breve{\theta}_N(x)$ and the momentum to $\breve{\Pi}_N(x)$. Here the normalization in Eqs. (7.34) and (7.35) necessitates the factor $1/\sqrt{2}$. It is also possible to choose $\theta_J(x)/\sqrt{2}$ as the canonical coordinate. Problem 7.4 deals with identification of the canonical momentum in this case. The Hamiltonian $H_{\rm sG}$ with general value of $a_N$ is called the sine-Gordon model. Using the equivalence of the sine-Gordon model with $a_N = 1$ to the free massive Thirring model, exact solution has been obtained for some interacting models. Sections 7.8 and 7.9 give such examples.

In the classical sine-Gordon model where $\theta_N$ and $\Pi_N$ are c-numbers, the spatial configuration of $\theta_N(x) = \pm\pi$ as $x \to \pm\infty$ is called a soliton. The reverse change of the phase is called the anti-soliton. After quantization, these phase configurations, as described by Eq. (7.47), correspond to fermion excitations. In contrast to common examples such as the Cooper pair where an even number of fermions constitute a bosonic entity, the sine-Gordon model provides a reverse example that the coherent bosonic excitations form a fermion.

## 7.4 Inclusion of Forward Scattering

Equipped with the formalism which may look too heavy for free fermions, we now consider two-body interactions in spinless systems. Let us first take the following form:

$$H_{\rm int} = \frac{1}{2L}\sum_{q,\alpha=\pm}\left[2g_2 : \rho_\alpha(q)\rho_{-\alpha}(-q) : + g_4 : \rho_\alpha(q)\rho_\alpha(-q) :\right], \tag{7.62}$$

where $g_2$ describes the interaction between R and L branches, while $g_4$ describes the one within each branch. In the total Hamiltonian $H_{\rm B} + H_{\rm int}$, the whole effect of $g_4$ is taken into account by the shift

$$v_F^* = v_F + g_4/(2\pi) \tag{7.63}$$

of the Fermi velocity. We assume that $g_4$ satisfies the property $v_F^* > 0$. On the other hand, $g_2$ gives rise to off-diagonal terms with respect to the indices $\pm$ (or R, L) of the density operators. Note that the $g_2$ term does not annihilate the non-interacting ground state even with normal ordering. Hence the ground state is reconstructed.

We shall diagonalize the total Hamiltonian $H = H_{\rm B} + H_{\rm int}$ by a kind of canonical transformation. It is convenient to introduce the positive parameter $K$ and the renormalized velocity $v_{\rm s}$ by the following relations:

$$\frac{v_{\rm s}}{K} = v_F^* + \frac{g_2}{2\pi}, \quad v_{\rm s}K = v_F^* - \frac{g_2}{2\pi}, \tag{7.64}$$

where $v_F^*$ is the shifted Fermi velocity defined by Eq. (7.63). In the weak-coupling limit $g_2 = 0$, the parameters become $K = 1$ and $v_s = v_F^*$. Assuming $|g_2| < 2\pi v_F^*$, which gives $v_s > 0$, we take the sum and difference of the two relations above to obtain

$$\frac{v_F^*}{v_s} = \frac{1}{2}\left(\frac{1}{K} + K\right) = \cosh 2\phi, \tag{7.65}$$

$$\frac{g_2}{2\pi v_s} = \frac{1}{2}\left(\frac{1}{K} - K\right) = \sinh 2\phi, \tag{7.66}$$

where $\phi$ and $K$ are related by $K = \exp(-2\phi)$. It is obvious that the dimensionless interaction $g_2/v_F^*$ is characterized by a single parameter. Eliminating $v_s$ in Eqs. (7.65) and (7.66), we indeed obtain such parameterization:

$$\frac{g_2}{2\pi v_F^*} = \tanh 2\phi = \frac{1-K^2}{1+K^2}, \tag{7.67}$$

which is equivalent to

$$K = \sqrt{\frac{2\pi v_F^* - g_2}{2\pi v_F^* + g_2}}. \tag{7.68}$$

Derivation of these results is the subject of Problem 7.6. As we show later in Eq. (7.73), $\phi$ corresponds to the (imaginary) rotation angle in the Bogoliubov transformation.

Using Eq. (7.64), and the density and current operators defined by Eqs. (7.30) and (7.31), we rewrite the total Hamiltonian including $H_{\text{int}}$ as

$$\begin{aligned} H &= \frac{\pi v_s}{L} \sum_{q>0} : \frac{1}{K} N(q)N(-q) + K J(q)J(-q) : \\ &= \frac{v_s}{2} \int_0^L \mathrm{d}x : \frac{1}{K}\left(\frac{\partial \Phi}{\partial x}\right)^2 + K \Pi(x)^2 : \end{aligned} \tag{7.69}$$

with $\Phi(x) = \theta_N(x)/\sqrt{8\pi}$ and $\Pi(x) = \sqrt{8\pi}\,\Pi_N(x)$. This form of $H$ motivates us to introduce the quantities

$$\tilde{\Phi} = \sqrt{K}\Phi, \quad \tilde{\Pi} = \Pi/\sqrt{K}, \tag{7.70}$$

which corresponds to renormalized coordinate and momentum of the bosonic field. The commutation relation $[\tilde{\Phi}, \tilde{\Pi}] = [\Phi, \Pi]$ shows that $\tilde{\Phi}$ and $\tilde{\Pi}$ are qualified as canonical variables for quantization. We further introduce new quantities by

$$\tilde{N}(q) = \tilde{\rho}_{\rm R}(q) + \tilde{\rho}_{\rm L}(q) = \sqrt{K}N(q), \tag{7.71}$$

$$\tilde{J}(q) = \tilde{\rho}_{\rm R}(q) - \tilde{\rho}_{\rm L}(q) = \frac{1}{\sqrt{K}}J(q), \tag{7.72}$$

which give the relation between the density operators as

$$\begin{pmatrix} \tilde{\rho}_{\rm R}(q) \\ \tilde{\rho}_{\rm L}(q) \end{pmatrix} = \begin{pmatrix} \cosh\phi, & -\sinh\phi \\ -\sinh\phi, & \cosh\phi \end{pmatrix} \begin{pmatrix} \rho_{\rm R}(q) \\ \rho_{\rm L}(q) \end{pmatrix}, \tag{7.73}$$

from the sum and difference of Eqs. (7.71) and (7.72). As we shall show shortly, the Hamiltonian is diagonalized in terms of the new variables $\tilde{\rho}_\alpha(q)$. The setup by Eq. (7.73) is an example of the Bogoliubov transformation for bosons. The imaginary angle in hyperbolic functions is contrasted with the case of fermions where a real angle specifies the transformation as discussed in Problem 5.2 around Eq. (5.95). The unitary nature of the transformation Eq. (7.73) appears in conservation of commutation relations for density operators. We note, however, that the $2 \times 2$ matrix in Eq. (7.73) is *not* unitary.

The unitary operator that represents this Bogoliubov transformation as $\tilde{\rho}_\alpha(q) = U^\dagger \rho_\alpha(q) U$ is given by

$$U = \exp\left[\sum_{q\neq 0} \frac{2\pi\phi}{Lq} \rho_{\rm R}(q)\rho_{\rm L}(-q)\right] \equiv \exp S. \tag{7.74}$$

The proof of equivalence to Eq. (7.73) is the subject of Problem 7.7. The unitarity of $U$ is recognized by the relation $S^\dagger = -S$. The transformed Hamiltonian reads

$$\tilde{H} \equiv U^\dagger H U = \frac{2\pi v_{\rm S}}{L} \sum_{q>0,\alpha} :\tilde{\rho}_\alpha(q)\tilde{\rho}_\alpha(-q):, \tag{7.75}$$

which is diagonal in $\alpha$. The form of $\tilde{H}$ is the same as that of free particles except for the replacement $v_F \to v_{\rm S}$. The normal ordering is taken for the original density operators. If one redefines the normal ordering in terms of $\tilde{\rho}$, the only difference is a constant term appearing in Eq. (7.75), which corresponds to the shift in the ground state energy.

## 7.5 Momentum Distribution Near the Fermi Level

A great advantage of the bosonization method is its ability to derive correlation functions of one-dimensional fermion systems. Let us first consider the right-going (R) free spinless fermions. The basic quantity is the phase correlation function

$$g_{\mathrm{R}}(x-y) = \langle \theta_{\mathrm{R}}(x)\theta_{\mathrm{R}}(y)\rangle, \tag{7.76}$$

which can be derived at $T = 0$ explicitly as

$$g_{\mathrm{R}}(x) = \sum_{q>0} \frac{2}{N_q} \exp(\mathrm{i}qx) = \ln \frac{L}{2\pi(\eta - \mathrm{i}x)}, \tag{7.77}$$

with use of Eqs. (7.24) and (7.52). Then the exponentiated correlation function can be derived as

$$\langle e^{\alpha\theta_{\mathrm{R}}(x)} e^{\beta\theta_{\mathrm{R}}(0)}\rangle = C \exp\left[\alpha\beta g_{\mathrm{R}}(x)\right] = C \left(\frac{L}{2\pi(\eta - \mathrm{i}x)}\right)^{\alpha\beta}, \tag{7.78}$$

$$C = \exp\left\langle \frac{1}{2}\left(\alpha^2 + \beta^2\right) \theta_{\mathrm{R}}(0)^2 \right\rangle \tag{7.79}$$

with the cumulant property Eq. (7.14).

Referring to Eq. (7.51), we can represent the fermion annihilation operator $\psi_{\mathrm{R}}(0)$ by putting $\beta = i$ in $\exp[\beta\theta_{\mathrm{R}}(0)]$. Similarly, putting $\alpha = -\mathrm{i}$ in $\exp[\alpha\theta_{\mathrm{R}}(x)]$ gives $\psi_{\mathrm{R}}^{\dagger}(x)\exp(\mathrm{i}k_F x)$. Hence for free fermions we find the correlation function, or the density matrix, as

$$\langle \psi_{\mathrm{R}}^{\dagger}(x)\psi_{\mathrm{R}}(0)\rangle_0 \sim (x + \mathrm{i}\eta)^{-1} \exp(-\mathrm{i}k_F x), \tag{7.80}$$

where we have discarded the numerical factor, and put the suffix 0 in the average to emphasize the non-interacting state. This result should agree with the one derived by the elementary method for right-going free fermions. In fact, the inverse Fourier transform of the momentum distribution:

$$n_{\alpha k} = \langle c_{\alpha k}^{\dagger} c_{\alpha k}\rangle_0 = \theta(-\alpha k), \tag{7.81}$$

which is the step function, leads to the result given by Eq. (7.80) for $\alpha = +1$ (R branch). Here $\eta$ $(> 0)$ works as the convergence factor in $\exp(-\mathrm{i}kx + \eta k)$ at $k \to -\infty$. When we consider finite $x$, we can neglect $\eta$ in Eq. (7.80).

We proceed to interacting spinless fermions where the unitary transformation $U = \exp S$ constructs the new ground state $|\tilde{0}\rangle = U|0\rangle$ from that of free fermions $|0\rangle$. The density matrix is now given by

$$\rho_{\mathrm{R}}(x,0) \equiv \left\langle \tilde{0}|\psi_{\mathrm{R}}^{\dagger}(x)\psi_{\mathrm{R}}(0)|\tilde{0}\right\rangle = \left\langle U^{\dagger}\psi_{\mathrm{R}}^{\dagger}(x)\psi_{\mathrm{R}}(0)U\right\rangle_0 . \tag{7.82}$$

In terms of the transformed variables with tilde such as $\tilde{\theta}_{\mathrm{R}} = U^{\dagger}\theta_{\mathrm{R}}U$ we obtain

$$U^{\dagger}\psi_{\mathrm{R}}(x)U \equiv \tilde{\psi}_{\mathrm{R}}(x) = \frac{1}{\sqrt{2\pi\eta}}\exp[\mathrm{i}\tilde{\theta}_{\mathrm{R}}(x) + \mathrm{i}k_F x]F_{\mathrm{R}}. \tag{7.83}$$

Explicit calculation is most concisely performed by going to density ($N$) and current ($J$) variables since the unitary transformation then amounts just to multiplication of $K^{\pm 1/2}$. Namely, we obtain

$$\begin{aligned}\tilde{\theta}_{\mathrm{R}}(x) &= \frac{1}{2}\left[\tilde{\theta}_N(x) + \tilde{\theta}_J(x)\right],\\ \tilde{\theta}_N(x) &= \sqrt{K}\theta_N(x),\;\; \tilde{\theta}_J(x) = \theta_J(x)/\sqrt{K},\end{aligned} \tag{7.84}$$

The phase correlation functions in the non-interacting case are given by

$$\langle\theta_N(x)\theta_N(0)\rangle_0 = \langle\theta_J(x)\theta_J(0)\rangle_0 = g_{\mathrm{R}}(x) + g_{\mathrm{L}}(x) = \ln\left(\frac{L}{2\pi x}\right)^2, \tag{7.85}$$

where $g_{\mathrm{L}}(x) = g_{\mathrm{R}}(x)^*$ represents the left-going branch.

Note that $\rho_{\mathrm{R}}(x,0)\exp(\mathrm{i}k_F x)$ is an odd function of $x$ as in the non-interacting case given by Eq. (7.80). This property is most easily seen by writing

$$\begin{aligned}\rho_{\mathrm{R}}(x,0) &= \frac{1}{2\pi\eta}\left\langle \exp\left(-\mathrm{i}\tilde{\theta}_{\mathrm{R}}(x) + \mathrm{i}\tilde{\theta}_{\mathrm{R}}(0) + \frac{1}{2}\left[\tilde{\theta}_{\mathrm{R}}(x), \tilde{\theta}_{\mathrm{R}}(0)\right] - \mathrm{i}k_F x\right)\right\rangle_0\\ &= \frac{1}{2\pi\eta}\exp\left(-\frac{1}{2}\left\langle\left[\tilde{\theta}_{\mathrm{R}}(x) - \tilde{\theta}_{\mathrm{R}}(0)\right]^2\right\rangle_0 + \frac{1}{2}\mathrm{i}\pi\,\mathrm{sgn}(x) - \mathrm{i}k_F x\right),\end{aligned} \tag{7.86}$$

where we have used Eq. (7.14) and the commutation rule Eq. (7.27). We now use Eqs. (7.84) and (7.85) to obtain

$$\begin{aligned}\langle\tilde{0}|\psi_{\mathrm{R}}^{\dagger}(x)\psi_{\mathrm{R}}(0)|\tilde{0}\rangle e^{\mathrm{i}k_F x} &\sim \mathrm{sign}(x)\exp\left[\frac{1}{4}\left(K + \frac{1}{K}\right)\ln\left(\frac{L}{2\pi x}\right)^2\right]\\ &\sim x^{-1}|x|^{-\gamma}\end{aligned} \tag{7.87}$$

with $\gamma = v_F^*/v_s - 1 = (K + K^{-1})/2 - 1 = 2\sinh^2\phi$ as deduced from Eqs. (7.65) and (7.66). Note that $\gamma$ vanishes without interactions. By Fourier transform we obtain the singular behavior at the Fermi points as

$$n_{\alpha k} = \langle\tilde{0}|c_{\alpha k}^\dagger c_{\alpha k}|\tilde{0}\rangle \sim -\mathrm{sign}(\alpha k)|k|^\gamma + \mathrm{const.} \tag{7.88}$$

Note that the physical momentum is given by $k + \alpha k_F$. Because of $\gamma > 0$, the discontinuity at the Fermi surface disappears by the interaction effect.

## 7.6 Separation of Spin and Charge

We now come to the point to deal with interacting fermions with the spin degrees of freedom. The kinetic energy is given by

$$\begin{aligned} H_0 &= v_\mathrm{F} \sum_{k,\pm,\sigma} \left[ (\pm k - k_\mathrm{F}) c_{\pm,k,\sigma}^\dagger c_{\pm,k,\sigma} \right] \\ \to \quad H_\mathrm{B} &= \frac{2\pi v_\mathrm{F}}{L} \sum_{q>0,\alpha=\pm,\sigma} : \rho_{\alpha,\sigma}(q)\rho_{\alpha,\sigma}(-q) :, \end{aligned} \tag{7.89}$$

where the density operators have the spin index $\sigma$ ($=\uparrow, \downarrow$), and are defined by

$$\rho_{\pm,\sigma}(q) = \sum_k c_{\pm,k,\sigma}^\dagger c_{\pm,k+q,\sigma}. \tag{7.90}$$

We combine the spin components to construct the quantities

$$\rho_\alpha(q) = [\rho_{\alpha\uparrow}(q) + \rho_{\alpha\downarrow}(q)]/\sqrt{2}, \tag{7.91}$$

$$\sigma_\alpha(q) = [\rho_{\alpha\uparrow}(q) - \rho_{\alpha\downarrow}(q)]/\sqrt{2}, \tag{7.92}$$

with $\alpha = \pm$ or, equivalently, R and L. Accordingly the kinetic energy is rewritten as

$$H_\mathrm{B} = \frac{2\pi v_\mathrm{F}}{L} \sum_{q>0,\alpha=\pm} : \rho_\alpha(q)\rho_\alpha(-q) + \sigma_\alpha(q)\sigma_\alpha(-q) :, \tag{7.93}$$

where the charge $\rho$ and the spin $\sigma$ degrees of freedom appear separately. If the interaction Hamiltonian also has separate components for spin and charge, the total Hamiltonian can be diagonalized by two Bogoliubov transformations: one for spin and the other for charge. The renormalized velocity of spin and charge excitations are in general different from each other. Namely, if the interaction constants $g_2$, $g_4$ are different for spin and charge components, the Bogoliubov angles are also different. In this case, it appears as if spin and charge move independently from each other, in strong contrast with the case in Fermi liquids. This situation is specific to one-dimensional systems, and is called the spin–charge separation.

The total Hamiltonian $H$ is decomposed into spin and charge parts as $H = H_\rho + H_\sigma$. Each part is given in the real space by

$$H_\rho = \frac{v_\rho}{2}\int_0^L \mathrm{d}x : \frac{1}{K_\rho}\left(\frac{\partial \Phi_\rho}{\partial x}\right)^2 + K_\rho \Pi_\rho(x)^2 : \tag{7.94}$$

$$H_\sigma = \frac{v_\sigma}{2}\int_0^L \mathrm{d}x : \frac{1}{K_\sigma}\left(\frac{\partial \Phi_s}{\partial x}\right)^2 + K_\sigma \Pi_s(x)^2 :, \tag{7.95}$$

where $\Phi_\xi, \Pi_\xi$ ($\xi = \rho, s$) are defined by analogy with the spinless case as in Eqs. (7.8), (7.91), and (7.92). Accordingly the phase field $\theta(x) = \sqrt{4\pi}\Phi(x)$ introduced in Eq. (7.8) for $N$and $J$ components is generalized as

$$\theta_{N\rho} = \frac{1}{2}\left(\theta_{\mathrm{R}\uparrow} + \theta_{\mathrm{L}\uparrow} + \theta_{\mathrm{R}\downarrow} + \theta_{\mathrm{L}\downarrow}\right), \tag{7.96}$$

$$\theta_{J\rho} = \frac{1}{2}\left(\theta_{\mathrm{R}\uparrow} - \theta_{\mathrm{L}\uparrow} + \theta_{\mathrm{R}\downarrow} - \theta_{\mathrm{L}\downarrow}\right), \tag{7.97}$$

$$\theta_{Ns} = \frac{1}{2}\left(\theta_{\mathrm{R}\uparrow} + \theta_{\mathrm{L}\uparrow} - \theta_{\mathrm{R}\downarrow} - \theta_{\mathrm{L}\downarrow}\right), \tag{7.98}$$

$$\theta_{Js} = \frac{1}{2}\left(\theta_{\mathrm{R}\uparrow} - \theta_{\mathrm{L}\uparrow} - \theta_{\mathrm{R}\downarrow} + \theta_{\mathrm{L}\downarrow}\right). \tag{7.99}$$

The dynamics of spin and charge are described by these four independent variables. These one-dimensional fermion systems are called Tomonaga–Luttinger liquids. The gapless excitation spectrum is common with the Fermi liquid in three dimensions, but the different velocities of spin and charge are the distinct feature in Tomonaga–Luttinger liquid.

Let us return to the momentum distribution function but including spins now. The right-going (R) fermions with spin down has the phase

$$\theta_{\mathrm{R}\downarrow} = \frac{1}{2}\left(\theta_{N\rho} + \theta_{J\rho} - \theta_{Ns} - \theta_{Js}\right) \tag{7.100}$$

from Eqs. (7.96) to (7.99). In the correlation functions of $N$ and $J$, the interaction effect enters through the factors $K$ and $1/K$, respectively. Hence by the same reasoning that leads to Eq. (7.88), we find that the momentum distribution $n_{k\downarrow}$ has the singular exponent

$$\gamma = \frac{v_F^*}{v_{\mathrm{s}}} - 1 = \frac{1}{4}\left(K_\rho + K_\rho^{-1} + K_\sigma + K_\sigma^{-1}\right) - 1 \tag{7.101}$$

which is positive as in Eq. (7.88), and vanishes without interactions. The same exponent applies to the up spin as well.

We can derive spin and charge correlation functions in a similar manner. For example, the transverse spin correlation function, which involves the components $s_{\pm} = s_x \pm i s_y$ and the wave number $2k_F$, are derived from

$$\langle U^{\dagger} \psi^{\dagger}_{\mathrm{L}\uparrow}(x) \psi_{\mathrm{R}\downarrow}(x) \psi^{\dagger}_{\mathrm{R}\downarrow}(0) \psi_{\mathrm{L}\uparrow}(0) U \rangle_0, \tag{7.102}$$

where the unitary transformation $U$ generalizes the one in Eq. (7.82) to systems with spin. The $x$-dependent phase $\theta_{\mathrm{R}\downarrow}(x) + \theta_{\mathrm{L}\uparrow}(x) + 2k_F x$ appears after bosonization. Another component with the wave number $-2k_F$ is derived by exchanging R and L in Eq. (7.102). We obtain from Eqs. (7.96) to (7.99)

$$\theta_{\mathrm{L}\uparrow} = \frac{1}{2}\left(\theta_{N\rho} - \theta_{J\rho} + \theta_{Ns} - \theta_{Js}\right). \tag{7.103}$$

Then together with Eq. (7.100) we can derive the relevant phase. Namely, after the unitary transformation by $U$, we obtain

$$\tilde{\theta}_{\mathrm{R}\downarrow} + \tilde{\theta}_{\mathrm{L}\uparrow} = \tilde{\theta}_{N\rho} - \tilde{\theta}_{Js} = \sqrt{K_\rho}\theta_{N\rho} - \frac{1}{\sqrt{K_\sigma}}\theta_{Ns}. \tag{7.104}$$

Following the same steps as in Eqs. (7.78) and (7.87), we obtain the exponent

$$\nu = K_\rho + K_\sigma^{-1}, \tag{7.105}$$

which appears in the transverse spin correlation function as

$$\langle s_-(x) s_+(0) \rangle \sim |x|^{-\nu} \cos(2k_F x). \tag{7.106}$$

On the other hand, the longitudinal correlation function, which involves the $z$ component of the spin, is contributed by $\psi^{\dagger}_{\mathrm{L}\sigma}(x)\psi_{\mathrm{R}\sigma}(x)$ with the same spin component and the wave number $2k_F$. The exchange of R and L branches gives the opposite wave number $-2k_F$. The bosonization accompanies the following phase, apart from $2k_F x$, as given by

$$\tilde{\theta}_{\mathrm{R}\sigma} + \tilde{\theta}_{\mathrm{L}\sigma} = \tilde{\theta}_{N\rho} + \sigma \tilde{\theta}_{Ns} = \sqrt{K_\rho}\theta_{N\rho} + \sigma\sqrt{K_\sigma}\theta_{Ns}, \tag{7.107}$$

where $\sigma$ in the RHS is interpreted as $\sigma$ as $\sigma = \pm 1$. Thus the exponent of the longitudinal correlation function is given by $K_\rho + K_\sigma$. Provided the rotational invariance is present for the spin direction, the longitudinal exponent must agree with the transverse one $K_\rho + K_\sigma^{-1}$. This requirement leads to the remarkable relation:

$$K_\sigma = K_\sigma^{-1} = 1, \tag{7.108}$$

which is the same as the non-interacting case. In the repulsive Hubbard model, for example, the rotational invariance is present. Then we have $K_\sigma = 1$, but the spin correlation function itself involves $K_\rho$ which is in general different from unity. Note that the algebraic decay of the correlation function is the necessary condition for $K_\sigma = 1$. As we shall discuss in Sect. 7.8, the exact solution with a superconducting ground state has a spin gap, and is characterized by $K_\sigma = 1/2$. Such superconducting state is realized in the attractive Hubbard model as an example of $K_\sigma \neq 1$ in spite of the rotational invariance.

Similar analysis can be performed for the charge correlation function. Problem 7.8 deals with its asymptotic behavior.

## 7.7 Backward and Umklapp Scatterings

In addition to the forward scattering, we now take into account scatterings from R to L branches, and vice versa. Such scatterings involve the change $\pm 2k_F$ of the momentum, as illustrated in Fig. 7.3. The backward scattering shown in (a) conserves the total momentum of two particles, while the Umklapp scattering shown in (b) accompanies the momentum change by $\pm 4k_F$. Such change is allowed if the crystal momentum is conserved, as will be discussed in more detail later. In the present section, we analyze effects of these scatterings with large momentum transfer.

The annihilation operators with momenta near the Fermi surfaces R and L are written as $\psi_{\mathrm{R}\sigma}(x)$, $\psi_{\mathrm{L}\sigma}(x)$ with spin $\sigma$. According to the uncertainty principle, the definite coordinate $x$ for a particle is incompatible with definite momentum $\pm k_F$. We assume therefore that the range of momenta superposed around $\pm k_F$ is much smaller than $2k_F$. The Hamiltonian for the backward scattering in Fig. 7.3a is described in the real space by

$$H_1 = \frac{1}{2}\int_0^L \mathrm{d}x \sum_\sigma \left[ g_{1\parallel}\left(\psi^\dagger_{\mathrm{R}\sigma}\psi^\dagger_{\mathrm{L}\sigma}\psi_{\mathrm{R}\sigma}\psi_{\mathrm{L}\sigma} + \mathrm{h.c.}\right)\right.$$
$$\left. + g_{1\perp}\left(\psi^\dagger_{\mathrm{R}\sigma}\psi^\dagger_{\mathrm{L}\bar{\sigma}}\psi_{\mathrm{R}\bar{\sigma}}\psi_{\mathrm{L}\sigma} + \mathrm{h.c.}\right)\right], \tag{7.109}$$

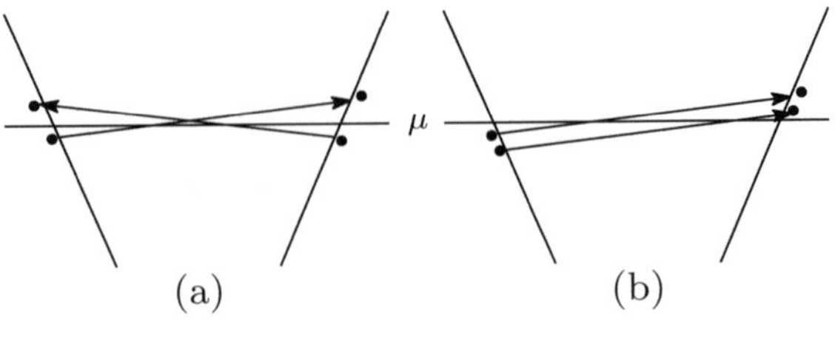


**Fig. 7.3** Scatterings with large momentum transfer: (**a**) backward scattering with coupling constant written as $g_1$; (**b**) Umklapp scattering with coupling constant $g_3$

with $\bar{\sigma} \equiv -\sigma$, and the argument $x$ has been omitted in all field operators. The term with $g_{1\parallel}$ can be rewritten also in the forward scattering form:

$$-g_{1\parallel}\psi^{\dagger}_{\mathrm{R}\sigma}\psi_{\mathrm{R}\sigma}\psi^{\dagger}_{\mathrm{L}\sigma}\psi_{\mathrm{L}\sigma}, \tag{7.110}$$

by exchanging two Fermi operators. The explicit form is given by

$$H_{1\parallel} = -\frac{g_{1\parallel}}{2L}\sum_{q,\alpha=\pm} :\rho_{\alpha}(q)\rho_{-\alpha}(-q) + \sigma_{\alpha}(q)\sigma_{-\alpha}(-q): \tag{7.111}$$

which indeed looks like a forward scattering.

Similar treatment for the $g_{1\perp}$ interaction results in forward scattering with spin-flip. Namely, $g_{1\perp}$ can be regarded as an exchange interaction. Using Eq. (7.51), we bosonize the $g_{1\perp}$ terms to obtain

$$g_{1\perp}\sum_{\sigma}\psi^{\dagger}_{\mathrm{R}\sigma}\psi^{\dagger}_{\mathrm{L}\bar{\sigma}}\psi_{\mathrm{R}\bar{\sigma}}\psi_{\mathrm{L}\sigma} + \mathrm{h.c.} = \frac{g_{1\perp}}{(\pi\eta)^2}\cos(2\theta_{Ns}), \tag{7.112}$$

where $\theta_{Ns}$ has been defined in Eq. (7.98). The Klein factors have been omitted since their product can be made unity.

Let us proceed to the Umklapp scattering illustrated in Fig. 7.3b. Here two fermions both of which are near the left Fermi surface (point) transfer to the region near the right Fermi surface, and vice versa. The Pauli principle allows Umklapp scattering only for different spins. The change of total momentum of two fermions is $\pm 4k_F$, and scattering is forbidden in general. However, if the density of fermions with spin 1/2 is such that the average number per site is unity, the Umklapp scattering is possible since it conserves the crystal momentum. Namely, with the lattice constant taken as unity, the density $n_\sigma$ per spin is given by $n_\sigma = 2k_F/(2\pi)$. Then $n_\sigma = 1/2$ corresponds to $4k_F = 2\pi$, which is the smallest reciprocal lattice. As mentioned in Sect. 2.3, the unit occupation per site is precisely the condition under which the Mott insulating state is realized. More details will be discussed in the next section.

The Hamiltonian for the Umklapp scattering is given by

$$H_3 = \frac{g_{3\perp}}{2}\int_0^L \mathrm{d}x\sum_{\sigma}\psi^{\dagger}_{\mathrm{R}\sigma}\psi^{\dagger}_{\mathrm{R}\bar{\sigma}}\psi_{\mathrm{L}\bar{\sigma}}\psi_{\mathrm{L}\sigma} + \mathrm{h.c.}. \tag{7.113}$$

Bosonization with use of Eq. (7.51) leads to the form

$$\begin{aligned} &g_{3\perp}\sum_{\sigma}\psi^{\dagger}_{\mathrm{R}\sigma}\psi^{\dagger}_{\mathrm{R}\bar{\sigma}}\psi_{\mathrm{L}\bar{\sigma}}\psi_{\mathrm{L}\sigma} + \mathrm{h.c.} \\ &\quad = \frac{g_{3\perp}}{(\pi\eta)^2}\cos\left[2\theta_{N\rho} + (4k_F - 2\pi)x\right], \end{aligned} \tag{7.114}$$

where $-2\pi x$ has no effect with the integer coordinate $x$. This form is convenient in the continuum approximation for $x$ since Eq. (7.113) is then finite only if $4k_F = 2\pi$. Otherwise Eq. (7.114) oscillates with continuum $x$, and Eq. (7.113) vanishes after integration over $x$.

We have seen that the Hamiltonian with large momentum transfer is still separated into spin and charge parts. In the spin part, the backward scattering acts like the exchange interaction. In the charge part, the Umklapp scattering becomes effective with the particular density $n = 1$. The important feature of large momentum transfer is the appearance of cosine terms with $\theta_{Ns}$ for spin, and $\theta_{N\rho}$ for charge. These terms give rise to excitation gaps as discussed below.

## 7.8 Superconductivity and Mott Insulator

In one-dimensional electron systems, the bosonization method makes it possible to take exact account of interaction phenomena such as superconductivity and Mott insulator. These topics are discussed in this section.

Let us start with superconductivity which has gapless charge excitations, but a finite gap for spin excitations. We have seen in Sect. 7.6 that the forward scattering does not lead to the excitation gap. Hence the backward scattering should be responsible for the spin gap in the Tomonaga–Luttinger liquid. It is known from another method of exact solution, which is called the Bethe ansatz [6], that the Hubbard model with negative $U$ has the superconducting ground state. By combining the idea of renormalization, we can make contact with exact solutions obtained by bosonization and Bethe ansatz.

We analyze in detail the bosonized form of the backscattering term

$$\frac{2g_{1\perp}}{(2\pi\eta)^2}\int_0^L \mathrm{d}x \cos\left[2\theta_{Ns}(x)\right], \tag{7.115}$$

which has appeared in Eq. (7.112). As we have discussed in Sect. 7.4, Bogoliubov transformation with the forward scattering parameter $K$ is performed by Eq. (7.70). For the spin liquid, this amounts to $\tilde{\theta}_{Ns} = \theta_{Ns}/\sqrt{K_\sigma}$. In the special case $K_\sigma = 1/2$, the cosine term in Eq. (7.115) becomes

$$\cos\sqrt{2}\tilde{\theta}_{Ns}(x). \tag{7.116}$$

Similar cosine form, with $\tilde{\theta}_{Ns}(x)$ replaced by $\check{\theta}_N(x) = \theta_N(x)/\sqrt{2}$, has appeared in Eq. (7.59), which describes hybridization between R and L branches of free fermions. Hence the spin liquid with $K_\sigma = 1/2$ behaves as non-interacting fermions with hybridization gap. Physically, the gap is for spin excitations. On the other hand, the charge excitation remains gapless unless the Umklapp scattering becomes effective. This contrasting situation for spin and charge excitations corresponds to the superconducting state.

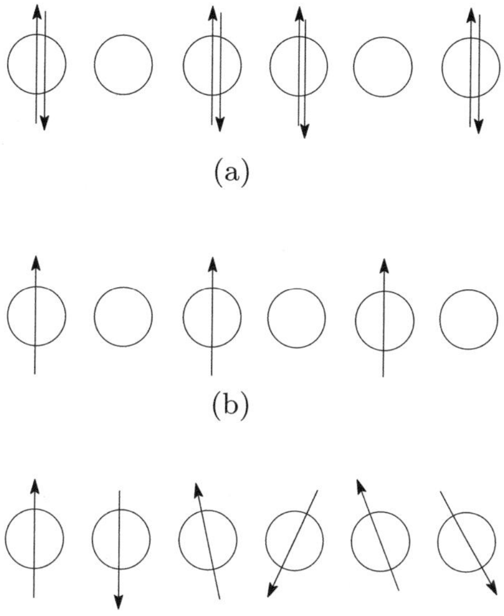


**Fig. 7.4** Illustration of strong-coupling limit for understanding nature of excitation gaps: (**a**) superconducting state with charge gap $\Delta_c = 0$, and spin gap $\Delta_s \neq 0$; (**b**) fermionic picture of the anisotropic XXZ spin model with vacant site corresponding to down spins. Repulsion among fermions leads to $\Delta_c \neq 0$, which means $\Delta_s \neq 0$ in the original spin model; (**c**) Mott insulating state with $\Delta_c \neq 0$ and $\Delta_s = 0$

Equations (7.67) and (7.111) combine to give the relation

$$\frac{g_{1\parallel}}{2\pi v_F^*} = -\frac{1-K_\sigma^2}{1+K_\sigma^2} = -\frac{3}{5}, \tag{7.117}$$

where the last equality follows from $K_\sigma = 1/2$ [7]. Since $g_{1\parallel} < 0$ is related to negative $U$, it is not surprising that superconductivity is realized in this case. Nevertheless, the exact result in the special case Eq. (7.117) demonstrates the power of bosonization method. Figure 7.4a illustrates a spin configuration that leads to gapless charge and gapped spin excitations in the strong-coupling limit.

We now go a bit aside to discuss another origin of the spin gap, which is due to anisotropy in the exchange interaction. In contrast with the bosonization so far discussed, we map here conversely the spin excitations to fermionic ones. Let us consider the following one-dimensional spin chain:

$$H = J\sum_i \left(S_i^x S_{i+1}^x + S_i^y S_{i+1}^y + \Delta_z S_i^z S_{i+1}^z\right), \tag{7.118}$$

with the anisotropy parameter $\Delta_z$ which is often called the XXZ model. The Heisenberg model corresponds to $\Delta_z = 1$, while the Ising model corresponds to the limit $\Delta_z = \infty$. The case $\Delta_z = 0$ is called the XY model, which can be solved exactly by introducing fictitious fermion operators. Namely, we define $\psi_i^\dagger$ and $\psi_i$

for each site $i$ as

$$S_i^z = \psi_i^\dagger \psi_i - 1/2 \equiv n_i - 1/2, \tag{7.119}$$

$$S_i^- = S_i^x - \mathrm{i} S_i^y = \psi_i \exp\left(\mathrm{i}\pi \sum_{j=1}^{i-1} n_j\right), \tag{7.120}$$

with $i = 1$ for the left end, and $i = N$ for the right end of the chain. We obtain the commutation relations

$$\{\psi_i, \psi_j\} = 0, \ \{\psi_i^\dagger, \psi_j\} = \delta_{ij}, \tag{7.121}$$

from the spin commutation rule. Hence $\psi^\dagger, \psi$ represents the creation and annihilation operators of fermions, and the corresponding transformation is called the Jordan–Wigner transformation. This is closely related to the Klein factor described in Eq. (7.55). In terms of the fictitious fermions, Eq. (7.118) is rewritten as

$$H = \frac{J}{2} \sum_i \left[ \psi_i^\dagger \psi_{i+1} + \psi_{i+1}^\dagger \psi_i + 2\Delta_z \left(n_i - \frac{1}{2}\right)\left(n_{i+1} - \frac{1}{2}\right)\right], \tag{7.122}$$

where the anisotropy parameter $\Delta_z$ represents the two-body interaction of fermions. With $\Delta_z = 0$, the system is a collection of free fermions with the spectrum

$$\epsilon_k = J \cos k,$$

with $-\pi < k \le \pi$. In the ground state, $N/2$ fermions occupy the negative energy states with the Fermi wave number $k_F = \pi/2$.

With $\Delta_z > 1$, the spin excitations have an energy gap. In the fermionic representation Eq. (7.122), the gap appears as the charge excitation gap. As illustrated in Fig. 7.4b, the charge gap in the strong-coupling limit corresponds to crystallization of fermions, which are originally up spins, due to the repulsion at nearest neighbors.

We now return to one-dimensional electron systems with spin 1/2, and focus on the charge gap caused by the Umklapp scattering. With special value of the interaction strength, which corresponds to the charge version of Eq. (7.117), exact solution is possible. We introduce the charge fermions with R and L branches. Then Eq. (7.114) for the Umklapp scattering is written in terms of R and L phase variables as

$$\theta_{N\rho} = (\theta_{\mathrm{R}\rho} + \theta_{\mathrm{L}\rho})/\sqrt{2}, \quad \theta_{NJ} = (\theta_{\mathrm{R}\rho} - \theta_{\mathrm{L}\rho})/\sqrt{2}. \tag{7.123}$$

By analogy with the spin fermion case, the system with $K_\rho = 1/2$ is mapped to free fermions with hybridization of R and L branches. This corresponds to the special value $g_{2\rho}/(2\pi v_F^*) = 3/5$ for the forward scattering of charge. The notation $g_2$ has been introduced in Eq. (7.62), for the spinless system, as the strength of the forward scattering involving R and L branches. Here we have added the index $\rho$ to represent the corresponding interaction that leads to renormalization of $v_F$ to $v_\rho$ as in Eq. (7.94). On the other hand, the spin part remains a gapless Tomonaga–Luttinger liquid. This state corresponds to the Mott insulating state, which is illustrated in Fig. 7.4c in the strong-coupling limit.

It is intuitively clear that the ground states for exactly solvable cases with $g_{1\|}/(2\pi v_F^*) = -3/5$ and $g_{2\rho}/(2\pi v_F^*) = 3/5$ are connected smoothly to the strong-coupling limit illustrated in Fig. 7.4. If we start from the weak-coupling limit, on the other hand, it is more subtle to discuss the formation of an energy gap. The relevant method in the latter case is the renormalization group for the sine-Gordon model, which we shall not discuss in this book because of its complexity [8]. It is known that in the case with $g_{1\|} < 0$ or $g_{2\rho} > 0$, the absolute value of the relevant coupling increases by renormalization. Namely, the Tomonaga–Luttinger liquid is unstable against gap formation no matter how small is the bare interactions.

## 7.9 Kondo Effect Revisited

We reconsider Kondo effect from the view point of bosonization. It turns out that a special value of the exchange interaction, which is called the Toulouse limit, makes the Kondo model equivalent to a one-dimensional free fermion model. The Toulouse limit corresponds to an impurity version of the free fermion model with hybridization of R and L branches, which has been discussed as the massive Thirring model in Sect. 7.3. Moreover we can derive exact solution for the two-channel Kondo model for another special value of the exchange interaction [9].

There are two ways to map the Kondo model to a one-dimensional impurity model. In the first way, one takes the positive spatial coordinate that corresponds to the radial coordinate $r$ in three dimensions. A positive wave number $k$ is assigned to the outgoing $s$-wave, while a negative $k$ is assigned to the incoming one. In the second way, the incoming $s$-wave with $\exp(-ikr)$ is interpreted as having the negative spatial coordinate $-r$. In this version the spatial coordinate has both positive and negative values, while the wave number is restricted to be positive. Namely, the Kondo model is mapped to an impurity model where the magnetic impurity at the origin interacts with right-going one-dimensional electrons. Here we adopt the latter mapping.

### 7.9.1 Bosonization and Fictitious Fermions in Kondo Model

The location of an impurity is at one of discrete lattice points with the lattice constant $a$. Hence the dimensionless field operator $\sqrt{a}\psi_\sigma(x)$ of conduction electron is interpreted as the annihilation operator of a Wannier state at site $x$. Then $H_{\rm exc}$ with the exchange anisotropy in Eq. (6.25) is written as

$$H_{\text{exc}} = \frac{1}{2} J_z S_z a \left[ \psi_\uparrow^\dagger(0)\psi_\uparrow(0) - \psi_\downarrow^\dagger(0)\psi_\downarrow(0) \right] + \frac{1}{2} J_\perp a \left[ S_+ \psi_\downarrow^\dagger(0)\psi_\uparrow(0) + S_- \psi_\uparrow^\dagger(0)\psi_\downarrow(0) \right] \equiv H_z + H_\perp. \tag{7.124}$$

On the other hand, the bosonization of conduction electrons is most conveniently performed for continuous space. We interpret the slowly varying wave function in the continuum space as the envelope function multiplying the Wannier orbitals at each lattice point. Since the restriction $|q| < \pi/a$ is imposed on the envelope function in the first Brillouin zone, we identify $\eta = a$ and regard that the high momentum region affects the low-energy physics only through renormalization of model parameters.

The kinetic energy of the right-going conduction electrons is given by

$$H_c = v_F \sum_{k\sigma} (k - k_F) c_{k\sigma}^\dagger c_{k\sigma}, \tag{7.125}$$

which is measured from the Fermi level. The bosonized expression reads

$$H_c = \frac{v_F}{4\pi} \int_0^L \mathrm{d}x \left[ \left( \frac{\partial \theta_c}{\partial x} \right)^2 + \left( \frac{\partial \theta_s}{\partial x} \right)^2 \right], \tag{7.126}$$

where the phase fields $\theta_c, \theta_s$ are defined by

$$\theta_c = 2^{-1/2}(\theta_\uparrow + \theta_\downarrow), \quad \theta_s = 2^{-1/2}(\theta_\uparrow - \theta_\downarrow), \tag{7.127}$$

$$\psi_\sigma(x) = \frac{F_\sigma}{\sqrt{2\pi a}} \exp[\mathrm{i}\theta_\sigma(x)], \tag{7.128}$$

with $F_\sigma$ being the Klein factor.

In terms of the phase field $\theta_s$, the spin density of conduction electrons is given by

$$\frac{1}{2} \sum_{\sigma \pm 1} \sigma \psi_\sigma^\dagger(x) \psi_\sigma(x) = \frac{1}{2\sqrt{2}\pi} \frac{\partial \theta_s(x)}{\partial x}, \tag{7.129}$$

with reference to Eq. (7.26). Then $H_{\text{exc}}$ is rewritten as

$$\frac{J_z a}{2\sqrt{2}\pi} S_z \frac{\partial \theta_s}{\partial x} + \frac{J_\perp}{4\pi} \left[ S_+ F_\downarrow^\dagger F_\uparrow \exp(\mathrm{i}\sqrt{2}\theta_s) + \text{h.c.} \right], \tag{7.130}$$

where $\theta_s$ is at $x = 0$. The charge bosons do not interact with the localized spin. Hence we discard $\theta_c$, and consider only the spin bosons hereafter.

Let us first consider the effect of $S_z$. By analogy with the phase shift in potential scattering, we account for the distortion of wave functions by a local unitary transformation

$$U_\gamma = \exp\left[i\gamma\theta_s(0)S_z\right]. \tag{7.131}$$

Then $\partial\theta_s/\partial x$ is transformed as

$$U_\gamma^\dagger \frac{\partial\theta_s}{\partial x} U_\gamma = \frac{\partial\theta_s}{\partial x} - \frac{2\pi}{a}\gamma S_z \delta(x,0), \tag{7.132}$$

with reference to Eqs. (7.10) and (7.27). Here $\delta(x,0)$ means the Kronecker delta. The spin bosons undergo a change in kinetic energy as

$$U_\gamma^\dagger \left(\frac{\partial\theta_s}{\partial x}\right)^2 U_\gamma = \left(\frac{\partial\theta_s}{\partial x}\right)^2 - \frac{4\pi}{a}\gamma S_z \frac{\partial\theta_s}{\partial x}\delta(x,0) + \left(\frac{2\pi\gamma}{a}\right)^2 \delta(x,0). \tag{7.133}$$

By a choice of $\gamma$ such that $\gamma = J_z\rho_c/\sqrt{2}$ with $\rho_c = a/(2\pi v_F)$, the second term in the RHS cancels with the first term in Eq. (7.130). More details of this cancellation is the subject of Problem 7.9. The third term in Eq. (7.133) is a constant, and can be neglected.

We shall show that the transverse exchange $H_\perp$ is transformed to a hybridization form of local and itinerant fermions which represent the spin degrees of freedom. In $U_\gamma^\dagger S_+ U_\gamma$, we can put $S_z = -1/2$ in $U_\gamma$, and $S_z = 1/2$ in $U_\gamma^\dagger$. Then we obtain

$$U_\gamma^\dagger S_\pm U_\gamma = \exp\left[\mp \mathrm{i}\gamma\theta_s(0)\right] S_+, \tag{7.134}$$

where the result for $S_-$ can be obtained from $(U_\gamma^\dagger S_+ U_\gamma)^\dagger$. Hence in the special case of $\gamma = \sqrt{2}-1$, which is equivalent to $J_z\rho_c = 2-\sqrt{2}$, we obtain

$$U_\gamma^\dagger H_\perp U_\gamma = \frac{1}{4\pi} J_\perp \left[S_+ F_\downarrow^\dagger F_\uparrow \exp(\mathrm{i}\theta_s) + \text{h.c.}\right] \tag{7.135}$$

from Eq. (7.130). Here $\exp(\mathrm{i}\theta_s)$ constitutes the annihilation operator of spin fermions if supplied with appropriate Klein factor $F_s$. It is reasonable to restrict $F_s$ as $F_\downarrow^\dagger F_\uparrow \propto F_s$ since $F_\downarrow^\dagger F_\uparrow$ decreases the spin of conduction electrons by unity, and leaves intact the total number. To preserve the bosonic commutation property of $\psi_\downarrow^\dagger(x)\psi_\uparrow(x)$, we choose for the proportionality constant another Klein factor $F_S^\dagger$ so that

$$F_\downarrow^\dagger F_\uparrow = F_S^\dagger F_s. \tag{7.136}$$

In this setting $S_+ F_S^\dagger$ acts on the local spin space, and causes the unit increase of the $z$-component. The increase is now interpreted as creation of a local spin fermion, so that the sum of conduction and local spin fermion numbers is conserved. We thus define the dimensionless spin fermion operators as

$$c_s(x) = F_s \exp[\mathrm{i}\theta_s(x)]/\sqrt{2\pi}, \quad S_+F_S^\dagger = d^\dagger, \quad F_S S_- = d. \tag{7.137}$$

It can be easily checked by use of Eq. (7.57) that $d$ and $c_s(x)$ indeed obey the anticommutation rule for fermions. Then we finally obtain the form:

$$U_\gamma^\dagger H_\perp U_\gamma = \frac{1}{\sqrt{8\pi}} J_\perp \left[d^\dagger c_s(0) + c_s^\dagger(0) d\right] \equiv H_{\mathrm{hyb}}, \tag{7.138}$$

which displays hybridization between local and itinerant fermions at the lattice site $x = 0$.

Thus the original Kondo exchange $H_{\mathrm{exc}}$ with the special value $J_z\rho_c = 2 - \sqrt{2} \sim 0.6$ is transformed to the hybridization form between spin fermions $d$ and $c_s$, without any further interaction. The kinetic energy of spin fermions is rewritten from the bosonic form as

$$H_{\mathrm{kin}} = v_F \sum_k kC_s(k)^\dagger C_s(k), \tag{7.139}$$

where

$$C_s(k) = \frac{1}{\sqrt{N}} \sum_x c_s(x) \exp(-ikx), \tag{7.140}$$

with $N$ being the number of lattice sites. Such bilinear form of the Hamiltonian $H_{\mathrm{kin}} + H_{\mathrm{hyb}}$ has been discussed already in Sect. 3.7, and the corresponding Green function has been obtained in Eq. (3.123). Then it is clear that the ground state is a resonant state of spin fermions without the energy gap. Since the weak-coupling result shows renormalization flow to larger $J_z$, the value of $J_z\rho_c$ should cross the special value $2-\sqrt{2}$ during the renormalization where the exact solution is possible. The ground state of the weak-coupling Kondo model is thus confirmed to be a local Fermi liquid as has been discussed in Chap. 6.

### 7.9.2 Two-Channel Kondo Model

We now turn to the two-channel Kondo model, which is the $n = 2$ version of Eq. (6.83). We demonstrate that the nontrivial fixed point is obtained exactly for the special value of the exchange interaction $J_z$ at $J_z\rho_c = 1$ [9]. We begin with the bosonized form of the conduction-electron operator as

$$\psi_{l\sigma}(x) = \frac{F_{l\sigma}}{\sqrt{2\pi a}} \exp[\mathrm{i}\theta_{l\sigma}(x)], \tag{7.141}$$

where $l = \pm 1$ describes the orbital degrees of freedom, which we now call "flavor", and $F_{l\sigma}$ is the corresponding Klein factor. In place of the phase fields $\theta_{l\sigma}(x)$, it is convenient to introduce new phase fields $\theta_c, \theta_s, \theta_f$ and $\theta_{sf}$ by

$$\theta_c = \frac{1}{2}(\theta_{+\uparrow} + \theta_{+\downarrow} + \theta_{-\uparrow} + \theta_{-\downarrow}) = \frac{1}{2}\sum_{l\sigma} \theta_{l\sigma}, \tag{7.142}$$

$$\theta_s = \frac{1}{2}(\theta_{+\uparrow} - \theta_{+\downarrow} + \theta_{-\uparrow} - \theta_{-\downarrow}) = \frac{1}{2}\sum_{l\sigma} \sigma\theta_{l\sigma}, \tag{7.143}$$

$$\theta_f = \frac{1}{2}(\theta_{+\uparrow} + \theta_{+\downarrow} - \theta_{-\uparrow} - \theta_{-\downarrow}) = \frac{1}{2}\sum_{l\sigma} l\theta_{l\sigma}, \tag{7.144}$$

$$\theta_{sf} = \frac{1}{2}(\theta_{+\uparrow} - \theta_{+\downarrow} - \theta_{-\uparrow} + \theta_{-\downarrow}) = \frac{1}{2}\sum_{l\sigma} l\sigma\theta_{l\sigma}. \tag{7.145}$$

The inverse transformation characterized by the same unitary matrix gives the original phase fields $\theta_{l\sigma}$ in terms of the new ones. Then the kinetic energy is given by

$$H_c = \frac{v_F}{4\pi}\int_0^L \mathrm{d}x \sum_{\alpha} \left(\frac{\partial \theta_\alpha}{\partial x}\right)^2, \tag{7.146}$$

with the index $\alpha$ indicating $c$, $s$, $f$, and $sf$. The spin density of conduction electrons is given by

$$\frac{1}{2}\sum_{l,\sigma\pm 1} \sigma \psi_{l\sigma}^\dagger(x)\psi_{l\sigma}(x) = \frac{1}{4\pi}\frac{\partial \theta_s(x)}{\partial x}, \tag{7.147}$$

to be compared with Eq. (7.129). Thus the exchange interaction is rewritten as

$$H_z = \frac{J_z a}{4\pi} S_z \frac{\partial \theta_s}{\partial x}, \tag{7.148}$$

$$H_\perp = \frac{J_\perp}{4\pi} S_+ \sum_l F_{l\downarrow}^\dagger F_{l\uparrow} \exp(\mathrm{i}\theta_s + \mathrm{i}_l\theta_{sf}) + \mathrm{h.c.} \tag{7.149}$$

We proceed to rewrite the products of Klein factors in Eq. (7.149) so as to represent the change $\Delta N_\alpha$ of new fermions. We have met a similar case with spin fermions in Eq. (7.137). Using Eqs. (7.142)–(7.145) which also determine linear relations between the changes $\Delta N_\alpha$ and $\Delta N_{l\sigma}$, we obtain

$$\Delta N_{+\uparrow} - \Delta N_{+\downarrow} = \Delta N_s + \Delta N_{sf}, \tag{7.150}$$

$$\Delta N_{-\uparrow} - \Delta N_{-\downarrow} = \Delta N_s - \Delta N_{sf}. \tag{7.151}$$

Then the corresponding Klein factors are converted as [10]

$$F_{+\uparrow}^{\dagger}F_{+\downarrow} = F_{sf}^{\dagger}F_{s}^{\dagger}, \quad F_{-\uparrow}^{\dagger}F_{-\downarrow} = F_{sf}F_{s}^{\dagger}, \tag{7.152}$$

$$F_{+\downarrow}^{\dagger}F_{+\uparrow} = F_{s}F_{sf}, \quad F_{-\downarrow}^{\dagger}F_{-\uparrow} = F_{s}F_{sf}^{\dagger}, \tag{7.153}$$

which are to be compared with Eq. (7.136) for the original Kondo model.

We now make the unitary transformation generated by $U_1$ which corresponds to $U_\gamma$ in Eq. (7.131) with $\gamma = 1$. We concentrate on the special case with $J_z\rho_c = 1$. Then $U_1^{\dagger}H_zU_1$ is canceled by the second term in the RHS of Eq. (7.133), and $\theta_s$ disappears from $U_1^{\dagger}H_{\perp}U_1$. Namely, we obtain

$$U_1^{\dagger}H_{\perp}U_1 = \frac{J_{\perp}}{\sqrt{8\pi}}S_+F_s\left[c_{sf} + c_{sf}^{\dagger}\right] + \text{h.c.}, \tag{7.154}$$

where new fermion operators $c_{sf}, c_{sf}^{\dagger}$ at $x = 0$ have been introduced. Note that Eq. (7.154) does not conserve the number of $sf$ fermions. For general sites $x$, we define

$$c_{sf}(x) = F_{sf}\exp[\mathrm{i}\theta_{sf}(x)]/\sqrt{2\pi}, \tag{7.155}$$

and its Hermitian conjugate. By using the local spin fermion operator defined by $d^{\dagger} \equiv S_+F_s$, and the anticommutation property of fermion operators, we obtain a peculiar form of hybridization:

$$U_1^{\dagger}H_{\perp}U_1 = \frac{1}{\sqrt{8\pi}}J_{\perp}(d^{\dagger} - d)(c_{sf} + c_{sf}^{\dagger}) \equiv H_{\text{hyb}}. \tag{7.156}$$

It is possible to diagonalize the Hamiltonian exactly with $J_z\rho_c = 1$ because of its bilinear form in fermion operators. In order to account for the mixing of particle and hole components in $H_{\text{hyb}}$, we employ the Bogoliubov transformation. Since only the species $sf$ couples with local fermions, we introduce the corresponding fermion operator in the momentum space by the first Bogoliubov transformation

$$\begin{pmatrix} C_+(k) \\ C_-(k) \end{pmatrix} = \frac{1}{\sqrt{2}}\begin{pmatrix} 1 & 1 \\ 1 & -1 \end{pmatrix}\begin{pmatrix} C_{sf}(k) \\ C_{sf}(-k)^{\dagger} \end{pmatrix} \equiv U\begin{pmatrix} C_{sf}(k) \\ C_{sf}(-k)^{\dagger} \end{pmatrix} \tag{7.157}$$

with $C_{sf}(k)$ being the Fourier transform of $c_{sf}(x)$ in Eq. (7.155), and the matrix $U = (\sigma_x + \sigma_z)/\sqrt{2}$ is unitary. The kinetic energy of the $sf$ fermions is given by

$$H_{\text{kin}} = \sum_k v_F k : C_{sf}(k)^{\dagger}C_{sf}(k) := \frac{1}{2}\sum_k\sum_{p=\pm} v_F k : C_p(k)^{\dagger}C_p(k) :, \tag{7.158}$$

with the particle–hole symmetry. The factor 1/2 in the rightmost side comes from double counting of the same state as a particle state and hole one. This is analogous to the case of the four-component theory of superconductivity with same factor 1/2 in Eq. (5.63). The details of calculation are the subject of Problem 7.10. Note that $d^\dagger - d$ couples only with $C_+(k)$, so that $C_-(k)$ fermions remain intact.

We further introduce the two-component field $d_\pm$ by the same Bogoliubov transformation:

$$\boldsymbol{\phi}_d \equiv \begin{pmatrix} d_+ \\ d_- \end{pmatrix} = U \begin{pmatrix} d \\ d^\dagger \end{pmatrix} = \frac{1}{\sqrt{2}}(1 + \mathrm{i}\sigma_y) \begin{pmatrix} d \\ d^\dagger \end{pmatrix}, \tag{7.159}$$

with the unitary matrix $U$ defined in Eq. (7.157). A closely related two-component field is defined by $\psi_1 = d_+$ and $\psi_2 = \mathrm{i}d_-$ both of which are real (Hermite) operators with $\psi_i^\dagger = \psi_i$. These real operators obey the anticommutation rule: $\{\psi_i, \psi_j\} = \delta_{ij}$ with $i, j = 1, 2$. Such fermionic objects are called Majorana particles, which make it explicit that a complex fermion is composed of two real objects. The transformation to the Majorana basis is written as

$$\begin{pmatrix} \psi_1 \\ \psi_2 \end{pmatrix} \equiv \boldsymbol{\psi} = \begin{pmatrix} 1 & 0 \\ 0 & \mathrm{i} \end{pmatrix} \boldsymbol{\phi}_d = \frac{1}{\sqrt{2}} \begin{pmatrix} 1 & 1 \\ -\mathrm{i} & \mathrm{i} \end{pmatrix} \begin{pmatrix} d \\ d^\dagger \end{pmatrix}. \tag{7.160}$$

It is instructive to derive the Green function of decoupled local fermions in various bases introduced above. Let us take the Hamiltonian $H_d = \epsilon_d S_z$ where the Zeeman splitting of the local spin state is written as $\epsilon_d$, which also corresponds to the energy level of the $d$-fermion as seen from the expressions:

$$S_z = \frac{1}{2}(d^\dagger \ d)\sigma_z \begin{pmatrix} d \\ d^\dagger \end{pmatrix} = \frac{1}{2}\boldsymbol{\psi}^\dagger \sigma_y \boldsymbol{\psi} = -\frac{1}{2}\boldsymbol{\phi}_d^\dagger \sigma_x \boldsymbol{\phi}_d. \tag{7.161}$$

Thus the Green function in the $(d, d^\dagger)$ basis is given by $(z - \epsilon_d\sigma_z/2)^{-1}$. Note that the hole component $1/(z + \epsilon_d/2)$ follows from the particle component $1/(z - \epsilon_d/2)$ by the replacement $\epsilon_d \to -\epsilon_d$. Namely, the same state appears twice as particle and hole states. Hence the weight factor 1/2 is necessary in the counting of fermionic states as in Eq. (7.158). In the $\boldsymbol{\phi}_d$ basis we obtain the Green function as

$$G_0(z) = (z + \epsilon_d\sigma_x/2)^{-1}, \tag{7.162}$$

which will be most conveniently used in the following calculation.

We now consider the transformed form of Eq. (7.156) as given by

$$H_{\rm hyb} = \frac{-1}{\sqrt{8\pi N}} J_\perp \sum_k d_-^\dagger C_+(k) + \text{h.c.}. \tag{7.163}$$

The bilinear form of the Hamiltonian composed of $H_{\rm kin}$ and $H_{\rm hyb}$ is characterized by the matrix $h$ as given by

$$h = \begin{pmatrix} -\epsilon_d\sigma_x/2 & \hat{V}^\dagger & 0 \\ \hat{V} & h_+ & 0 \\ 0 & 0 & h_- \end{pmatrix}, \tag{7.164}$$

where $h_\pm$ are both $N \times N$ diagonal matrices with elements $\epsilon_i = v_F k_i$. The momentum $k_i$ runs from $-\pi/a$ to $\pi/a$ with the interval $2\pi/(Na)$, where $a$ is the lattice constant. The hybridization enters through $\hat{V}$ which is the $N \times 2$ matrix with the first column being all zero, and all the $N$ elements in the second column are the same: $V = -J_\perp/(\sqrt{8\pi N})$. Hence the $2 \times 2$ Green function $G_2(z)$ including the hybridization effect is given by

$$G_2(z)^{-1} = G_0(z)^{-1} - \hat{V}^\dagger(z - h_+)^{-1}\hat{V} \equiv G_0(z)^{-1} - \Sigma(z), \tag{7.165}$$

where $\Sigma(z)$ is a $2\times2$ matrix in which the only nonzero element is $\Sigma(z)_{22}$. Following the same procedure as in Eq. (3.120), we obtain for Im $z > 0$ and $|z| \ll D$,

$$\Sigma(z)_{22} = \mathrm{i}\,(J_\perp)^2\,\rho_c/8 \equiv i\Delta_2 \tag{7.166}$$

with $\rho_c = a/(2\pi v_F)$. Hence $G(z)$ is explicitly obtained as

$$G_2(z)^{-1} = \begin{pmatrix} z & \epsilon_d/2 \\ \epsilon_d/2 & z + \mathrm{i}\Delta_2 \end{pmatrix} = z + \frac{1}{2}\epsilon_d\sigma_x - \frac{\mathrm{i}}{2}(1 - \sigma_z)\Delta_2. \tag{7.167}$$

This concise result contains essential features of the two-channel Kondo model, including the nontrivial fixed point discussed in Chap. 6.

To appreciate the implication of Eq. (7.167) we first consider the most interesting case $\epsilon_d = 0$, where $G_2(z)$ becomes diagonal. The spectral intensity is given by

$$\rho_+(\epsilon) = -\frac{1}{\pi}\mathrm{Im}G_{11}(\epsilon + \mathrm{i}0_+) = \delta(\epsilon), \tag{7.168}$$

$$\rho_-(\epsilon) = -\frac{1}{\pi}\mathrm{Im}G_{22}(\epsilon + \mathrm{i}0_+) = \frac{1}{\pi}\frac{\Delta_2}{\epsilon^2 + \Delta_2^2}. \tag{7.169}$$

Namely, the $d_+$ particle is decoupled from other fermions, and form the sharp level at the Fermi level.

The impurity susceptibility at temperature $T \ll \Delta_2$ is now derived as

$$\begin{aligned}\chi_z(T) &= \int_0^\beta \mathrm{d}\tau\,\langle S_z(\tau)S_z\rangle = \frac{1}{2}\int \mathrm{d}\epsilon \int \mathrm{d}\epsilon_1 \frac{f(\epsilon_1) - f(\epsilon)}{\epsilon - \epsilon_1}\rho_-(\epsilon)\rho_+(\epsilon_1) \\ &= \int \frac{d\epsilon}{4\pi}\frac{\tanh(\beta\epsilon/2)}{\epsilon}\cdot\frac{\Delta_2}{\epsilon^2 + \Delta_2^2} \sim \frac{1}{2\pi\Delta_2}\ln\frac{\Delta_2}{T},\end{aligned} \tag{7.170}$$

with use of Eq. (7.161). The logarithmic $T$-dependence is one of the typical signatures of the non-Fermi liquid state. Here and in the following, the integration over $\epsilon$ is, unless otherwise specified, always limited within the range of fermion spectrum $\pm v_F\pi/a$.

We proceed to derivation of the entropy. The thermodynamic potential $\Omega$ of the system is given by

$$\begin{aligned}\Omega &= -\frac{1}{2}T\,\mathrm{Tr}\,\ln(1+e^{-\beta h}) = -\frac{1}{2}T\int d\epsilon\,\ln(1+e^{-\beta\epsilon})\,\mathrm{Tr}\,\delta(\epsilon-h)\\ &= -\frac{1}{2}\int d\epsilon f(\epsilon)\frac{1}{\pi}\mathrm{Im}\,\mathrm{Tr}\,\ln(\epsilon+\mathrm{i}0_+-h),\end{aligned} \tag{7.171}$$

where $f(\epsilon) = (e^{\beta\epsilon}+1)^{-1}$, and the overall factor 1/2 compensates the double counting of each state as particle state and hole one. In order to separate the impurity contribution, we resort to the matrix identity:

$$\begin{aligned}\mathrm{Tr}\,\ln(z-h) &= \ln\det(z-h)\\ &= -\ln\det G_2(z) - \ln\det G_+(z) - \ln\det G_-(z),\end{aligned} \tag{7.172}$$

with $G_\pm(z) = (z-h_\pm)^{-1}$. The impurity contribution $\Delta\Omega$ is obtained from Eq. (7.171) by subtraction of the $C_\pm$-fermion parts associated with $G_\pm(z)$. In the most interesting case $\epsilon_d = 0$, the result becomes particularly simple, as given by

$$\Delta\Omega = -\int\frac{d\epsilon}{2\pi}f(\epsilon)\mathrm{Im}\left(\ln\frac{1}{\epsilon+\mathrm{i}0_+}+\ln\frac{1}{\epsilon+\mathrm{i}\Delta_2}\right). \tag{7.173}$$

We thus obtain for $T \ll \Delta_2$,

$$\Delta\Omega = -\frac{T}{2}\ln 2 - \frac{\pi^2}{12}\cdot\frac{T^2}{\Delta_2} + \text{const.} \tag{7.174}$$

with use of partial integration to obtain the first term in the RHS. For the second term in the RHS we have used the Sommerfeld expansion formula:

$$\int_{-\infty}^{\infty} d\epsilon f(\epsilon)g(\epsilon) = \int_{-\infty}^{0} d\epsilon g(\epsilon) + \frac{\pi^2}{6}T^2 g'(0) + O(T^4), \tag{7.175}$$

with a smooth function $g(\epsilon)$. Thus the extra entropy $\Delta S$ due to the impurity is derived for $T \ll \Delta_2$ as

$$\Delta S = -\frac{\partial}{\partial T}\Delta\Omega = \frac{1}{2}\ln 2 + \frac{\pi^2}{6}\cdot\frac{T}{\Delta_2}. \tag{7.176}$$

In the opposite limit $T \gg \Delta_2$, we obtain $\Delta S = \ln 2$ immediately from Eq. (7.173), as expected for a fermionic impurity state.

With $\epsilon_d = 0$, the ground state is degenerate as signaled by Eq. (7.168); the energy is independent of the occupancy of the level. One may naively expect that the corresponding entropy $\Delta S$ is $\ln 2$. However, we have obtained $\Delta S = \ln\sqrt{2} + O(T/\Delta_2)$. In the thermodynamic limit $N \to \infty$, the fractional entropy persists down to $T = 0$. This strange result is naturally interpreted as due to the decoupled Majorana particle $\psi_1 = (d + d^\dagger)/\sqrt{2}$ [9]. Since the Majorana particle is half of the ordinary fermion, the associated entropy is also half of the fermionic one. The specific heat derived from Eq. (7.176) is linear in $T$ as in the Fermi liquid. On the other hand, perturbation theory in $J_z\rho_c - 1$ shows that the specific heat behaves as $T \ln T$ once $J_z$ deviates from the special value $J_z\rho_c = 1$ [9].

Now we consider the case of $\epsilon_d \neq 0$, where the ground state is no longer degenerate. We obtain

$$\Delta\Omega = -\int \frac{d\epsilon}{2\pi} f(\epsilon) \tan^{-1}\frac{\epsilon\Delta_2}{\epsilon^2 - \epsilon_d^2/4} \sim -\frac{\pi\Delta_2}{48\epsilon_d^2}T^2 + O(T^4) + \text{const.}. \tag{7.177}$$

The rightmost result is valid at low temperatures $T \ll |\epsilon_d|$. Note that the $T$-linear term in the entropy (or specific heat), which is given by $-\partial\Delta\Omega/\partial T$, grows endlessly as $|\epsilon_d|$ approaches to zero. In the opposite range $T \gg |\epsilon_d|$ of temperature, $\Delta\Omega$ is dominated by $O(T^4)$ terms, and depends hardly on $\epsilon_d$. This is understood easily from the integral with $f(\epsilon)$ in Eq. (7.177).

Let us summarize how the Zeeman splitting $\epsilon_d$ controls the behavior of the system. With $\epsilon_d \neq 0$, $\psi_1$ and $\psi_2$ are never decoupled, and the Majorana particle does not show up in the ground state. Namely, there is no fractional entropy and no divergence in $\chi_z$ in contrast to that in Eq. (7.170). In terms of the renormalization group, $\epsilon_d$ is a relevant parameter that drives the system away from the nontrivial fixed point. However, in the temperature range $|\epsilon_d| \ll T \ll \Delta_2$, the system cannot feel the Zeeman splitting, and behaves as if it is in the non-Fermi liquid state described by Eq. (7.176).

## Problems

**7.1** With quantization of $\Phi(x)$ and $\Pi(x)$, $H_{\text{string}}$ in Eq. (7.4) appears as a bosonic system. Demonstrate the equivalence to $H_{\text{B}}$ in Eq. (7.1).

**7.2** Prove the BCH formula given by Eq. (7.12).

**7.3*** Prove the expectation value for the exponentiated bosonic operators as given by Eq. (7.13).

**7.4** Represent the Hamiltonian $H_{\mathrm{B}}$ in terms of $\theta_J$ and $\nabla\theta_N$ as proportional to the field coordinate and the momentum, respectively.

**7.5*** Demonstrate the equivalence between the free massive Thirring model and the sine-Gordon model.

**7.6** Derive the results given by Eqs. (7.67) and (7.68).

**7.7** Show that the operator defined by Eq. (7.74) generates the same transformation as given by Eq. (7.73).

**7.8** Derive the exponent controlling the asymptotic behavior of the charge correlation function.

**7.9** Derive the condition for $\gamma$ in Eq. (7.131) so that the new Hamiltonian $U^\dagger(H + H_{\mathrm{exc}})U$ does not contain the interaction term with $J_z$.

**7.10** Derive the kinetic energy of $sf$ fermions in the rightmost side of Eq. (7.158)

## Solutions to Problems

**Problem 7.1**
By using the Fourier decomposition we obtain

$$\nabla\Phi(x) = \sum_q \frac{q}{\sqrt{2|q|L}} \left(b_q + b_{-q}^\dagger\right) \exp(iqx), \tag{7.178}$$

which leads to the integral

$$\int_0^L \mathrm{d}x\, (\nabla\Phi)^2 = \sum_q \frac{|q|}{2} (b_q + b_{-q}^\dagger) \left(b_{-q} + b_q^\dagger\right). \tag{7.179}$$

Similar calculation using Eq. (7.7) gives another integral:

$$\int_0^L \mathrm{d}x\, \Pi(x)^2 = \sum_q \frac{|q|}{2} (b_q - b_{-q}^\dagger) \left(b_{-q} - b_q^\dagger\right). \tag{7.180}$$

Hence the sum of Eqs. (7.179) and (7.180) gives the bosonic form $H_{\mathrm{B}}$ in Eq. (7.1) with $\omega_q = v|q|$.

**Problem 7.2**
We shall first show the equality

$$e^{A+B} = e^A e^B e^{-[A,B]/2}, \tag{7.181}$$

with the assumption that $[A, B]$ is a c-number. It is convenient to introduce an auxiliary function $F(g) = e^{gA} B e^{-gA}$ with the parameter $g$. The derivative with respect to $g$ is given by

$$\frac{\mathrm{d}F}{\mathrm{d}g} = e^{gA}[A, B]e^{-gA} = [A, B], \tag{7.182}$$

where we have used the condition that $[A, B]$ commutes with any operator. Considering the initial condition $F(0) = B$, we obtain immediately $F(g) = B + g[A, B]$. This result is equivalent to

$$e^{gA} B = (B + g[A, B])\, e^{gA}. \tag{7.183}$$

On the other hand, another auxiliary function $J(g) = e^{gA} e^{gB} e^{-g^2[A,B]/2}$ has the $g$-derivative

$$\frac{\mathrm{d}J}{\mathrm{d}g} = e^{gA}\,(A + B - g[A, B])\, e^{gB} e^{-g^2[A,B]/2} = (A + B)J(g), \tag{7.184}$$

where the last equality follows from exchange between $e^{gA}$ and $B$ with use of Eq. (7.183), and resulting cancellation of $g[A, B]$. Then integration with the initial condition $J(0) = 1$ leads to $J(g) = e^{g(A+B)}$. By putting $g = 1$, we arrive at Eq. (7.12).

**Problem 7.3***

We first consider the special case $\alpha = \gamma$. Putting $\phi = b + b^\dagger$, we derive the generating function

$$G(\alpha) = \langle \exp(\alpha\phi) \rangle = \sum_{n=0}^{\infty} \frac{\alpha^n}{n!} \langle \phi^n \rangle, \tag{7.185}$$

which corresponds to the LHS of Eq. (7.13). The moments $\langle \phi^n \rangle$ in the generating function can be derived by considering the harmonic oscillator under a fictitious external field $x$, which is defined by

$$H(x) = \omega b^\dagger b + x\phi = \omega \left(b^\dagger + \frac{x}{\omega}\right)\left(b + \frac{x}{\omega}\right) - \frac{x^2}{\omega}. \tag{7.186}$$

Here $b + x/\omega$ is interpreted as the annihilation operator in a harmonic oscillator with its origin shifted. Since the commutation relation is independent of $x$, the spectrum of $H(x)$ is the same as that of $H(0)$ except for the uniform shift $-x^2/\omega$. Therefore the partition function is derived as

$$Z(x) \equiv \mathrm{Tr}e^{-\beta H(x)} = Z(0)\exp(\beta x^2/\omega). \tag{7.187}$$

From the form of $\exp[-\beta H(x)]$ as given by Eq. (7.186), the moments are derived as

$$(-\beta)^n \langle \phi^n \rangle = \frac{1}{Z(0)} \frac{\partial^n Z(x)}{\partial x^n}\bigg|_{x=0}, \tag{7.188}$$

where the $x$-dependence of $Z(x)$ is given explicitly by Eq. (7.187). Hence we obtain

$$\langle \phi^{2m} \rangle = \frac{(2m)!}{m!} \left( \frac{1}{\beta\omega} \right)^m, \tag{7.189}$$

while the moments with odd power are all zero. Substituting the result into the RHS of Eq. (7.185), we obtain

$$G(\alpha) = \sum_{m=0}^{\infty} \frac{\alpha^{2m}}{m!} (\beta\omega)^{-m} = \exp\left( \frac{\alpha^2}{\beta\omega} \right) = \exp\left( \frac{\alpha^2}{2} \langle \phi^2 \rangle \right), \tag{7.190}$$

where $\langle \phi^2 \rangle = 2/(\beta\omega)$ has been used. Considering $\langle \phi^2 \rangle = 2\langle b^\dagger b \rangle + 1$, we obtain Eq. (7.13) in the case of $\alpha = \gamma$.

We recall that cumulants $\langle \phi^n \rangle_c$ are defined by

$$\ln G(\alpha) = \sum_{n=1}^{\infty} \frac{\alpha^n}{n!} \langle \phi^n \rangle_c. \tag{7.191}$$

Comparison with Eq. (7.190) shows that only the second-order cumulant $\langle \phi^2 \rangle_c = \langle \phi^2 \rangle$ is nonzero. In other words, the set of operators $\phi^n$ have expectation values analogous to the Gaussian distribution of $\phi$. We refer to Problem 3.9 for the classical Gaussian distribution.

In the general case $\alpha \neq \gamma$ in Eq. (7.13), expansion of the LHS reveals that only those terms survive which have equal powers of $\alpha$ and $\gamma$. Namely, all nonzero terms are obtained by replacing $\alpha^2$ in the previous case by $\alpha\gamma$. Hence the replacement $\alpha^2 \to \alpha\gamma$ in Eq. (7.190) gives Eq. (7.13) in the general case.

**Problem 7.4**

From Eqs. (7.30)–(7.35), we can represent quantities with indices R, L in terms of $N$, $J$. Namely, we obtain

$$H_{\rm B} = \frac{\pi v_F}{L} \sum_{q>0} : N(q)N(-q) + J(q)J(-q) : \tag{7.192}$$

$$= \frac{v_F}{16\pi} \int_0^L {\rm d}x : (\nabla\theta_N)^2 + (\nabla\theta_J)^2 : . \tag{7.193}$$

The relation

$$\frac{1}{2}[\theta_a(x), \rho_b(y)] = -\mathrm{i}\delta_{ab}\delta(x-y), \tag{7.194}$$

with $a, b = N, J$ is regarded as the canonical commutation rule of field coordinate and momentum, as described by Eq. (7.5). If one regards $\check{\theta}_N(x) = \theta_N(x)/\sqrt{2}$ as the canonical field coordinate, then the canonical momentum $-\check{\Pi}_N(y)$ is identified as $\rho_J(y)/\sqrt{2} = (2\sqrt{2}\pi)^{-1}\nabla\theta_J(y)$. It is also possible to regard $\check{\theta}_J(x) = \theta_J(x)/\sqrt{2}$ as the field coordinate. Then the corresponding momentum $-\check{\Pi}_J(y)$ is identified as $\rho_N(y)/\sqrt{2} = (2\sqrt{2}\pi)^{-1}\nabla\theta_N(y)$. In this way we obtain the two equivalent canonical sets such as

$$H_{\mathrm{B}} = \frac{v_F}{2}\int_0^L \mathrm{d}x : \frac{1}{4\pi}\left(\nabla\check{\theta}_N\right)^2 + 4\pi\check{\Pi}_N^2 : \tag{7.195}$$

$$= \frac{v_F}{2}\int_0^L \mathrm{d}x : \frac{1}{4\pi}\left(\nabla\check{\theta}_J\right)^2 + 4\pi\check{\Pi}_J^2 : . \tag{7.196}$$

With use of the scaling described by Eq. (7.8), we recover the harmonic oscillator form Eq. (7.4) where $v$ is understood as $v_F$.

**Problem 7.5***
Let us choose $C_{\mathrm{R}} = 1$ and $C_{\mathrm{L}} = (-1)^{N+1}$ for the Klein factor, and bosonize the fermion operators as

$$\psi_{\mathrm{R}}(x) = \frac{1}{\sqrt{L}} : \exp[\mathrm{i}\hat{\theta}_{\mathrm{R}}(x)] :, \tag{7.197}$$

$$\psi_{\mathrm{L}}(x) = (-1)^{N+1}\frac{1}{\sqrt{L}} : \exp[-\mathrm{i}\hat{\theta}_{\mathrm{L}}(x)] :, \tag{7.198}$$

with $\hat{\theta}_\alpha(x) = \theta_\alpha(x) + \bar{\theta}_\alpha$ for $\alpha = \mathrm{R}, \mathrm{L}$. Then we obtain

$$\psi_{\mathrm{R}}^\dagger(x)\psi_{\mathrm{L}}(x) = \frac{(-1)^N}{L} : \exp[-\mathrm{i}\hat{\theta}_N(x)] :, \tag{7.199}$$

$$\psi_{\mathrm{L}}^\dagger(x)\psi_{\mathrm{R}}(x) = \frac{(-1)^N}{L} : \exp[\mathrm{i}\hat{\theta}_N(x)] :, \tag{7.200}$$

In deriving Eq. (7.200) we have used the BCH formula in the form

$$: e^A :: e^B : = e^{A_+}e^{A_-}e^{B_+}e^{B_-} = e^{A_+}e^{B_+}e^{A_-}e^{B_-}\exp[A_-, B_+]$$
$$=: \exp(A+B) :, \tag{7.201}$$

where $A_\pm$ and $B_\pm$ represent the boson creation (+) and annihilation (-) parts, and the c-number $[A_-, B_+]$ corresponds to $[\theta_{L-}(x), \theta_{R+}(x)] = 0$. Thus the final result is given by

$$\psi_{\mathrm{R}}^{\dagger}(x)\psi_{\mathrm{L}}(x) + \psi_{\mathrm{L}}^{\dagger}(x)\psi_{\mathrm{R}}(x) = (-1)^{N}\frac{2}{L} : \cos[\theta_N(x)] : . \tag{7.202}$$

In the absence of the current $J = N_{\mathrm{R}} - N_{\mathrm{L}}$, we obtain the even integer for $N = 2N_{\mathrm{R}}$. Adding the fermionic kinetic energy as described by $H_{\mathrm{B}}$, we recognize that the massive Thirring model in the $J = 0$ sector is mapped to the sine-Gordon model given by Eq. (7.60).

**Problem 7.6**

We obtain Eq. (7.67) by taking the ratio of Eqs. (7.65) and (7.66). Next we utilize Eq. (7.67) to obtain:

$$1 \pm \tanh 2\phi = \frac{2\pi v_F^* \pm g_2}{2\pi v_F^*}. \tag{7.203}$$

The ratio of two quantities with $\pm$ leads to

$$\frac{1 - \tanh 2\phi}{1 + \tanh 2\phi} = e^{-4\phi} = K^2 = \frac{2\pi v_F^* - g_2}{2\pi v_F^* + g_2}. \tag{7.204}$$

Alternatively, the ratio of two relations in Eq. (7.64) immediately gives Eq. (7.68).

**Problem 7.7**

For any pair of operators $X$, $Y$ we have the relation

$$e^Y X e^{-Y} = X + [Y, X] + \frac{1}{2}[Y, [Y, X]] + \cdots \equiv (\exp Y_L)X, \tag{7.205}$$

where $Y_L X \equiv [Y, X]$ is defined by analogy with the Liouville operator used in Sect. 3.6. The first equality is confirmed by comparing the term $Y^n X Y^m$ in the Taylor expansion of $\exp(\pm Y)$, and in the multiple commutators. The last concise expression is confirmed by the Taylor expansion of $\exp(Y_L)$ and the definition of $Y_L$.

Let us consider the particular case: $X = b_q^{\dagger},\ Y = -S$ with

$$S = \sum_{q>0} \phi\left(b_q b_{-q} - b_q^{\dagger} b_{-q}^{\dagger}\right), \tag{7.206}$$

with reference to Eq. (7.23). Then with $U = \exp S$ as defined by Eq. (7.74), we obtain

$$\begin{aligned} U^{\dagger} b_q U &= b_q\left(1 + \frac{1}{2}\phi^2 + \cdots\right) - b_{-q}^{\dagger}\left(\phi + \frac{1}{6}\phi^3 + \cdots\right) \\ &= b_q \cosh\phi - b_{-q}^{\dagger} \sinh\phi. \end{aligned} \tag{7.207}$$

In a similar manner we obtain $U^\dagger b_{-q}^\dagger U = b_{-q}^\dagger \cosh\phi - b_q \sinh\phi$. Rewriting the relation in terms of $\rho_\alpha(q)$, we obtain Eq. (7.73).

**Problem 7.8**
The component in the density operator $\rho(x)$ which contribute around the wave number $2k_F$ is given by

$$\rho(x) \to \sum_\sigma \psi_{\mathrm{L}\sigma}^\dagger(x)\psi_{\mathrm{R}\sigma}(x). \tag{7.208}$$

The phase for each spin component after bosonization is given by Eq. (7.107), which is the same as the phase for the longitudinal ($z$) component of the spin correlation. Hence the exponent of the charge correlation function becomes $K_\rho + K_\sigma$ which agrees with the exponent of the longitudinal spin correlation.

**Problem 7.9**
The second term in the RHS of Eq. (7.133) takes the same form as the interaction term with $J_z$ given by Eq. (7.130). Taking account of the coefficient in the kinetic energy in Eq. (7.126), we obtain the zero-sum condition:

$$\frac{a}{2\sqrt{2}\pi}J_z - \frac{v_F}{4\pi}4\pi\gamma = 0, \tag{7.209}$$

which leads to $J_z\rho_c = \sqrt{2}\gamma$. Here $\rho_c = a/(2\pi v_F)$ is the density of states for conduction electrons with the $s$-wave.

**Problem 7.10**
The kinetic energy in Eq. (7.158) is rewritten as

$$\begin{aligned} H_{\mathrm{kin}} &= \frac{1}{2}\sum_k v_F k : C_{sf}(k)^\dagger C_{sf}(k) - C_{sf}(-k)^\dagger C_{sf}(-k) : \\ &= \frac{1}{2}\sum_k v_F k : \left(C_{sf}(k)^\dagger,\, C_{sf}(-k)\right)\begin{pmatrix} C_{sf}(k) \\ C_{sf}(-k)^\dagger \end{pmatrix} :, \end{aligned} \tag{7.210}$$

where the factor 1/2 compensates the double counting in summation over $k$. The second line follows from the anticommutation property of fermions together with the normal ordering. Using the unitary matrix $U$ given by Eq. (7.157), we obtain

$$\left(C_{sf}(k)^\dagger,\, C_{sf}(-k)\right)U^\dagger U\begin{pmatrix} C_{sf}(k) \\ C_{sf}(-k)^\dagger \end{pmatrix} = \left(C_+(k)^\dagger,\, C_-(k)^\dagger\right)\begin{pmatrix} C_+(k) \\ C_-(k) \end{pmatrix}, \tag{7.211}$$

which reproduces the form in the rightmost side of Eq. (7.158).

## References


1. Tomonaga, S.: Prog. Theor. Phys. **5**, 349 (1950)
2. Luttinger, J.M.: J. Math. Phys. **4**, 1154 (1963)
3. Alvarez-Gaume, L., Vazquez-Mozo, M.A.: An Invitation to Quantum Field Theory. Springer, Berlin (2012)
4. Mattis, D.C., Lieb, E.H.: J. Math. Phys. **6**, 304 (1965)
5. Haldane, F.D.M.: J. Phys. C **14**, 2585 (1981)
6. Bethe, H.: Z. Phys. **71**, 205 (1931)
7. Luther, A., Emery, V.J.: Phys. Rev. Lett. **33**, 389 (1974)
8. Coleman, S.: Phys. Rev. D **11**, 2088 (1975)
9. Emery, V.J., Kivelson, S.: Phys. Rev. B **46**, 10812 (1992)
10. Zarand, G., von Delft, J.: Phys. Rev. B **61**, 6918 (2000)

# Chapter 8
# Fractionalization of Charge and Statistics

**Abstract** In this chapter we discuss most drastic non-perturbative effects of interactions. Namely, mutual interactions may give rise to new quasi-particles which are neither fermions nor bosons. We begin with the best-known example of such quasi-particles with a fractional charge in two-dimensional electron systems. Then we proceed to a simpler case in one dimension, where exact solution including such exotic quasi-particles is available. The bosonic description discussed in the previous chapter is regarded as the long-wavelength limit in the more general framework for non-perturbative effects. We emphasize that the fractional charge does not contradict with indivisible nature of electron, since such quasi-particle emerges as a collective phenomenon out of a non-perturbative ground state. All the treatment in this chapter use the first quantization formalism.

## 8.1 Magnetic Flux and Geometric Phase

Electron is regarded as indivisible. Protons and neutrons, which used to be believed indivisible, turn out to be composite of quarks. However, electron is still regarded as indivisible like other leptons such as neutrino. In condensed matter, on the other hand, this viewpoint is not always valid. Great varieties arise in condensed matter because the ground state, or vacuum in condensed matter, depends on systems.

The view on physical vacuum has changed with time. For a long time until 1930, there was little doubt that vacuum is vacant with no structure. By his association of negative energy states with the antiparticle [1], Dirac put forward a groundbreaking view on vacuum.[1] The Dirac picture was, however, modest as compared with the picture of vacuum after 1960. Namely, by analogy with the BCS theory of superconductivity, Nambu [2] proposed that the physical vacuum undergoes spontaneous symmetry breaking. In this way, the current picture of

[1] Dirac interpreted proton as the antiparticle of electron [1]. After the experimental discovery in 1932, the antiparticle of electron has been identified as positron.



Y. Kuramoto, *Quantum Many-Body Physics*, Lecture Notes in Physics 934,
https://doi.org/10.1007/978-4-431-55393-9_8

physical vacuum has become closer to an ordered ground state in condensed matter; the vacuum is full of structures.

One cannot observe the ground state by experiment, but only excitations out of it. Hence if the ground state of a many-particle system is connected continuously from the non-interacting counterpart, the Landau picture of quasi-particles, or minor modification thereof, will apply. However, if the ground state with interactions is separated from the non-interacting state, the observable excitations are also different from the Landau quasi-particles. The most striking example of the ground state distinct from the non-interacting counterpart has been found in two-dimensional electrons in strong magnetic fields [3]. From the magnitude of the Hall effect in GaAs-$Ga_xAl_{1-x}As$ heterostructures, it has been deduced that the particles responsible for the transport are fractionally charged. Hence the experimental result has been called the fractional quantum Hall effect (FQHE). Shortly after the observation of FQHE, convincing interpretation was offered by Laughlin [4], who proposed a trial wave function for the ground state of the system.

Before discussing the non-perturbative interaction effects, it is necessary to be acquainted with some elements in quantum mechanics for a charged particle without spin. The first one is the Aharonov–Bohm (AB) effect [5] which shows that the vector potential influences the observables even though there is no magnetic field. Let us consider the simplest case illustrated in Fig. 8.1 where the conducting ring in the $xy$-plane is threaded by a magnetic flux $\phi_{\rm AB}$ along the $z$-axis. The flux is infinitesimally thin so that there is no magnetic field except at $x = y = 0$ in the plane:

$$B(\boldsymbol{r}) = \phi_{\rm AB}\delta(\boldsymbol{r}), \tag{8.1}$$

where $\boldsymbol{r} = (x, y)$. The corresponding vector potential $\boldsymbol{A}(\boldsymbol{r})$ is given by

$$\boldsymbol{A} = \frac{\phi_{\rm AB}}{2\pi r^2}\hat{z} \times \boldsymbol{r}, \tag{8.2}$$

with the use of the unit vector $\hat{z}$ along the $z$-axis. In terms of the polar coordinate $(r, \theta)$, the components are written as

$$A_r = 0, \quad A_\theta = \frac{\phi_{\rm AB}}{2\pi r}. \tag{8.3}$$

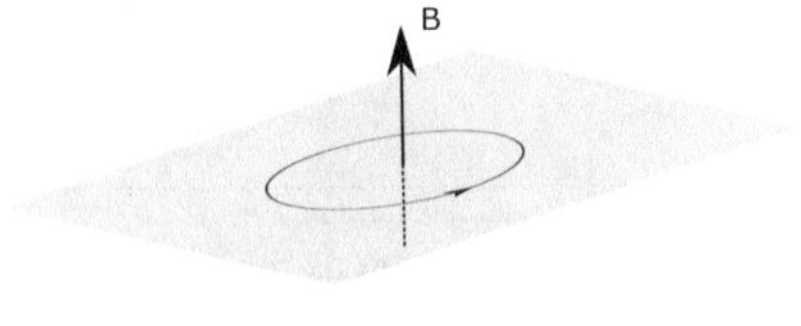


**Fig. 8.1** System of conducting ring on the $xy$-plane with a thin magnetic flux along the $z$-direction threading the plane

The Hamiltonian of an electron confined in the ring with radius $R$ is given by

$$H = \frac{1}{2m}\left(p_\theta + eA_\theta\right)^2, \tag{8.4}$$

with $p_\theta = (\mathrm{i}R)^{-1}\partial/\partial\theta$. In the case of $\phi_{\mathrm{AB}} = 0$, eigenfunctions are of the form $\psi_k(\theta) = \exp(\mathrm{i}k\theta)$ with $k$ integers. Accordingly the eigenvalues are given by $E_k = k^2/(2mR^2)$. There are two-fold degeneracies with $\pm k$, except for the ground state $k = 0$. For nonzero $\phi_{\mathrm{AB}}$, we introduce the ratio $\alpha = \phi_{\mathrm{AB}}/\phi_1$ where $\phi_1 = 2\pi/e$ is the unit flux. Note that the flux quantum $\phi_0$ is defined by $\phi_0 \equiv h/(2e) = \phi_1/2$ with the Planck constant $h = 2\pi\hbar$ recovered. The factor $1/2$ comes from the charge $2e$ of a Cooper pair. With the gauge transformation

$$\psi_{\mathrm{AB}}(\theta) = \exp\left(-\mathrm{i}\alpha\theta\right)\psi_k(\theta), \tag{8.5}$$

we obtain

$$e^{i\alpha\theta}\left(p_\theta + eA_\theta\right)^2\psi_{\mathrm{AB}}(\theta) = p_\theta^2\psi_k(\theta) = (k/R)^2\psi_k(\theta). \tag{8.6}$$

Thus the vector potential $A_\theta$ has been absorbed in the phase factor. It may then appear that the magnetic flux does not play any role. However, we have to recall that physical wave functions must be single-valued. Thus the new periodicity condition for $\theta$ requires $n \equiv k - \alpha$ be an integer, which leads to the spectrum

$$E_n = \frac{1}{2mR^2}\left(n + \alpha\right)^2, \tag{8.7}$$

with integer $n$. Namely, the ground state energy does depend on $\phi_{\mathrm{AB}}$, and oscillates between 0 and $1/(2mR^2)$ with the period $\Delta\alpha = 1$.

The origin of observable effect by the vector potential, which has been gauged away, is its singularity at $\boldsymbol{r} = 0$. Accordingly such gauge transformation $\exp[\mathrm{i}\alpha\theta(\boldsymbol{r})]$ for general coordinate $\boldsymbol{r}$ is ill-defined at $\boldsymbol{r} = 0$, and is called the singular gauge transformation. As a consequence, the phase factor in Eq. (8.5) does not have the periodicity $2\pi$ for $\theta$ with non-integer $\alpha$. We emphasize that the spatial singularity matters even if the wave function vanishes there. In contrast, the spectrum remains invariant against the ordinary gauge transformation, which is well defined at all spatial positions. Let us confirm the invariance by employing a new vector potential:

$$\boldsymbol{A} \to \boldsymbol{A} + \nabla\chi \tag{8.8}$$

with $\chi(\boldsymbol{r})$ being a real and regular function of $\boldsymbol{r}$. We make the Fourier decomposition in the angular coordinate:

$$\chi(R,\theta) = \sum_{n=-\infty}^{\infty}\chi_n(R)\exp(\mathrm{i}n\theta), \tag{8.9}$$

with $\chi_{-n} = \chi_n^*$. Then we recognize that the new phase factor $\exp(-ieR\chi)$ multiplying $\psi_{\rm AB}$ in Eq. (8.5) has the proper periodicity $2\pi$ in $\theta$, and absorbs the change of the vector potential without any physical effect.

The second element in one-body quantum mechanics is called the Berry (or geometric) phase [6, 7], which generalizes the AB phase associated with the vector potential. In general, nonzero phase may appear as a result of infinitely slow (adiabatic) change of a system parameter that takes a loop trajectory in real or hypothetical space. Specifically we consider a variant of the system in Fig. 8.1, removing the confining potential along the ring but adding a smooth potential well with the minimum at $\boldsymbol{\rho}$. Let $\psi(\boldsymbol{r}-\boldsymbol{\rho})$ be the (real) wave function of the groundstate with $\phi_{\rm AB}=0$. We assume that $\boldsymbol{\rho}$ is initially at $\boldsymbol{\rho}=(R,0)$, and moves slowly along the ring where the spread of $\psi(\boldsymbol{r}-\boldsymbol{\rho})$ is much smaller than $R$. With $\phi_{\rm AB}\neq 0$, the eigenfunction $\psi_{\boldsymbol{\rho}}(\boldsymbol{r})$ is given by

$$\psi_{\boldsymbol{\rho}}(\boldsymbol{r}) = \exp\left[ie\int_{\boldsymbol{\rho}}^{\boldsymbol{r}} \mathrm{d}\boldsymbol{r}'\cdot\boldsymbol{A}\left(\boldsymbol{r}'\right)\right]\psi(\boldsymbol{r}-\boldsymbol{\rho}), \tag{8.10}$$

with the use of the gauge transformation. The integration path of $\boldsymbol{r}'$ through $[\boldsymbol{\rho},\boldsymbol{r}]$ does not make a loop around the flux so that $\psi_{\boldsymbol{\rho}}(\boldsymbol{r})$ be single-valued.

Let us derive the phase factor acquired by $\psi_{\boldsymbol{\rho}}(\boldsymbol{r})$ after the loop-wise move of $\boldsymbol{\rho}$. It is obvious that an infinitesimal move of $\boldsymbol{\rho}$ leads to the additional phase $\delta\theta$ in $\psi_{\boldsymbol{\rho}}(\boldsymbol{r})$ as given by

$$\delta\theta = -e\delta\boldsymbol{\rho}\cdot\boldsymbol{A}(\boldsymbol{\rho}). \tag{8.11}$$

Hence we obtain the net change $\Delta\theta$ of the phase as

$$\Delta\theta = e\oint \mathrm{d}\boldsymbol{\rho}\cdot\boldsymbol{A}(\boldsymbol{\rho}) = \pm 2\pi\alpha, \tag{8.12}$$

where $\pm$ depends on the direction of rotation. The result is the same as in Eq. (8.5), except that the groundstate energy is now independent of $\alpha$. On the other hand, if the loop-wise move of $\boldsymbol{\rho}$ does not include the origin $\boldsymbol{r}=0$, there is no phase factor acquired. The extra phase depends only on whether the path includes the origin, and is independent of the details. Furthermore, the phase is independent of the potential that makes $\psi(\boldsymbol{r}-\boldsymbol{\rho})$ the groundstate. In this sense the extra phase characterizes the topology of the surface $S$.

Accumulation of the phase factor along the loop integral may happen under more general conditions. Let $|\psi_{\boldsymbol{\lambda}}\rangle$ be the groundstate of some Hamiltonian with $\boldsymbol{\lambda}$ being a slowly varying vector which generalizes the position $\boldsymbol{\rho}$. The infinitesimal change $\delta\theta$ of the phase of $|\psi_{\boldsymbol{\lambda}}\rangle$ is given by

$$\delta\theta = -\mathrm{Im}\ln\langle\psi_{\boldsymbol{\lambda}}|\psi_{\boldsymbol{\lambda}+\delta\boldsymbol{\lambda}}\rangle = -\mathrm{Im}\left\langle\psi_{\boldsymbol{\lambda}}|\frac{\partial}{\partial\boldsymbol{\lambda}}\psi_{\boldsymbol{\lambda}}\right\rangle\cdot\delta\boldsymbol{\lambda} \equiv \boldsymbol{A}_{\rm B}(\boldsymbol{\lambda})\cdot\delta\boldsymbol{\lambda}, \tag{8.13}$$

which defines the fictitious vector (Berry) potential $\boldsymbol{A}_{\mathrm{B}}(\boldsymbol{\lambda})$. We have here used the property $\delta\langle\psi_{\boldsymbol{\lambda}}|\psi_{\boldsymbol{\lambda}}\rangle = 0$ by the normalization of $\psi_{\boldsymbol{\lambda}}$. The net change of $\theta$ after the move of $\boldsymbol{\lambda}$ along the loop $C$ is given by the Stokes theorem as

$$\Delta\theta = \oint_C \boldsymbol{A}_{\mathrm{B}}(\boldsymbol{\lambda}) \cdot \mathrm{d}\boldsymbol{\lambda} = \int_S \nabla \times \boldsymbol{A}_{\mathrm{B}} \cdot \mathrm{d}\boldsymbol{S}, \tag{8.14}$$

where the surface $S$ is enclosed by the loop $C$, and $\nabla \times \boldsymbol{A}_{\mathrm{B}}$ is the fictitious magnetic field. Note that $\Delta\theta$ is independent of the gauge transformation of $\boldsymbol{A}_{\mathrm{B}}$ as in Eq. (8.8). In a general case of $n$-dimensional parameter $\lambda_i$, the terminology of differential geometry [8] is properly used: each component $A_{\mathrm{B}i}$ of the $n$-dimensional Berry potential is called the connection along the direction $i$, and the surface ($S$) integral in Eq. (8.14) involves $\Omega_{ij} \equiv \partial_i A_{\mathrm{B}j} - \partial_j A_{\mathrm{B}i}$ which is called the Berry curvature. It has thus become clear that the Berry phase is a proper generalization of the AB phase. We discuss in the next section how the Berry phase is utilized in fractional statistics of quasi-particles. The Berry phase appears in a variety of phenomena in condensed matter [7].

As the third and final element in quantum mechanics, we survey the one-electron problem in magnetic field $\boldsymbol{B} = (0, 0, B)$. Neglecting the spin degrees of freedom for simplicity, the Hamiltonian of an electron confined in the two-dimensional plane is given by

$$H = \frac{1}{2m}\left(\pi_x^2 + \pi_y^2\right), \tag{8.15}$$

where $(\pi_x, \pi_y) = \boldsymbol{\pi} = \boldsymbol{p} + e\boldsymbol{A}$. Taking the vector potential in the symmetric gauge such as $\boldsymbol{A} = \boldsymbol{B} \times \boldsymbol{r}/2$, we obtain the coordinate representation:

$$\pi_x = -\mathrm{i}\frac{\partial}{\partial x} + \frac{1}{2\ell_B^2}y, \quad \pi_y = -\mathrm{i}\frac{\partial}{\partial y} - \frac{1}{2\ell_B^2}x, \tag{8.16}$$

where $\ell_B = \sqrt{c\hbar/eB}$ is called the cyclotron radius, with $\hbar$ and the light velocity $c$ written explicitly. We obtain the estimate $\ell \sim 81$ Å for $B \sim 10$ T.

It is convenient to introduce the operators

$$a \equiv \frac{\ell_B}{\sqrt{2}}\left(\pi_x + \mathrm{i}\pi_y\right), \quad a^\dagger \equiv \frac{\ell_B}{\sqrt{2}}\left(\pi_x - \mathrm{i}\pi_y\right), \tag{8.17}$$

which satisfy the bosonic commutation rule: $[a, a^\dagger] = 1$. Then the Hamiltonian is rewritten as

$$\hat{H} = \left(a^\dagger a + \frac{1}{2}\right)\hbar\omega_{\mathrm{c}} \tag{8.18}$$

with

$$\omega_c = \frac{\hbar}{m\ell_B^2} = \frac{eB}{mc}. \tag{8.19}$$

The magnitude of $\hbar\omega_c$ is about 1 meV with $B \sim 10$ T. Since $a^\dagger a$ is a non-negative operator, the groundstate energy is given by $\hbar\omega_c/2$. Hereafter we put $\hbar = c = 1$ again.

We next introduce the dimensionless complex coordinates $z$ and $z^*$ by

$$z = \frac{1}{2\ell_B}(x + \mathrm{i}y), \qquad z^* = \frac{1}{2\ell_B}(x - \mathrm{i}y), \tag{8.20}$$

which give the derivatives as

$$\frac{\partial}{\partial z} = \ell_B \left( \frac{\partial}{\partial x} - \mathrm{i}\frac{\partial}{\partial y} \right), \qquad \frac{\partial}{\partial z^*} = \ell_B \left( \frac{\partial}{\partial x} + \mathrm{i}\frac{\partial}{\partial y} \right). \tag{8.21}$$

Then we obtain useful representations:

$$a = -\frac{\mathrm{i}}{\sqrt{2}} \left( z + \frac{\partial}{\partial z^*} \right) = -\frac{\mathrm{i}}{\sqrt{2}} e^{-|z|^2} \frac{\partial}{\partial z^*} e^{|z|^2}, \tag{8.22}$$

$$a^\dagger = \frac{\mathrm{i}}{\sqrt{2}} \left( z^* - \frac{\partial}{\partial z} \right) = -\frac{\mathrm{i}}{\sqrt{2}} e^{|z|^2} \frac{\partial}{\partial z} e^{-|z|^2}. \tag{8.23}$$

From Eq. (8.22) we recognize that wave functions of the form

$$\phi(z) = f(z) \exp\left( -|z|^2 \right), \tag{8.24}$$

with $f(z)$ being any polynomial of $z$, give the groundstate energy $\omega_c/2$, since it is annihilated by $a$, namely

$$a^\dagger a|\phi\rangle = 0. \tag{8.25}$$

Thus we choose the unnormalized set of wave functions for the degenerate ground state as

$$\phi_m(z) = z^m \exp\left( -|z|^2 \right), \tag{8.26}$$

with $m$ being non-negative integers.

The functions $\phi_m$ form a complete set for the ground Landau level. The degeneracy $n_B$ per unit area increases with increasing $B$, which we now derive. The maximum of $|\phi_m(z)|^2$ corresponds to $|z|^2 = m/2$, and the width of the charge distribution decreases as $1/\sqrt{m}$. Thus with $m \gg 1$ the charge density takes the circular shape with the diameter $R_m$ given by

$$R_m^2 = 2\ell_B^2 m. \tag{8.27}$$

Namely, the system with diameter $R_N$ can accommodate $N + 1$ states from $m = 0$ to $N$. Thus the degeneracy per unit area is given by

$$n_B = \frac{N+1}{\pi R_N^2} = \frac{1}{2\pi \ell_B^2} + O\left(\frac{1}{N}\right). \tag{8.28}$$

We shall give another derivation later.

## 8.2 Fractional Charge in Two Dimensions

With these preliminaries of quantum mechanics in the previous section, we can now proceed to many-electron systems. We begin with $N$ non-interacting electrons in strong magnetic field. The ground state with the most compact charge distribution is given by the Slater determinant:

$$\Psi_1(z_1, z_2, \ldots, z_N) = \det \begin{vmatrix} \phi_0(z_1) & \phi_0(z_2) & \cdots \phi_0(z_N) \\ \phi_1(z_1) & \phi_1(z_2) & \cdots \phi_1(z_N) \\ \cdots & \cdots & \cdots \\ \phi_{N-1}(z_1) & \phi_{N-1}(z_2) & \cdots \phi_{N-1}(z_N) \end{vmatrix}, \tag{8.29}$$

which is equivalent to

$$\Psi_1 = \prod_{1 \le i < j \le N} (z_i - z_j) \prod_{i=1}^{N} \exp\left(-|z_i|^2\right). \tag{8.30}$$

It is obvious that $\Psi_1(z_1, z_2, \ldots, z_N)$ is antisymmetric against interchange of any two coordinates. Thus $\Psi_1$ satisfies the basic property for fermions. In the thermodynamic limit $N \to \infty$, $\Psi_1(z_1, z_2, \ldots, z_N)$ gives the fully occupied Landau level with the electron density $n = n_B$.

Now we can turn to our primary interest: the system of $N$ interacting electrons. Experimentally, the Coulomb repulsion among electrons is the most relevant interaction. However, we do not use the explicit form, and simply assume that the relevant energy scale due to the interaction is much smaller than $\omega_c$. For FQHE systems, Laughlin has constructed a simple but highly nontrivial trial wave function that takes advantage of repulsive interaction between electrons. The trial function reads

$$\Psi_q(z_1, z_2, \ldots, z_N) = \prod_{i<j}(z_i - z_j)^q \prod_{i=1}^{N} \exp\left(-|z_i|^2\right), \tag{8.31}$$

with $q$ being odd integer. Evidently $\Psi_q$ satisfies the antisymmetric property for fermions. As compared with $\Psi_1$ with the same $N$, electrons in $\Psi_q$ with $q \geq 3$ are more separate from one another.

There are a variety of ways to see that $\Psi_q$ describes a state with $1/q$ occupation of the ground Landau level in the thermodynamic limit. We shall use the analogy to classical plasma [4], which also demonstrates the fractional charge most transparently. The analogy begins with the norm in the form:

$$\langle \Psi_q | \Psi_q \rangle \equiv \int \mathrm{d}z_1 \mathrm{d}z_2 \cdots \mathrm{d}z_N \exp(-\beta H_c), \tag{8.32}$$

$$\frac{1}{2}\beta H_c(z_1, z_2, \ldots, z_N) = \sum_i |z_i|^2 - q \sum_{i<j} \ln\left|z_i - z_j\right|. \tag{8.33}$$

The quantity $H_c$ corresponds to the Hamiltonian of the two-dimensional classical plasma at fictitious temperature $T = 1/\beta$. In the thermodynamic limit, only such configuration that gives the minimum of $\beta H_c$ needs to be considered. The Coulomb potential between two-dimensional particles with unit charge $Q$ is given by

$$V_2(z_i - z_j) = -\frac{Q^2}{2\pi} \ln|z_i - z_j|, \tag{8.34}$$

the derivation of which is discussed in Problem 8.1. Since $V_2(z_i - z_j)$ has the dimension of energy, the dimension of $Q$ is different from that of ordinary charge in three-dimensional systems. On the other hand, the neutralizing background with the charge density $-Q\rho$ gives the harmonic potential as given by

$$V_1(z_i) = Q^2 \ell_B^2 \rho |z_i|^2. \tag{8.35}$$

Problem 8.1 also discusses how this potential comes out. The classical plasma must be homogeneous in the thermodynamic limit. Otherwise the lack of charge neutrality causes the logarithmically divergent potential to break up the system. Comparison of $V_1$ and $V_2$ with Eq. (8.33) gives the correspondence

$$\frac{2q}{\beta} = \frac{Q^2}{2\pi}, \quad \frac{2}{\beta} = Q^2 \ell_B^2 \rho. \tag{8.36}$$

With the charge neutrality, which is necessary for the minimum energy, the electron density is the same as $\rho$ as given by

$$\rho = \frac{1}{2\pi\ell_B^2 q} = \frac{n_B}{q} \equiv \nu n_B. \tag{8.37}$$

Thus $\Psi_q$ indeed corresponds to the occupation $\nu = 1/q$ of the ground Landau level. The special case $q = 1$ reproduces the degeneracy given by Eq. (8.28).

We proceed to discuss the fractional charge by modifying $H_c$ as

$$\frac{\beta}{2} H_{\mathrm{ch}} \equiv \frac{\beta}{2} H_c - \sum_{i=1}^{N} \ln |z_i - w|, \tag{8.38}$$

which corresponds to inserting an external charge at $w$ into the plasma. The magnitude of the charge must be $Q/q$ since the mutual interaction term in $H_c$ has the coefficient $q$. The Hamiltonian $H_{\mathrm{ch}}$ is realized by the wave function:

$$\Psi_{q\mathrm{h}}(z_1, z_2, \ldots, z_N; w) = \prod_{i=1}^{N} (z_i - w) \Psi_q(z_1, z_2, \ldots, z_N). \tag{8.39}$$

Thus $\Psi_{q\mathrm{h}}$ describes an excited state with a fractionally charged ($= e/q$) entity at $w$, which is called the quasi-hole. In the special case $w = 0$, it is obvious that $\Psi_{q\mathrm{h}}$ is expanded a little more than $\Psi_q$ due to the increase by $N$ of the total angular momentum. Thus the overall density is less than that of $\Psi_q$ by $O(1/N)$.

Similarly, we can modify $H_c$ in another way as

$$\frac{\beta}{2} H_{\mathrm{cp}} \equiv \frac{\beta}{2} H_c + \sum_{i=1}^{N} \ln |z_i - w|, \tag{8.40}$$

which amounts to inserting an external fractional ($1/q$) charge with the sign opposite from that of $N$ particles. In the special case $w = 0$, the associated factor

$$\exp\left( -\sum_i \ln z_i \right) = \prod_i z_i^{-1} \tag{8.41}$$

is reflected in the corresponding wave function $\Psi_{q\mathrm{p}}$ as reduction of the power of each $z_i$ in the polynomial part of $\Psi_q$. However, any negative power should not appear in the final result. Thus the reduction is accomplished by $\partial/\partial z_i$ rather than by $1/z_i$. Namely, we obtain

$$\Psi_{q\mathrm{p}}(z_1, z_2, \ldots, z_N; w = 0) = \exp\left( -\sum_{i=1}^{N} |z_i|^2 \right) \prod_{k=1}^{N} \frac{\partial}{\partial z_k} \prod_{i<j} (z_i - z_j)^q . \tag{8.42}$$

Since the derivatives decrease the total angular momentum by $N$ from that of $\Psi_q$, the overall density of electrons increases by $O(1/N)$. The change is ascribed to the fractionally charged quasi-electron at the origin $w = 0$. In a similar manner, the case with $w \neq 0$ can be treated with a little more complication [4].

The plasma analogy makes it clear how the charge of a quasi-particle acquires a fractional magnitude. We now ask how the quasi-particle affects the phase of the wave function. It is convenient to consider the case with two quasi-particles. We derive the change of the phase of wave function after one of them encircles the other. The resultant change corresponds to the Berry phase, and the half of the change, which corresponds to the interchange of quasi-particles, describes their statistics. Namely, they are regarded as bosons if there is no change, and as fermions if the sign reverses. It is convenient to allow for different species of quasi-particles with the wave function:

$$\Psi_{qhp}(z_1, z_2, \ldots, z_N; w, v) \equiv \prod_{i=1}^{N}(z_i - w)(z_i - v)^p \Psi_q(z_1, z_2, \ldots, z_N) \tag{8.43}$$

with $p$ being an integer. If we set $p = 1$, $\Psi_{qh1}$ describes a state with two identical holes present, and $\Psi_{qhq}$ with $p = q$ describes another state with a hole at $v$ and an extra electron present at $v$. Although $\Psi_{qh1}$ is symmetric against interchange of $w$ and $v$, this does not mean their bosonic property since $w$ and $v$ are just parameters in the $N$-electron system. On the other hand, the Berry phase as a result of adiabatic move $w$ and/or $v$ reflects the property of quasi-particles in the system with $N$ electrons.

We shall inspect evolution of the phase of $\Psi_{qhp}$ as $w$ encircles adiabatically around the origin with radius $R$ ($\gg \ell_B$) [9]. Let us consider the two cases: (1) $|v| \gg R$, and (2) $|v| \ll R$ with resulting phase changes $\theta_1$ and $\theta_2$. We regard $\Delta\theta = \theta_1 - \theta_2$ as the Berry phase of the hole in the presence of another object at $v$. Thus $\Delta\theta/2$ corresponds to the interchange of two objects at $w$ and $v$. The manipulation of closed loop encircling or avoiding the other object is in line with derivation of the Berry phase in Sect. 8.1. In the present case, however, it is essential that we integrate over the position $z_i$ of a large number of electrons.

In the case (1), the factor $(z_i - v)^p \sim (-v)^p$ influences only normalization, and we may take $\Psi_{qh}$ in Eq. (8.39) in considering the variation of $w = Re^{i\phi}$. Since the norm $\langle\Psi_{qh}|\Psi_{qh}\rangle$ does not change by $\delta\phi$, we may use the unnormalized form for variation. Thus according to Eq. (8.13) we need to evaluate

$$\delta\theta_1 = \frac{-1}{\langle\Psi_{qh}|\Psi_{qh}\rangle}\mathrm{Im}\langle\Psi_{qh}|\delta\Psi_{qh}\rangle. \tag{8.44}$$

Using the relation

$$\delta \ln \Psi_{qh} = \sum_i \delta \ln(z_i - w), \tag{8.45}$$

we obtain

$$\begin{aligned}\delta\theta_1 &= -\frac{1}{\langle\Psi_{q\mathrm{h}}|\Psi_{q\mathrm{h}}\rangle}\int \mathrm{d}z_1,\ldots,\mathrm{d}z_N|\Psi_{q\mathrm{h}}|^2 \mathrm{Im}\,\delta\sum_i \ln(z_i-w)\\ &= -\int \mathrm{d}z\rho(z)\mathrm{Im}\,\delta\ln(z-w),\end{aligned} \tag{8.46}$$

where we have used the electron density:

$$\rho(z)=\int \mathrm{d}z_1,\ldots,\mathrm{d}z_N\frac{|\Psi_{q\mathrm{h}}|^2}{\langle\Psi_{q\mathrm{h}}|\Psi_{q\mathrm{h}}\rangle}\sum_i\delta(z-z_i). \tag{8.47}$$

As $w$ encircles the contour with radius $R$, only the region $|z|<R$ contributes to $\oint dw\mathrm{Im}\ln(z-w)\neq 0$. In this region we have $\rho(z)=\rho_0=\nu/(2\pi\ell_B^2)$ in the thermodynamic limit $N\to\infty$. We thus obtain the simple result

$$\theta_1=\rho_0\int_{|z|\le R}\mathrm{d}z\int_0^{2\pi}\mathrm{d}(\arg w)=2\pi N_R \tag{8.48}$$

with $N_R=\rho_0\pi R^2=\nu(R/\ell_B)^2$.

In the case (2), we use $\Psi_{q\mathrm{h}p}$ with $v=0$ for evaluation of $\theta_2$. By similar reasoning to the case of (1) we obtain

$$\delta\theta_2=-\int \mathrm{d}z\rho_p(z)\mathrm{Im}\,\delta\ln(z-w), \tag{8.49}$$

$$\rho_p(z)=\int \mathrm{d}z_1,\ldots,\mathrm{d}z_N\frac{|\Psi_{q\mathrm{h}p}|^2}{\langle\Psi_{q\mathrm{h}p}|\Psi_{q\mathrm{h}p}\rangle}\sum_i\delta(z-z_i). \tag{8.50}$$

The density $\rho_p(z)$ includes a deficit region due to the factor $\prod_i(z_i-v)^p$ in $\Psi_{q\mathrm{h}p}$. As a result the integration gives

$$\int_{|z|\le R}\mathrm{d}z\rho_p(z)=N_R-\frac{p}{q}=N_R-p\nu. \tag{8.51}$$

The case $p=q$ gives $N_R-1$ in the RHS, which is understood naturally, since the object at $v$ has the same role as electrons but is not counted as such. In another case $p=1$ the deficit of the density is less than the case $p=q$ $(\ge 3)$, and gives the fractional statistics. In fact, the phase that amounts to exchange of two identical holes is given by

$$\frac{1}{2}(\theta_1-\theta_2)=\pi\nu. \tag{8.52}$$

Actually, the sign reverses depending on the direction of rotation in the exchange. The complete filling with $\nu = 1$ recovers the Fermi statistics. Otherwise, the statistics is neither Fermi nor Bose, but fractional.

We may consider the move of $v$, instead of $w$, along the circle with radius $R$. Correspondingly we derive the phase $\theta_3$ for the case $|w| \gg R$, and another phase $\theta_4$ where $|w| \ll R$. Following the procedure parallel to the previous cases (1) and (2), we obtain

$$\theta_3 = -\int dz\rho(z)\text{Im}\ \ln(z - v)^p = 2\pi p N_R, \tag{8.53}$$

$$\theta_4 = -\int dz\rho_1(z)\text{Im}\ \ln(z - v)^p = 2\pi p \left(N_R - \frac{1}{q}\right), \tag{8.54}$$

where the electron density $\rho_1(z)$ includes a deficit region due to the factor $\prod_i (z_i - w)$ in $\Psi$. Thus we obtain with $p = 1$

$$\frac{1}{2}(\theta_3 - \theta_4) = \pi p/q = \pi\nu, \tag{8.55}$$

which is the same as the result in Eq. (8.52). The equivalence persists even if $p \neq 1$. Since the exchange of two objects gives identical results whether $w$ or $v$ makes a move, the exchange statistics is well defined. In the special case of $p = 1$, the fractional statistics of quasi-holes is characterized by $\nu = 1/q$.

As presented above, the wave functions constructed by Laughlin have convincingly demonstrated the emergence of fractional charge and exotic statistics. In this scheme, however, it is difficult to analyze effects of finite temperature, and dynamical property of the system exhibiting the FQHE. We shall thus turn to a one-dimensional system where fractional particles also emerge. The great advantage of the one-dimensional system is that exact analytic solution is available not only for thermodynamics but also for dynamics [10]. We shall see that the mechanism for the fractional charge shares something in common with the FQHE system, but there are also different aspects. Especially there is no symmetry between quasi-holes and quasi-electrons in the one-dimensional system as described in detail below. Comparison of both systems will give deeper insight into such non-perturbative effects as charge fractionalization.

## 8.3 Sutherland Model and Its Exact Eigenvalues

In the rest of this chapter we consider mostly a one-dimensional system of spinless $N$ particles. We take the Hamiltonian:

$$H = -\sum_{i=1}^{N} \frac{\partial^2}{\partial x_i^2} + \frac{\pi^2}{L^2} \sum_{i \neq j} \frac{\lambda(\lambda - 1)}{\sin^2\left[\pi\left(x_i - x_j\right)/L\right]}, \tag{8.56}$$

where $x_i$ is the coordinate of $i$-th particle, and $L$ is the length of the system with the periodic boundary condition. The denominator in the interaction term corresponds to the chord distance squared for a ring with circumference $L$, and behaves as $r^2$ for the interparticle distance $r = |x_i - x_j|$ with $r/L \ll 1$. The coupling parameter $\lambda$ is taken to be a positive integer for simplicity. We have taken units such that all variables are dimensionless. This model is known as the Sutherland model [11], which realizes fractionally charged quasi-particles in its exact solution. The model may appear very special because of its interaction form. In fact, the Sutherland model corresponds to the fixed point model in the sense of renormalization toward a non-Fermi liquid ground state. In other words, the model includes the Tomonaga–Luttinger liquid for density excitations in the limit of long wavelength.

We take the following form of $N$-particle wave function:

$$\Psi_g(x_1, x_2, \ldots, x_N) = \prod_{j<k} \sin^\lambda \frac{\pi\left(x_j - x_k\right)}{L}, \tag{8.57}$$

which turns out to describe the ground state of Eq. (8.56). With an odd number for $\lambda$, the wave function is antisymmetric against exchange of two coordinates, which corresponds to the property of fermions. In the particular case of $\lambda = 1$, $\Psi_g$ reduces to the Slater determinant for free fermions, which is consistent with the vanishing interaction $\lambda(\lambda - 1) = 0$ in Eq. (8.56). The reduction to the Slater determinant is the subject of Problem 8.2. On the other hand, with an even integer for $\lambda$, the exchange of two coordinates in $\Psi_g(x_1, x_2, \ldots, x_N)$ leaves the wave function invariant, which corresponds to the property of bosons.

We now demonstrate that Eq. (8.57) describes the ground state of Eq. (8.56) for any positive value of $\lambda$. The argument is actually not restricted to the case of integer $\lambda$, provided the power of the sine function is well-defined. We first take the derivative of $\Psi_g$ to obtain

$$\frac{\partial}{\partial x_j}\Psi_g = \sum_{i(\neq j)} \frac{\lambda\pi}{L} \cot \frac{\pi(x_j - x_i)}{L}\Psi_g. \tag{8.58}$$

Further derivative $\Psi_g$ leads to the result

$$\begin{aligned}\left(\frac{L}{\pi}\right)^2 \Psi_g^{-1} \sum_{i=1}^N \frac{\partial^2}{\partial x_i^2}\Psi_g = \lambda^2 \sum_{i=1}^N \left[\sum_{j(\neq i)} \cot \frac{\pi(x_i - x_j)}{L}\right]^2 \\ - \lambda \sum_{i\neq j} \sin^{-2} \frac{\pi\left(x_i - x_j\right)}{L},\end{aligned} \tag{8.59}$$

where the first summand in the RHS is separated into the two-site contribution

$$\sum_{i\neq j}\cot^2\frac{\pi\left(x_i-x_j\right)}{L}=\sum_{i\neq j}\sin^{-2}\frac{\pi\left(x_i-x_j\right)}{L}-N(N-1), \tag{8.60}$$

and the three-site one

$$\sum_i\sum_{j(\neq i)}\sum_{k(\neq i,j)}\cot\frac{\pi\left(x_i-x_j\right)}{L}\cot\frac{\pi\left(x_i-x_k\right)}{L}. \tag{8.61}$$

Surprisingly, Eq. (8.61) gives a constant $N(N-1)(N-2)/3$ independent of the coordinates! This fact results from the identity

$$\cot(ij)\cot(ik)+\cot(ji)\cot(jk)+\cot(ki)\cot(kj)=-1, \tag{8.62}$$

with the abbreviated notation: $\cot(ij)\equiv\cot\left[\pi\left(x_i-x_j\right)/L\right]$. The proof of this identity is the subject of Problem 8.3. Consequently, the RHS of Eq. (8.59) is given by

$$\left(\frac{\pi}{L}\right)^2\sum_{i\neq j}\frac{\lambda(\lambda-1)}{\sin^2\left[\pi\left(x_i-x_j\right)/L\right]}-E_{0,N}, \tag{8.63}$$

where the first term corresponds to the interaction term in $H$ given by Eq. (8.56), while the constant

$$E_{0,N}=(\pi\lambda/L)^2N(N^2-1)/3 \tag{8.64}$$

gives the eigenenergy associated with $\Psi_g$. Thus by arranging terms in Eq. (8.59) we obtain $H\Psi_g=E_{0,N}\Psi_g$. We shall show later that $E_{0,N}$ is indeed the lowest eigenvalue of $H$.

In addition to the ground state energy, we can derive all eigenenergies of the Sutherland model without much effort. For this purpose we take any eigenfunction $\Psi$ in the form $\Psi=\Psi_g\Phi$. Then we consider a new Schrödinger equation for $\Phi(x_1,\ldots,x_N)$ with the condition that $\Phi(x_1,\ldots,x_N)$ be invariant against interchange of coordinates. For an odd integer $\lambda$, this construction makes $\Psi$ antisymmetric against coordinate exchange, as an inheritance from $\Psi_g$. Thus $\Psi$ is capable of describing identical fermions. From the original eigenvalue equation $H\Psi_g\Phi=E\Psi_g\Phi$, we obtain a new equation for $\Phi$ as

$$\left(-\sum_i\frac{\partial^2}{\partial x_i^2}+\sum_{i\neq j}\frac{2\pi\lambda}{L}\cot\frac{\pi(x_i-x_j)}{L}\frac{\partial}{\partial x_i}\right)\Phi=\left(E-E_{0,N}\right)\Phi. \tag{8.65}$$

In the following we make extensive use of the complex coordinate $z_j = \exp(2\pi \mathrm{i} x_j/L)$, which leads to the following expressions:

$$\frac{\partial}{\partial x_j} = \frac{2\pi \mathrm{i} z_j}{L} \frac{\partial}{\partial z_j}, \qquad \cot \frac{\pi(x_l - x_j)}{L} = \mathrm{i} \frac{z_l + z_j}{z_l - z_j}. \tag{8.66}$$

In the solution of Problem 8.2, Eq. (8.128) utilizes the complex coordinates. The Hamiltonian can also be represented in terms of $z_j$. Namely, Eq. (8.65) can be written as

$$\left(h^{(1)} + \lambda h^{(2)}\right) \Phi = \mathcal{E} \Phi, \tag{8.67}$$

with $\mathcal{E} = [L/(2\pi)]^2 \left(E - E_{0,N}\right)$ and

$$h^{(1)} = \sum_i \left(z_i \frac{\partial}{\partial z_i}\right)^2, \quad h^{(2)} = \sum_{i<j} \left(\frac{z_i + z_j}{z_i - z_j}\right) \left(z_i \frac{\partial}{\partial z_i} - z_j \frac{\partial}{\partial z_j}\right). \tag{8.68}$$

Note that $h^{(1,2)}$ are not Hermitian, but $h = h^{(1)} + \lambda h^{(2)}$ have real eigenvalues because the original Hamiltonian is Hermitian. The transformation $H \to h = \Psi_\mathrm{g}^{-1} H \Psi_\mathrm{g}$ is an example of similarity transformations which are not necessarily unitary, but keep the eigenvalues of $H$.

Let us proceed to derive all eigenenergies. We begin with the two-body ($N = 2$) problem in order to understand the strategy which is valid for any value of $N$. Consider the symmetrized function $\Phi(z_1, z_2) = z_1^{\kappa_1} z_2^{\kappa_2} + z_1^{\kappa_2} z_2^{\kappa_1}$. By operating $h^{(1)}$ upon $\Phi$ we immediately obtain the eigenvalue $\kappa_1^2 + \kappa_2^2$. We next operate $h^{(2)}$ upon $\Phi$. The result is 0 in the case of $\kappa_1 = \kappa_2$, while in the case of $\kappa_1 > \kappa_2$, we obtain the following result:

$$\begin{aligned}
&\left(\frac{z_1 + z_2}{z_1 - z_2}\right) \left(z_1 \frac{\partial}{\partial z_1} - z_2 \frac{\partial}{\partial z_2}\right) \left(z_1^{\kappa_1} z_2^{\kappa_2} + z_1^{\kappa_2} z_2^{\kappa_1}\right) \\
&= (\kappa_1 - \kappa_2) \left(\frac{z_1 + z_2}{z_1 - z_2}\right) \left(z_1^{\kappa_1} z_2^{\kappa_2} - z_1^{\kappa_2} z_2^{\kappa_1}\right) \\
&= (\kappa_1 - \kappa_2)(z_1 + z_2) \left(z_1^{\kappa_1 - 1} z_2^{\kappa_2} + \cdots + z_1^{\kappa_2} z_2^{\kappa_1 - 1}\right) \\
&= (\kappa_1 - \kappa_2) \left(z_1^{\kappa_1} z_2^{\kappa_2} + 2 z_1^{\kappa_1 - 1} z_2^{\kappa_2 + 1} + \cdots + 2 z_1^{\kappa_2 + 1} z_2^{\kappa_1 - 1} + z_1^{\kappa_2} z_2^{\kappa_1}\right).
\end{aligned} \tag{8.69}$$

It is clear from Eq. (8.69) that application of $h$ keeps the degree $\kappa = \kappa_1 + \kappa_2$ of homogeneous polynomials in $z_1$ and $z_2$. Then the set of homogeneous polynomials is arranged into the decreasing order of the maximum power of $z_1$. Namely, we obtain $\Phi_1 = z_1^\kappa + z_2^\kappa$, $\Phi_2 = z_1^{\kappa-1} z_2 + z_1 z_2^{\kappa-1}, \ldots$ and so on. Here we do not care about normalization of $\Phi_i$. With this definition of the basis set, the result $(h^{(1)} + \lambda h^{(2)})\Phi_i$ is expressed as a linear combination of $\Phi_j$ with $j \geq i$. This means that the Hamiltonian matrix with this ordered basis has a triangular form with zero components in the lower triangular part. Since the eigenvalues of a triangular matrix are exhausted by its diagonal elements, we obtain

$$\mathcal{E}(\kappa_1, \kappa_2) = \kappa_1^2 + \kappa_2^2 + \lambda|\kappa_1 - \kappa_2| \tag{8.70}$$

for the two-body eigenvalues. Here the condition $\kappa_1 \geq \kappa_2$ is no longer necessary because of the absolute value.

For the $N$-body wave function $\Phi(z_1, z_2, \ldots, z_N)$ with the total momentum $\kappa$, we take the same ordering of the homogeneous polynomials. Namely, in the set $(\kappa_1, \kappa_2, \ldots, \kappa_N)$ with $\kappa = \sum_i \kappa_i$, the first priority is a larger value of $\kappa_1$. If this is the same in the two polynomials, the second priority is a larger value of $\kappa_2$, and so on. In this way the set of basis functions are generated. For example, we obtain

$$\begin{aligned} \Phi_1 &= \sum_i z_i^\kappa, \quad \Phi_2 = \sum_{i \neq j} z_i^{\kappa-1} z_j, \\ \Phi_3 &= \sum_{i \neq j} z_i^{\kappa-2} z_j^2, \quad \Phi_4 = \sum_{i \neq j \neq k \neq i} z_i^{\kappa-2} z_j z_k. \end{aligned} \tag{8.71}$$

Note that $h^{(2)}$ acts on a pair of coordinates, say $z_i, z_j$. Then application on $\Phi_l$ results in the diagonal element $|\kappa_i - \kappa_j|$ and a linear combination of $\Phi_m$ with $m > l$. Hence the Hamiltonian is upper triangular, and the eigenvalues are given by its diagonal elements, just as in the case of $N = 2$. Thus Eq. (8.70) is generalized as

$$\mathcal{E}(\kappa_1, \ldots, \kappa_N) = \sum_{i=1}^{N} \kappa_i^2 + \frac{\lambda}{2} \sum_{i,j=1}^{N} \left|\kappa_i - \kappa_j\right|. \tag{8.72}$$

In this way we have obtained all eigenenergies of the $N$-particle system. The ground state energy is given by $\mathcal{E}(0, 0, \ldots, 0) = 0$ which corresponds to Eq. (8.64). The interaction effect enters only in the diagonal element in Eq. (8.72), which may look like the mean field theory. Nevertheless, the present result is exact. It is remarkable that a many-body problem can be solved exactly in such a simple manner. The eigenfunctions can also be derived exactly with a little more effort. We have already derived the ground state in Eq. (8.57) which corresponds to $\Phi = 1$. It is known that the excited states with nontrivial $\Phi$ are derived in the form called the Jack polynomials [10].

## 8.4 Rapidity of Quasi-Particles

We shall show in this section that the $\lambda$-dependent interaction term in Eq. (8.72) is absorbed into the new quantity called the rapidity, which in the present case refers to a modification of the momentum. Explicit construction of the rapidity is the subject of Problem 8.4. Namely, in terms of the rapidity, all eigenvalues are expressed as the sum of free quasi-particle energies. To proceed we first utilize the following relation:

$$\sum_{i<j}(\kappa_i - \kappa_j) = \sum_{i=1}^{N}(N+1-2i)\,\kappa_i, \tag{8.73}$$

the proof of which is also the subject of Problem 8.4. Then Eq. (8.72) is rewritten as

$$\mathcal{E} = \sum_{i=1}^{N}\left[\kappa_i^2 + \lambda(N+1-2i)\kappa_i\right] = \sum_{i=1}^{N}\left(\tilde{\kappa}_i^2 - \tilde{\kappa}_{i,0}^2\right), \tag{8.74}$$

where we have introduced the rapidity $\tilde{\kappa}_i$ by

$$\tilde{\kappa}_i = \kappa_i + \frac{\lambda}{2}(N+1-2i), \quad \tilde{\kappa}_{i,0} = \frac{\lambda}{2}(N+1-2i), \tag{8.75}$$

with $\tilde{\kappa}_{i,0}$ relevant to the ground state. It is remarkable that Eq. (8.74) has the form of difference between $\tilde{\kappa}_i^2$ and $\tilde{\kappa}_{i,0}^2$ both of which represent energies of free particles. The ground state has $\mathcal{E} = 0$ which corresponds to $\tilde{\kappa}_i = \tilde{\kappa}_{i,0}$, or equivalently $\kappa_i = 0$ for all $i$. In the special case with $\lambda = 1$, the rapidity $\tilde{\kappa}_i$ is the same as that of free fermions.

Alternative representation of the rapidity is given by

$$\tilde{\kappa}_i = \kappa_i + \frac{\lambda}{2}\sum_{j(\neq i)} \operatorname{sgn}\left(\tilde{\kappa}_i - \tilde{\kappa}_j\right), \tag{8.76}$$

because of the relation

$$\sum_{j(\neq i)} \operatorname{sgn}\left(\tilde{\kappa}_i - \tilde{\kappa}_j\right) = \sum_{j=1}^{i-1}(-1) + \sum_{j=i+1}^{N}(+1) = N - 2i + 1. \tag{8.77}$$

The form of Eq. (8.76) will be used conveniently in Sect. 8.6.

The quantity $p_i \equiv 2\pi\tilde{\kappa}_i/L$ has the dimension of momentum, and is also called the rapidity. The main motivation to introduce $p_i$ is to use it as a continuous variable in the thermodynamic limit $L \to \infty$. Then the total energy $E$ as defined by $E - E_{0,N} = (2\pi/L)^2\mathcal{E}$ is given for a general value of $\lambda$ by

$$E = \sum_{i=1}^{N} (2\pi\tilde{\kappa}_i/L)^2 \equiv \sum_{i=1}^{N} p_i^2. \tag{8.78}$$

Thus the rapidity $p_i$ (or $\tilde{\kappa}_i$) has the meaning of renormalized momentum including the interaction effect. Problem 8.4 deals with derivation of Eq. (8.78).

Figure 8.2 illustrates the distribution of $\tilde{\kappa}_i$ in the case of $\lambda = 2$ and $N = 5$. The ground state is shown in (a) with $\tilde{\kappa}_1 = 4, \tilde{\kappa}_2 = 2, \ldots, \tilde{\kappa}_5 = -4$. Thus the quasi-particles with $\lambda = 2$ have an exclusion property stronger than that of free fermions ($\lambda = 1$) where the rapidity takes successive integers. On the other hand, in an excited state shown in (b), $\tilde{\kappa}_1$ has increased to 6 from 4. We regard this as a particle excitation, where $\tilde{\kappa}_1$ can take any integer larger than 4. In another excited state shown in (c), both $\tilde{\kappa}_1$ and $\tilde{\kappa}_2$ have increased by 1 from the ground state. As a result, the difference $\tilde{\kappa}_2 - \tilde{\kappa}_3 = 3$ is larger than the minimum difference 2, which is the case for all the other momenta. Such state is called a hole. More generally, a hole is present at $\tilde{\kappa}_i$ if all $\tilde{\kappa}_j$'s with $j < i$ have increased by 1.

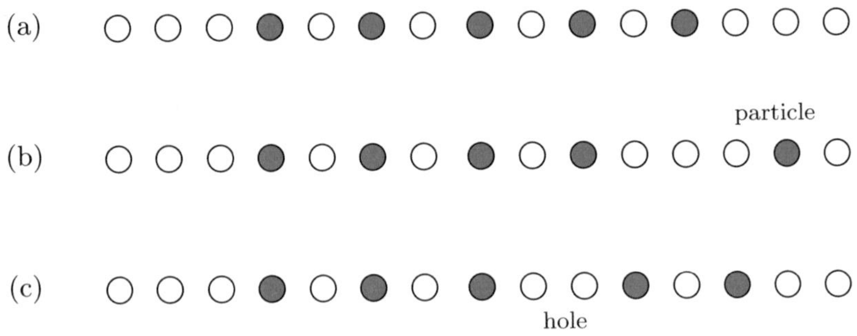


**Fig. 8.2** Distributions of $\tilde{\kappa}_i$ for $N = 5, \lambda = 2$. Filled circles represent occupied states, while the vacant ones are empty states. The center corresponds to the value $\tilde{\kappa}_i = 0$. (**a**) the ground state; (**b**) a particle excitation with the rapidity $\tilde{\kappa} = 6$; (**c**) a hole excitation whose location is between $\tilde{\kappa}_2$ and $\tilde{\kappa}_3$

## 8.5 Distribution Function and Entropy

In terms of the distribution function $\nu(\kappa)$ for the momentum $\kappa$, the excitation energy of Eq. (8.72) is written as

$$\Delta E[\nu] = \left(\frac{2\pi}{L}\right)^2 \left[\sum_{\kappa} \kappa^2 \nu(\kappa) + \frac{\lambda}{2} \sum_{\kappa} \sum_{\kappa'} \left|\kappa - \kappa'\right| \nu(\kappa) \nu\left(\kappa'\right)\right]. \tag{8.79}$$

Thermodynamics is derived by combining the entropy of the system. Let $G$ be the number of one-particle states in a given many-particle system. A block $\alpha$ in the momentum space contains $G_\alpha$ single-particle states specified by $\kappa_i$, and we assume $1 \ll G_\alpha \ll G$. The contribution to energy from each block is accounted for by a representative value $\kappa_\alpha$ inside the block. Then the averaged distribution function is defined by

$$\bar{\nu}_\alpha = \frac{N_\alpha}{G_\alpha}, \quad N_\alpha = \sum_{\kappa \in \alpha} \nu(\kappa). \tag{8.80}$$

The excitation energy is rewritten as the sum of block contributions as

$$\Delta \bar{E} = \left(\frac{2\pi}{L}\right)^2 \left[\sum_{\alpha} \kappa_\alpha^2 N_\alpha + \frac{\lambda}{2} \sum_{\alpha} \sum_{\alpha'} |\kappa_\alpha - \kappa_{\alpha'}| N_\alpha N_{\alpha'}\right]. \tag{8.81}$$

With this setting we derive the number of states $W_\alpha$ in each block assuming bosonic particles. Since identical bosons have no restriction on occupation of $\kappa_\alpha$ inside the block, we obtain

$$W_\alpha = \frac{(G_\alpha + N_\alpha - 1)!}{(G_\alpha - 1)! N_\alpha!}. \tag{8.82}$$

The entropy $S$ of the whole system is the sum of block contributions. Using the Stirling formula $\ln M! \sim M \ln M$ for a large number $M$, we obtain the standard formula for bosons:

$$S = \sum_{\alpha} \left[(\bar{\nu}_\alpha + 1) \ln(\bar{\nu}_\alpha + 1) - \bar{\nu}_\alpha \ln \bar{\nu}_\alpha\right] G_\alpha, \tag{8.83}$$

which will be extended later to general statistics. Combining with the internal energy given by Eq. (8.81) we can derive the free energy $F = E - TS$. It turns out that the thermodynamic potential $\Omega = E - TS - \mu N$ is more convenient for applying the variational (stationary) condition to determine $\nu(\kappa)$.

In the thermodynamic limit $N \to \infty$, the block quantities such as $2\pi\kappa_\alpha/L \to k$, $2\pi G_\alpha/L \to \mathrm{d}k$, $\bar{\nu}_\alpha \to \nu(k)$ are regarded as continuous variables. Then the thermodynamic potential is given by

$$\begin{aligned}
&\Omega[\nu(k)]/L - \pi^2\lambda^2 n^3/3 \\
&= \frac{1}{2\pi}\int_{-\infty}^{\infty} \mathrm{d}k\,(k^2-\mu)\nu(k) + \frac{\lambda}{4\pi}\int_{-\infty}^{\infty}\mathrm{d}k\int_{-\infty}^{\infty}\mathrm{d}k'\,|k-k'|\nu(k)\nu(k') \\
&- \frac{T}{2\pi}\int_{-\infty}^{\infty}\mathrm{d}k\,[(\nu(k)+1)\ln(\nu(k)+1) - \nu(k)\ln\nu(k)], \qquad (8.84)
\end{aligned}$$

where the second term in the LHS is the contribution from the ground state. The density $n$ is given by

$$n = \int \frac{\mathrm{d}k}{2\pi}\nu(k).$$

The equilibrium distribution function $\nu(k)$ is determined by the stationary condition for $\Omega$, which requires $\delta\Omega/\delta\nu(k) = 0$. The solution is given by

$$\nu(k) = \left\{\exp\left[\left(\tilde{\epsilon}(k)-\mu\right)/T\right]-1\right\}^{-1}, \qquad (8.85)$$

$$\tilde{\epsilon}(k) = k^2 + \lambda\int_{-\infty}^{\infty}\mathrm{d}k'|k-k'|\,\nu\left(k'\right). \qquad (8.86)$$

Equation (8.85) takes the form of the Bose distribution. However, the energy $\tilde{\epsilon}(k)$ includes the interaction effect, and is determined self-consistently by Eq. (8.86). Hence we regard $\tilde{\epsilon}(k)$ as an effective one-particle energy analogous to the quasi-particle energy in the Fermi liquid theory. We emphasize that the present scheme is exact for all energies, which is in contrast with the Fermi liquid theory. In this way, thermodynamic quantities can in principle be derived. Actually, the treatment is much more simplified with the idea of exclusion statistics which is explained in the next section.

## 8.6 Exclusion Statistics

We have indexed the neighboring momenta as $\kappa_i \geq \kappa_{i+1}$. As a result the rapidities $\tilde{\kappa}_i$ are constrained as

$$\tilde{\kappa}_i - \tilde{\kappa}_{i+1} = \kappa_i - \kappa_{i+1} + \lambda \geq \lambda. \qquad (8.87)$$

This restriction is regarded as a generalization of the Pauli exclusion principle, and is called the exclusion statistics [12]. We note here the difference from the exchange statistics including the fractional one as discussed in Sect. 8.2. In one-dimensional system, any spatial exchange avoiding the collision is not possible. Hence the statistics on this basis does not have a meaning, in contrast with higher dimensions. Hence the distinction including the one between ideal Fermi and Bose particles is made most naturally in terms of the degree of exclusion.

From now on we use another notation of the rapidity $p_i = 2\pi\tilde{\kappa}_i/L$ and the momentum $k_i = 2\pi\kappa_i/L$, both of which are regarded as continuous variables. Then Eq. (8.76) is written as

$$p_i = k_i + \frac{\pi\lambda}{L}\sum_{j(\neq i)} \mathrm{sgn}(p_i - p_j). \tag{8.88}$$

We introduce the rapidity distribution function by

$$\rho(p) = \frac{2\pi}{L}\sum_i \delta(p - p_i). \tag{8.89}$$

Then Eq. (8.88) can also be expressed as

$$p_i = k_i + \frac{\lambda}{2}\int \mathrm{d}p'\mathrm{sgn}\left(p_i - p'\right)\rho\left(p'\right). \tag{8.90}$$

Using the relation

$$\rho(p)\mathrm{d}p = \nu(k)\mathrm{d}k = (2\pi/L)\mathrm{d}N, \tag{8.91}$$

we take the $p$-derivative of both sides of Eq. (8.90) to obtain

$$1 = \frac{\mathrm{d}k}{\mathrm{d}p} + \lambda\rho(p) = \frac{\rho(p)}{\nu(k)} + \lambda\rho(p), \tag{8.92}$$

which is equivalently written as

$$\rho(p)^{-1} = \nu(k)^{-1} + \lambda. \tag{8.93}$$

By use of Eq. (8.93), $\nu(k)$ and $\rho(p)$ are expressed in terms of each other as

$$\nu(k) = \frac{\rho(p)}{1 - \lambda\rho(p)}, \quad \rho(p) = \frac{\nu(k)}{1 + \lambda\nu(k)}. \tag{8.94}$$

Making use of the rapidity distribution function, the energy is given in the free-particle form:

$$E = \sum_i p_i^2 = L\int \frac{\mathrm{d}p}{2\pi}p^2\rho(p). \tag{8.95}$$

Furthermore the entropy derived in Eq. (8.83) is equivalently written as

$$S = L \int_{-\infty}^{\infty} \frac{dp}{2\pi} \left[ (\rho + \rho^*) \ln (\rho + \rho^*) - \rho \ln \rho - \rho^* \ln \rho^* \right], \tag{8.96}$$

where the argument $p$ has been omitted in the integrand, and $\rho^*$ is given by

$$\rho^* = 1 - \lambda \rho, \tag{8.97}$$

which has the meaning of the hole distribution function. Derivation of Eq. (8.96), starting from Eq. (8.83) for bosons, is the subject of Problem 8.5. In the special case $\lambda = 1$, we obtain $\rho^* = 1 - \rho$, and the entropy reduces to that of fermions.

It is possible to formulate the exclusion statistics in a way independent of a particular model. We use $G$ defined as the number of one-particle states in a given many-particle system. We consider the influence of occupation of certain states on further occupation of particles. The number of available states is written as $D$. In the case of bosons, any occupation does not influence $D$, so that we have $D = G$. For fermions, the available number $D$ under $m$ occupied states is given by $D = G - m$ according to the Pauli principle. Hence in the $N$-particle systems, the available number for the $N$-th particle is given by

$$D = \begin{cases} G & \text{(bosons)}, \\ G - N + 1 & \text{(fermions)}. \end{cases} \tag{8.98}$$

These relations can be expressed in a unified manner with the use of a statistical parameter $g$ as

$$\Delta D = -g \Delta N, \tag{8.99}$$

where the change $\Delta D$ of available states is proportional to the change $\Delta N$ of the particles. The bosons have $g = 0$, while fermions have $g = 1$. In the Sutherland model, the exclusion statistics with $g = \lambda$ is realized. Here we extend the range of $g$ to be any non-negative number.

Let us proceed to determine the entropy associated with exclusion statistics. We consider a block $\alpha$ in the *rapidity space* which contains many ($G_\alpha \gg 1$) rapidities, but $G_\alpha$ is still much smaller than the number $G$ of the whole system. We have used similar idea to derive the bosonic entropy working in the *momentum space* with the use of $\nu(\kappa)$ as in Eq. (8.80). We now prefer rapidity to momentum and accordingly use $\rho(p)$. The contribution to energy from each block is accounted for by a representative value of rapidities inside the block. Then the averaged distribution function $\rho_\alpha$ is defined by

$$\rho_\alpha = \frac{N_\alpha}{G_\alpha}, \quad N_\alpha = \sum_{p \in \alpha} \rho(p). \tag{8.100}$$

We define the hole distribution function $\rho_\alpha^*$ by

$$\rho_\alpha^* = \frac{D_\alpha}{G_\alpha} = 1 - g\rho_\alpha, \tag{8.101}$$

with use of Eq. (8.99). It is consistent with Eq. (8.97) defined from a different route for $g = \lambda$. The number of states $W_\alpha$ in each block is given by

$$W_\alpha = \frac{(D_\alpha + N_\alpha - 1)!}{(D_\alpha - 1)!N_\alpha!} = \exp(S_\alpha), \tag{8.102}$$

where $S_\alpha$ is the entropy of the block. The entropy $S$ of the whole system is the sum of block contributions. Using the Stirling formula we obtain the result

$$S = \sum_\alpha G_\alpha \left[ (\rho_\alpha + \rho_\alpha^*) \ln (\rho_\alpha + \rho_\alpha^*) - \rho_\alpha \ln \rho_\alpha - \rho_\alpha^* \ln \rho_\alpha^* \right], \tag{8.103}$$

which is consistent with Eq. (8.96) obtained for a specific model.

## 8.7 Thermodynamics with Exclusion Statistics

We now derive thermodynamics with exclusion statistics without referring to a specific Hamiltonian such as the Sutherland model. The thermodynamic potential $\Omega$ of free particles with the statistical parameter $g$ is given by

$$\Omega/L = \int \frac{\mathrm{d}p}{2\pi}(\epsilon - \mu)\rho - T \int_{-\infty}^{\infty} \frac{\mathrm{d}p}{2\pi} \left[ (\rho + \rho^*) \ln (\rho + \rho^*) - \rho \ln \rho - \rho^* \ln \rho^* \right], \tag{8.104}$$

with $\rho^* = 1 - g\rho$. In the Sutherland model, the corresponding parameters are given by $\epsilon = p^2$ and $g = \lambda$.

The distribution function $\rho(p)$ is determined by the stationary condition $\delta\Omega/\delta\rho(p) = 0$, which leads to

$$\ln (1 + w) - g \ln \left(1 + w^{-1}\right) = \frac{\epsilon - \mu}{T}, \tag{8.105}$$

with $w \equiv \rho^*/\rho$. Equation (8.105) is also written as

$$\exp\left[(\epsilon - \mu)/T\right] = w^g (1 + w)^{1-g}. \tag{8.106}$$

The distribution functions are expressed in terms of $w$ and $g$ as

$$\rho = 1/(w + g), \quad \rho^* = w/(w + g), \tag{8.107}$$

with use of $\rho^* = 1 - g\rho$. Then $\Omega$ is compactly written as

$$\Omega/L = -T \int \frac{\mathrm{d}p}{2\pi} \ln\left[1 + w^{-1}\right]. \tag{8.108}$$

Let us check Eq. (8.108) for some special cases. In the case $g = 0$, Eq. (8.106) gives

$$w = \exp\left[(\epsilon - \mu)/T\right] - 1, \tag{8.109}$$

which leads to the bosonic distribution function

$$\rho = \frac{1}{\exp\left[(\epsilon - \mu)/T\right] - 1}, \quad \rho^* = 1 \tag{8.110}$$

and the thermodynamic potential

$$\Omega = T \int \frac{\mathrm{d}p}{2\pi} \ln\left[1 - \exp[-(\epsilon - \mu)/T]\right]. \tag{8.111}$$

In another special case $g = 1$, we use the relation $w = \exp\left[(\epsilon - \mu)/T\right]$. Then Eq. (8.106) leads to

$$\rho = \frac{1}{\exp\left[(\epsilon - \mu)/T\right] + 1}, \quad \rho^* = \frac{1}{\exp\left[-(\epsilon - \mu)/T\right] + 1}, \tag{8.112}$$

and

$$\Omega = -T \int \frac{\mathrm{d}p}{2\pi} \ln\left[1 + \exp[-(\epsilon - \mu)/T\right], \tag{8.113}$$

which obviously describes free fermions.

For other special cases of $g = 1/2$ and 2, we can derive $w$ as the solution of the quadratic equation. Moreover, these particular cases exhibit some typical characteristics of the exclusion statistics. With $g = 2$, we obtain the solution of Eq. (8.106) as

$$w = \frac{1}{2} e^{(\epsilon-\mu)/T} \left[1 + \sqrt{1 + 4e^{-(\epsilon-\mu)/T}}\right] \equiv \frac{1}{2} e^{(\epsilon-\mu)/T} \left(1 + R_2\right), \tag{8.114}$$

$$\rho = \frac{1}{2}\left(1 - \frac{1}{R_2}\right), \quad \rho^* = \frac{1}{R_2}. \tag{8.115}$$

Note that the restriction $\rho \leq 1/2$ with $R_2 \geq 1$ is stronger than the restriction $\rho \leq 1$ for fermions. In this sense such particles with $g > 1$ are sometimes called ultrafermions.

On the other hand, with $g = 1/2$ we obtain from Eq. (8.106)

$$w = \frac{1}{2}\left[-1 + \sqrt{1 + 4e^{2(\epsilon-\mu)/T}}\right] \equiv \frac{1}{2}(-1 + R), \tag{8.116}$$

$$\rho = 2/R, \quad \rho^* = 1 - 1/R. \tag{8.117}$$

Now the restriction $\rho \le 2$ with $R \ge 1$ is weaker than $\rho \le 1$ for fermions. Hence these particles with $g = 1/2$ have an exclusion property between fermions and bosons. They are sometimes called semions.

For general values of $g$, it is hard to obtain distribution functions in a closed form. However, there is a remarkable correspondence between $\rho$ and $\rho^*$ as

$$\rho \to g\rho^*, \quad w \to 1/w, \quad g \to 1/g, \quad \epsilon - \mu \to -(\epsilon - \mu)/g, \tag{8.118}$$

which may be called the particle-hole duality. The derivation is the subject of Problem 8.6.

## 8.8 Quasi-Particles and Quasi-Holes in Thermodynamics

We shall gain another insight into the quasi-particle picture with exclusion statistics. It is convenient to introduce the quantity $p_F$ by analogy with the Fermi momentum. The distribution functions at $T = 0$ are given by

$$\begin{cases} \rho(p) = 1/g, \ \rho^*(p) = 0, \ \left(|p| < p_F \equiv \mu^{1/2}\right) \\ \rho(p) = 0, \quad \rho^*(p) = 1, \ (|p| > p_F) \end{cases} \tag{8.119}$$

which are step functions like those of particles and holes with the Fermi statistics. The chemical potential is related to the density by $\mu = p_F^2 = (\pi n g)^2$.

The particle and hole excitations correspond to $|p| > p_F$ and $|p| < p_F$, respectively. Accordingly for general temperature we introduce the parameters as

$$w_p \equiv w, \ (\text{for } |p| > p_F), \quad w_h \equiv w^{-1}, \ (\text{for } |p| < p_F). \tag{8.120}$$

Then Eq. (8.105) is rewritten as

$$\epsilon_p(p)/T \equiv \left(p^2 - \mu\right)/T = \log\left(1 + w_p\right) - g_p \log\left(1 + w_p^{-1}\right), \tag{8.121}$$

$$\epsilon_h(p)/T \equiv \left(\mu - p^2\right)/(gT) = \log\left(1 + w_h\right) - g_h \log\left(1 + w_h^{-1}\right), \tag{8.122}$$

where we have distinguished exclusion properties of particles and holes by defining $g_\mathrm{p} = g$ and $g_\mathrm{h} = 1/g$. Using Eqs. (8.121) and (8.122), we write $\Omega$ as

$$\frac{\Omega}{L} = -T \int_{|p|>p_\mathrm{F}} \frac{\mathrm{d}p}{2\pi} \ln\left(1 + w_\mathrm{p}^{-1}\right) - \frac{T}{g} \int_{|p|<p_\mathrm{F}} \frac{\mathrm{d}p}{2\pi} \ln\left(1 + w_\mathrm{h}^{-1}\right), \tag{8.123}$$

where the first term in the RHS represents particle excitations, while the second term represents hole excitations. It is instructive to compare with Eq. (8.108) which does not divide the rapidity regions at $p_\mathrm{F}$. In Eq. (8.123), the hole part has the factor $1/g$ which reflects the scaling of rapidities.

We next derive the effective charge of particles and holes. This is simply achieved by looking at the coefficient to the chemical potential $\mu$. The effective charge $\mathrm{e_p}$ of a particle excitation is $\mathrm{e_p} = 1$ independent of the statistical parameter, since the coefficient of $\epsilon_\mathrm{p}$ is always $-1$. On the other hand, the coefficient of $\epsilon_\mathrm{h}$ is $1/g$, which implies the effective charge $\mathrm{e_h} = -1/g$ for holes. This result generalizes the particle–hole symmetry for $g = 1$ for fermions. Namely, with an integer $g \neq 1$, the corresponding holes have a fractional charge.

In closing this chapter, we summarize the main difference between the Sutherland model and the FQHE systems in two dimensions. In the Sutherland model, and more generally in systems with exclusion statistics, there is duality between particles and holes, but no particle–hole symmetry. The absence is easily understood from the parabolic single-particle spectrum that has no particle–hole symmetry. On the contrary, in the FQHE systems, quasi-particles obey the fractional *exchange* statistics, where the phase of the wave function by the interchange of two quasi-particles changes by $\pm\pi/q$, rather than $\pm\pi$ for electrons. The quasi-electrons and quasi-holes in Laughlin wave functions have the same fractional charge with magnitude $e/q$, which is not the case in exclusion statistics. This particle–hole symmetry is associated with the absence of kinetic energy in the ground Landau level. Provided higher Landau levels can be neglected, the ground Landau level has the particle–hole symmetry in the form of complete degeneracy.

From another viewpoint, one can compare construction of wave functions in the excited states with only quasi-particles or only quasi-holes. We have seen that constructions for quasi-holes are similar in both FQHE and Sutherland systems, as given in Eqs. (8.39) and $\Phi_\mathrm{g}\Phi$ in p. 198, respectively. However, for the particle excitation in the Sutherland model, there is no construction like Eq. (8.42). Namely, both particle and hole excitations are described in the form $\Phi_\mathrm{g}\Phi$. Hence fractional statistics is a kind of family where each member has something in common, but also has a distinct character.

## Problems

**8.1** Using the Gauss law for two-dimensional systems, derive Eqs. (8.34) and (8.35).

**8.2** Show that Eq. (8.57) describes the Slater determinant in the case of $\lambda = 1$.

**8.3** Prove the identity involving cotangent functions as given by Eq. (8.62).

**8.4** Prove the rapidity relation given by Eq. (8.73). With the use of this relation, show that the ground state energy is given by Eq. (8.78), as the sum of free quasi-particle contributions.

**8.5** Derive the entropy given by Eq. (8.96) starting from the expression in Eq. (8.84) for bosons.

**8.6** Express the duality relation between the distribution functions $\rho$ and $\rho^*$ for a general value of $g$.

## Solutions to Problems

**Problem 8.1**
With the unit charge $Q$ at the origin, the electric field $\boldsymbol{E}(\boldsymbol{r})$ at $\boldsymbol{r}$ is pointing radially and its magnitude depends only on $r = |\boldsymbol{r}|$. Hence the Gauss law takes the form:

$$2\pi r E(r) = Q, \tag{8.124}$$

where we have chosen the unit so that the coefficient in the RHS does not involve $\pi$. We obtain the scalar potential $\varphi(r)$ as

$$\varphi(r) = -\int_a^r \mathrm{d}r\, E(r) = -\frac{Q}{2\pi} \ln\left(\frac{r}{a}\right), \tag{8.125}$$

where $a$ is the length at which $\varphi(r) = 0$. Choosing $a = 2\ell_B$, we obtain $V_2$ given by Eq. (8.34).

To obtain the potential due to the neutralizing background with homogeneous charge density $-\rho Q$, we consider the integrated charge inside the circle with radius $r$. Then the Gauss law gives

$$2\pi r E(r) = -\pi r^2 \rho Q. \tag{8.126}$$

The potential now is given by

$$\varphi(r) = -\int_0^r \mathrm{d}r\, E(r) = \frac{Q}{4}\rho r^2. \tag{8.127}$$

Putting $r = 2\ell_B|z|$ we obtain $V_1$ in Eq. (8.35).

**Problem 8.2**
It is convenient to introduce the complex coordinate $z_j = \exp(2\pi \mathrm{i} x_j/L)$, which maps the original coordinate $x_j$ to a point $z_j$ on the unit circle in the complex plane.

The periodic boundary condition is now built in. Using the expression

$$\sin\left[\pi\left(x_j - x_k\right)/L\right] = \frac{1}{2\mathrm{i}\sqrt{z_j z_k}}\left(z_j - z_k\right), \tag{8.128}$$

the wave function is rewritten as

$$\Psi_{\mathrm{g}} = (2\mathrm{i})^{\lambda N(N-1)} \prod_j z_j^{\lambda N(N-1)} \prod_{j<k} \left(z_j - z_k\right)^{\lambda}. \tag{8.129}$$

In the case of $\lambda = 1$, the product of $z_j - z_k$ reduces to the Vandermonde determinant. With a shift of the total momentum by $\prod_j z_j^{\lambda N(N-1)}$, the determinant is equivalent to the Slater determinant where the plane-wave states are occupied up to the Fermi momentum.

**Problem 8.3**
The identity is equivalent to another expression

$$\cot\alpha \cot\beta + \cot\beta \cot\gamma + \cot\gamma \cot\alpha = 1 \tag{8.130}$$

with $\alpha + \beta + \gamma = 0$. The two terms on the LHS sum up into

$$(\cot\alpha + \cot\gamma)\cot\beta = (\cot\alpha + \cot\gamma)\frac{1 - \cot\alpha\cot\gamma}{\cot\alpha + \cot\gamma}, \tag{8.131}$$

where the addition formula for the cotangent function is used together with $\beta = -\alpha - \gamma$. By factoring out $\cot\alpha + \cot\gamma$, we obtain Eq. (8.130).

**Problem 8.4**
In the LHS of Eq. (8.73), the number of times for $\kappa_i$ to appear with the plus sign is $N - i$, from $\kappa_i - \kappa_{i+1}$ to $\kappa_i - \kappa_N$ for any $i$. On the other hand, the negative sign for $\kappa_i$ appears $i - 1$ times, from $\kappa_1 - \kappa_i$ to $\kappa_{i-1} - \kappa_i$. By combining both cases we obtain the number $N - i - (i - 1) = N + 1 - 2i$ for the coefficient in the RHS of Eq. (8.73). The same argument applies to Eq. (8.77). We then use the formula

$$\sum_{i=1}^{N}\left(\frac{N+1}{2} - \mathrm{i}\right)^2 = \frac{N}{12}\left(N^2 - 1\right), \tag{8.132}$$

which corresponds to the summation of $\tilde{\kappa}_{i,0}^2$. With multiplication of $(2\pi/L)^2$ and shifting the origin of energy by $E_{0,N}$ we recover Eq. (8.78).

Note that the rapidity $p$ and the momentum $k$ in the thermodynamic limit are related by

$$p = \frac{1}{2}\frac{\partial \epsilon(k)}{\partial k}, \tag{8.133}$$

$$\epsilon(k) = k^2 + \lambda \int dk' |k - k'| \nu(k'), \tag{8.134}$$

which shows the character of $p$ as the velocity of a quasi-particle.

**Problem 8.5**

In order to eliminate $\nu$ in favor of $\rho$, we find it most convenient to use Eq. (8.93). As a preliminary we arrange the bosonic entropy as

$$(\nu + 1)\ln(\nu + 1) - \nu \ln \nu = \nu \left[ \left(1 + \frac{1}{\nu}\right) \ln \left(1 + \frac{1}{\nu}\right) - \frac{1}{\nu} \ln \frac{1}{\nu} \right]. \tag{8.135}$$

Furthermore the factor $dk/dp = \rho(p)/\nu(k)$ enters through the change $k \to p$ of the integration variable. As a result, the first factor $\nu$ in Eq. (8.135) is replaced by $\rho$. Then we use the relation $\nu(k)^{-1} = \rho(p)^{-1} - g$ repeatedly, and use $\rho^* = 1 - g\rho$ after all $\nu$'s are eliminated. In this way we obtain Eq. (8.96) after a little rearrangement of terms.

**Problem 8.6**

We rewrite the relation of distribution functions so that the particles and holes exchange their roles. Dividing both sides of Eq. (8.105) by $-g$, we obtain

$$-\frac{\epsilon - \mu}{gT} = g^{-1} \ln w^{-1} + \left(1 - g^{-1}\right) \ln \left(1 + w^{-1}\right). \tag{8.136}$$

Furthermore Eq. (8.107) leads to the relation

$$g\rho^* = \frac{gw}{w + g} = \frac{1}{w^{-1} + g^{-1}}. \tag{8.137}$$

Hence we recognize the correspondence

$$\rho \to g\rho^*, \quad w \to 1/w, \quad g \to 1/g, \quad \epsilon - \mu \to -(\epsilon - \mu)/g \tag{8.138}$$

between the particles and holes in their distribution functions. Equivalently this relation is summarized as

$$\rho^* \left( \frac{\epsilon - \mu}{T}; g \right) = \frac{1}{g} \rho \left( \frac{\mu - \epsilon}{gT}; \frac{1}{g} \right), \tag{8.139}$$

as a kind of duality.

## References


1. Dirac, P.A.M.: Proc. R. Soc. Lond. A **126**, 360 (1930)
2. Nambu, Y., Jona-Lasinio, G.: Phys. Rev. **122**, 345 (1961)
3. Tsui, D.C., Stormer, H.L., Gossard, A.C., Laughlin, R.B.: Phys. Rev. Lett. **48**, 1559 (1982)
4. Laughlin, R.B.: Phys. Rev. Lett. **50**, 1395 (1983)
5. Aharonov, Y., Bohm, D.: Phys. Rev. **115**, 485 (1959)
6. Berry, M.V.: Proc. R. Soc. Lond. Ser. A **392**, 45 (1984)
7. Vanderbilt, D.: Berry Phases in Electronic Structure Theory. Cambridge University Press, Cambridge (2018)
8. Arfken, G., Frank, H.W., Harris, E.: Mathematical Methods for Physicists, 7th edn. Academic, Cambridge (2012)
9. Arovas, D., Schrieffer, J.R., Wilczek, F.: Phys. Rev. Lett. **53**, 722 (1984)
10. Kuramoto, Y., Kato, Y.: Dynamics of One-Dimensional Quantum Systems: Inverse-Square Interaction Models. Cambridge University Press, Cambridge (2009)
11. Sutherland, B.: Phys. Rev. A **4**, 2019 (1971)
12. Haldane, F.D.M.: Phys. Rev. Lett. **67**, 937 (1991)

# Chapter 9
# Many-Body Perturbation Theory

**Abstract** This chapter deals with the field theoretical method to many-body systems, especially the path-integral formalism for fermions with use of Grassmann numbers. Green function appears as a natural element in Feynman diagram that represents a group of terms in many-body perturbation theory. The graphical method appeals to intuition, and is convenient to reorganize the perturbation series to infinite order. The discussion in this chapter is rather technical, but is most useful in understanding physical concepts including the dynamical mean field theory to be explained in the next chapter.

## 9.1 Grassmann Numbers

We have discussed the coherent state already in Chap. 5 as a superposition of states with different number $n$ of bosonic particles. The phase factor $\exp(in\theta)$ in superposition has the fixed phase $\theta$, while $n$ taking positive integers. As shown in Eq. (5.4), the coherent state is the eigenstate of the annihilation operator that changes the particle number $n$ to $n-1$. Likewise, we can construct the fermionic coherent state by superposition of occupation numbers 0 and 1 for each state. The resultant state is very convenient for dealing with many fermion states.

It is necessary to introduce a special quantity called the Grassmann number that anticommutes with another Grassmann number. Namely, the Grassmann numbers $\eta_1$ and $\eta_2$ satisfy

$$\eta_1\eta_2 = -\eta_2\eta_1. \tag{9.1}$$

The anticommutation property holds also in the product with fermion creation or annihilation operator. On the other hand, a Grassmann number commutes with a c-number.



Y. Kuramoto, *Quantum Many-Body Physics*, Lecture Notes in Physics 934,
https://doi.org/10.1007/978-4-431-55393-9_9

We can define the complex conjugate so as to satisfy $(\eta^*)^* = \eta$ and $(\eta_1^*\eta_1)^* = \eta_1^*\eta_1$, as in the case of c-numbers. The construction of $(\eta_1\eta_2)^*$ is the subject of Problem 9.1. The derivative of the Grassmann number is defined as

$$\frac{\partial}{\partial \eta_i}\eta_j = \delta_{ij}, \tag{9.2}$$

which looks similar to ordinary derivative, except for $\partial\eta_i$ being also a Grassmann number.

Let us proceed to the integral of a function of Grassmann numbers. It suffices to define the integrals of 1 and $\eta$ since second and higher order terms of $\eta$ vanish. We impose the condition that the integral remains invariant against the linear transformation $\eta \to \eta + \xi$ with use of a constant Grassmann number $\xi$. This condition is satisfied by the definitions

$$\int \mathrm{d}\eta\eta = 1, \quad \int \mathrm{d}\eta 1 = 0, \tag{9.3}$$

where the first definition determines the norm of the integral, which is a c-number. The 0 in the second integral is a Grassmann number, but is indistinguishable from the ordinary 0. The second definition guarantees the invariance against linear transformation. Comparison of Eqs. (9.2) and (9.3) shows that derivative and integral of Grassmann numbers give identical results.

As the most important case, we take the integral of Gaussian functions of Grassmann numbers. We start with the identity: $\exp(-\eta^* A\eta) = 1 - \eta^* A\eta$, where $A$ is an arbitrary c-number. This identity can be confirmed by expanding the exponential. Thus we immediately obtain

$$\int \mathrm{d}\eta^*\mathrm{d}\eta \exp(-\eta^* A\eta) = A, \tag{9.4}$$

where we have used Eq. (9.3). The Gaussian integration is generalized as

$$\int \mathrm{d}\eta^*\mathrm{d}\eta \exp\left(-\eta^* A\eta + \xi^*\eta + \eta^*\xi\right) = A \exp\left(\xi^* A^{-1}\xi\right), \tag{9.5}$$

with $\xi$ being any Grassmann number. This result is obtained either by expanding the exponential or by the transformation $\eta \to \eta - A^{-1}\xi,\ \eta^* \to \eta - \xi^* A^{-1}$.

Let us derive the multi-variable version of the Gaussian integration. With $A$ being an $N \times N$ Hermitian matrix of c-number elements, the following result holds:

$$\int \prod_i \mathrm{d}\eta_i^*\mathrm{d}\eta_i \exp\left(-\sum_{ij}\eta_i^* A_{ij}\eta_j\right) = \det A. \tag{9.6}$$

The result is obvious if $A$ is a diagonal matrix, since the product of eigenvalues is just $\det A$. In a general case, we take a suitable unitary transformation to diagonalize $A$, and use the new integral variables which result from the same unitary transformation. Similarly, the multi-variable version of Eq. (9.5) is given by

$$\int \prod_i \mathrm{d}\eta_i^* \mathrm{d}\eta_i \exp\Big[-\sum_{ij} \eta_i^* A_{ij} \eta_j + \sum_i \left(\xi_i^* \eta_i + \eta_i^* \xi_i\right)\Big]$$

$$= \det A \exp\left(-\sum_{ij} \xi_i^* G_{ij} \xi_j\right) \equiv Z\left(\xi^*, \xi\right), \tag{9.7}$$

where we have introduced the $N \times N$ matrix $G = -A^{-1}$.

The average of fluctuating Grassmann variables is defined by

$$\langle \eta_i^* \eta_j \rangle_0 \equiv \int \prod_m \mathrm{d}\eta_m^* \mathrm{d}\eta_m \, \eta_i^* \eta_j \exp\left(-\sum_{ij} \eta_i^* A_{ij} \eta_j\right) / \det A. \tag{9.8}$$

This quantity is evaluated in terms of a second derivative of $Z(\xi^*, \xi)$ as

$$\langle \eta_i^* \eta_j \rangle_0 \equiv -\lim_{\xi^*, \xi \to 0} \frac{1}{Z} \frac{\partial^2}{\partial \xi_i \partial \xi_j^*} Z(\xi^*, \xi) = G_{ji}. \tag{9.9}$$

Hence we recognize the correspondence $\eta_j \to \partial/\partial \xi_j^*$, $\eta_i^* \to -\partial/\partial \xi_i$. With use of Eq. (9.7) for $\xi$-derivatives of higher orders, the average is generalized to the case of $2n$ Grassmann numbers. Namely, we obtain

$$\langle \eta_1^* \eta_{n+1} \eta_2^* \eta_{n+2}, \ldots, \eta_n^* \eta_{2n} \rangle_0 = \sum_P \operatorname{sgn} P \; G_{n+1,P(1)} G_{n+2,P(2)} G_{2n,P(n)}, \tag{9.10}$$

where $P$ represents the permutation of $n$ ($\leq N$) variables $(\eta_1^*, \eta_2^*, \ldots, \eta_n^*)$, and the sign factor $\operatorname{sgn} P = \pm 1$ follows according to whether the permutation is even or odd.

It is instructive to compare with the c-number Gaussian distribution. Let $\phi_i$ and $\nu_i$ with $i = 1, \ldots, N$ be a set of complex numbers. Then the Gaussian integration corresponding to Eq. (9.7) is given by

$$\int \prod_i \mathrm{d}\phi_i^* \mathrm{d}\phi_i \exp\Big[-\sum_{ij} \phi_i^* A_{ij} \phi_j + \sum_i \left(\nu_i^* \phi_i + \phi_i^* \nu_i\right)\Big]$$

$$= \pi^N \det A^{-1} \exp\left(-\sum_{ij} \nu_i^* G_{ij} \nu_j\right) \equiv Z_c(\nu^*, \nu), \tag{9.11}$$

with $G = -A^{-1}$ as in the Grassmann case. In the c-number case, we have to assume the convergence of the integral. The average is now given by

$$\langle \phi_i^* \phi_j \rangle_0 \equiv \lim_{\nu^*,\nu \to 0} \frac{1}{Z_c} \frac{\partial^2}{\partial \nu_i \partial \nu_j^*} Z_c(\nu^*, \nu) = G_{ji}. \tag{9.12}$$

The multi-variable correlation is given by

$$\langle \phi_1^* \phi_{n+1} \phi_2^* \phi_{n+2}, \ldots, \phi_n^* \phi_{2n} \rangle_0 = \sum_P G_{n+1,P(1)} G_{n+2,P(2)} G_{2n,P(n)}, \tag{9.13}$$

without the sign factor sgn $P$ in contrast to the Grassmann case. Later in Sect. 9.3, we shall use Eqs. (9.10) and (9.13) in proving the Wick's theorem, which plays an important role in many-body perturbation theory.

## 9.2 Coherent States for Fermions

We are now in a position to construct the coherent state for fermions. Let $f^\dagger$ the fermion creation operator of a certain state. The occupied state $|1\rangle$ and the vacant state $|0\rangle$ are related as $|1\rangle = f^\dagger|0\rangle$. The coherent state $|\eta\rangle$ is characterized by a Grassmann number $\eta$ and is defined by

$$|\eta\rangle \equiv \exp\left(f^\dagger \eta\right)|0\rangle = \left(1 + f^\dagger \eta\right)|0\rangle = |0\rangle - \eta|1\rangle. \tag{9.14}$$

It is clear from the definition that the coherent sate is a superposition of vacant and occupied states. Moreover, it is an eigenstate of the annihilation operator $f$ as indicated by

$$f|\eta\rangle = \eta|\eta\rangle. \tag{9.15}$$

The conjugate of this relation corresponds to $\langle \eta | = \langle \eta | f^\dagger$.

The inner product of two coherent states is given by

$$\langle \eta_1 | \eta_2 \rangle = \exp\left(\eta_1^* \eta_2\right). \tag{9.16}$$

Hence any function $F(f^\dagger, f) = af + bf^\dagger + cf^\dagger f$, including $f$ and $f^\dagger$, satisfies the relation:

$$\langle \eta_1 | F\left(f^\dagger, f\right) | \eta_2 \rangle = F\left(\eta_1^*, \eta_2\right) \exp\left(\eta_1^* \eta_2\right), \tag{9.17}$$

where the constants $a, b, c$ can be either c- or Grassmann numbers.

It is most important for practical purpose to have the completeness condition, or the representation of the unit operator in terms of coherent states. The set of coherent states forms a overcomplete set for the Fock space spanned by $|0\rangle$ and $|1\rangle$. The definition Eq. (9.14) leads to the completeness relation:

$$\int \mathrm{d}\eta^* \mathrm{d}\eta |\eta\rangle\langle\eta| \exp\left(-\eta^*\eta\right) = |0\rangle\langle 0| + |1\rangle\langle 1| = \mathbf{1}, \tag{9.18}$$

which is indispensable for representing the partition function in terms of path integrals to be discussed shortly.

As the simplest case of using Grassmann numbers, we take the Hamiltonian $H(f^\dagger, f) = Ef^\dagger f$. This trivial model most clearly demonstrates the characteristics of Grassmann numbers. The grand partition function $Z$ is given by $Z = 1 + \exp(-\beta E)$ with the origin of energy taken at the chemical potential. The same partition function is written in terms of Grassmann numbers by

$$Z = \int \mathrm{d}\eta_1^* \mathrm{d}\eta_1 \left\langle \eta_1 \left| \mathrm{e}^{-\beta H} \right| - \eta_1 \right\rangle \exp\left(-\eta_1^*\eta_1\right), \tag{9.19}$$

where we have used the completeness condition Eq. (9.18), together with

$$\langle n|\eta\rangle \left\langle \eta | e^{-\beta H} |n \right\rangle = \left\langle \eta | e^{-\beta H} |n \right\rangle \langle n| - \eta\rangle \tag{9.20}$$

for $n = 0, 1$. The minus sign in the state $|-\eta\rangle$ follows from the anticommutation property of $\eta$ and $\eta^*$. It will be shown below that the minus sign is also related to the antiperiodicity of the Green function with respect to the imaginary time $\beta$.

We can rewrite Eq. (9.19) in the form of path integral of the action, which is in fact valid for any Hamiltonian. For this purpose we take a large number $M$ ($\gg 1$) and put $\delta\tau = \beta/M$. Then we use the approximation $\exp(-\beta H) \sim (1 - \delta\tau H/M)^M$, which becomes exact in the limit of $M \to \infty$. This operation is called the Trotter (or Suzuki–Trotter) decomposition. We then insert the unit operator given by Eq. (9.18) between all adjacent pairs of $(1 - \delta\tau H/M)$, in total $(M - 1)$ times. The partition function is now given by

$$Z = \int \prod_{i=1}^{M} \mathrm{d}\eta_i^* \mathrm{d}\eta_i \, R\left(\eta_1^*, \eta_2\right) \exp\left[-\eta_1^*\left(\eta_1 - \eta_2\right)\right] R\left(\eta_2^*, \eta_3\right)$$
$$\times \cdots R\left(\eta_M^*, \eta_{M+1}\right) \exp\left[-\eta_M^*\left(\eta_M - \eta_{M+1}\right)\right], \tag{9.21}$$

where $R(\eta_1^*, \eta_2) = 1 - \delta\tau H(\eta_1^*, \eta_2) \sim \exp[-\delta\tau H(\eta_1^*, \eta_2)]$ and $\eta_{M+1} \equiv -\eta_1$ from Eq. (9.19). In the limit of large $M$, we may regard $\eta_i$ as $\eta(\tau)$ which is a function of the continuous variable $\tau$ in the range $0 < \tau < \beta$ with the boundary condition $\eta(\beta) = -\eta(0)$. Then $\eta(\tau)$ can be Fourier decomposed with odd Matsubara frequencies. Note that the maximum of the relevant Matsubara frequency should be smaller than the resolution $(\delta\tau)^{-1}$ of the Trotter decomposition. In other words, we may replace $\eta(\tau_i) - \eta(\tau_{i+1})$ by $\delta\tau\partial\eta(\tau)/\partial\tau$ provided $\eta(\tau)$ is slowly varying in the scale of $\delta\tau$. Under this condition the first part of the integrand in Eq. (9.21) is expressed as

$$R\left(\eta_1^*, \eta_2\right) \exp\left[-\eta_1^*\left(\eta_1 - \eta_2\right)\right] \sim \exp\left[-\delta\tau\, \eta^*(\tau)\left(\frac{\partial}{\partial\tau} + E\right)\eta(\tau)\right], \tag{9.22}$$

with $\tau \sim \tau_1 \sim \tau_2$ and similar expressions for general $\eta(\tau_i)$. The integration in Eq. (9.21) can be carried out to with use of Eq. (9.6) as

$$Z = \det_\tau\left[\delta\tau\left(\frac{\partial}{\partial\tau} + E\right)\right], \tag{9.23}$$

where the set of variables $\eta_i$ $(i = 1, 2, \ldots, M)$ is regarded as an $M$-dimensional vector, and $\det_\tau$ denotes the determinant of the $M \times M$ matrix in the $\tau$-space. The meaning of $\partial/\partial\tau$ will be made clear shortly by the Fourier transform of $\tau$ to Matsubara frequencies.

After taking the logarithm of both sides, we obtain trace of the matrix, instead of determinant, as given by

$$\ln Z = \mathrm{Tr}_\tau \ln\left[\frac{\beta}{M}\left(\frac{\partial}{\partial\tau} + E\right)\right] = -\beta\Omega, \tag{9.24}$$

where $\Omega$ is the thermodynamic potential. The trace operation is performed in the limit $M \to \infty$ with the basis set which is anti-periodic in the region $\tau \in [0, \beta]$. The orthonormal complete set $\exp(-\mathrm{i}\epsilon_n\tau)/\sqrt{\beta}$ with odd Matsubara frequencies $\epsilon_n = (2n+1)\pi T$ serves best for the basis set. The result is given, apart from a constant term, by

$$\beta\Omega = -\sum_n \ln\left(-\mathrm{i}\epsilon_n + E\right), \tag{9.25}$$

where $\partial/\partial\tau$ is replaced by its eigenvalue $-\mathrm{i}\epsilon_n$. The summation in Eq. (9.25) actually leads to a divergent result. The divergence originates from the replacement $\eta(\tau_i) - \eta(\tau_{i+1})$ in Eq. (9.21) by $\delta\tau\partial\eta(\tau)/\partial\tau$. As we have discussed between Eqs. (9.21) and (9.22), this approximation breaks down in the limit of large Matsubara frequency. Fortunately, this divergence can be managed in practical cases. Namely, if we take the second derivative of Eq. (9.25) with respect to $E$, we obtain the convergent result:

$$\frac{\partial^2\Omega}{\partial E^2} = \sum_n \frac{1}{(\mathrm{i}\epsilon_n - E)^2} = \frac{\beta^2 e^{\beta E}}{\left(e^{\beta E} + 1\right)^2}. \tag{9.26}$$

By two-fold integration with respect to $E$, we recover the physically meaningful part of Eq. (9.25). The actual procedure is the subject of Problem 9.2.

Generalizing the variable $\eta(\tau)$ used above, we introduce the Grassmann variable $\eta_\alpha(\tau)$ for each one-particle state $\alpha$. Note that $\alpha$ can be a continuous index such as the spatial coordinate in field theory. Since the integrand contains the functions $\eta_\alpha(\tau)$, the corresponding multiple $\eta$-integrals are called the functional integral, which is written symbolically as $\mathcal{D}\eta^*\mathcal{D}\eta$. By analogy with the path-integral theory for one-body quantum mechanics [1], the present version with Grassmann numbers is also called the path integral.

We are in the position to describe the procedure for the path-integral representation for a general Hamiltonian $H$ which may include mutual interactions. Let $F(\{f_\alpha^\dagger\},\{f_\alpha\})$ be a function of fermion operators where the creation operators always stand on the left of annihilation operators in each term. Such ordering is called the $N$-product as used in Eq. (7.21). As a generalization of Eq. (9.17), we obtain the matrix element

$$\left\langle\{\eta_\alpha\}\left|F\left(\left\{f_\alpha^\dagger\right\},\{f_\alpha\}\right)\right|\{\eta'_\alpha\}\right\rangle = F\left(\{\eta_\alpha^*\},\{\eta'_\alpha\}\right)\exp\left(\sum_\alpha \eta_\alpha^*\eta'_\alpha\right). \tag{9.27}$$

Then, as a generalization of Eq. (9.21), the partition function is written as

$$Z = \int \mathcal{D}\eta^*\mathcal{D}\eta \exp(-S) \tag{9.28}$$

$$S = \int_0^\beta \mathrm{d}\tau \left[\sum_\alpha \eta_\alpha^* \frac{\partial}{\partial\tau}\eta_\alpha + H(\tau)\right] \equiv \int_0^\beta \mathrm{d}\tau \mathcal{L}(\tau), \tag{9.29}$$

where $H(\tau)$ is the Hamiltonian in terms of Grassmann variables at $\tau$, in place of fermion operators. By analogy with the path-integral theory of quantum mechanics in real time, $S$ is called the action, and $\mathcal{L}(\tau)$ is called the Lagrangian.

## 9.3 Wick's Theorem

The Grassmann path integral is possible not only for the partition function but also for correlation functions including the Matsubara Green function. For a general Hamiltonian the Green function

$$G_{\beta\alpha}(\tau_2-\tau_1) = -\left\langle T_\tau f_\beta(\tau_2) f_\alpha^\dagger(\tau_1)\right\rangle = \left\langle T_\tau f_\alpha^\dagger(\tau_1) f_\beta(\tau_2)\right\rangle \tag{9.30}$$

with use of the Matsubara representation $f_\alpha(\tau) = \exp(\tau H) f_\alpha \exp(-\tau H)$ is equivalently given by

$$G_{\beta\alpha}(\tau_2-\tau_1) = \int \mathcal{D}\eta^* \mathcal{D}\eta\, \eta^*_\alpha(\tau_1)\eta_\beta(\tau_2) \exp(-S)/Z \equiv \langle \eta^*_\alpha(\tau_1)\eta_\beta(\tau_2)\rangle. \tag{9.31}$$

Note that the time ordering indicated by $T_\tau$ is absent in the Grassmann integral. In order to prove Eq. (9.31), we first assume $\tau_1 > \tau_2$ and rewrite $\langle f^\dagger_\alpha(\tau_1) f_\beta(\tau_2)\rangle$ as

$$Z^{-1}\mathrm{Tr}\exp[-(\beta-\tau_1)H]\, f^\dagger_\alpha \exp[-(\tau_1-\tau_2)H]\, f_\beta \exp[-\tau_2 H]. \tag{9.32}$$

Now we perform Tr operation using the completeness relation Eq. (9.18) and the Trotter decomposition for exponentials including $H$ as in Eq. (9.21). We thus obtain the first equality in Eq. (9.31). In the case of $\tau_1 < \tau_2$, we change the order of the fermion operators, and proceed in the same way to use the completeness relation and Trotter decomposition. The anticommutation property $\eta_\beta(\tau_2)\eta^*_\alpha(\tau_1) = -\eta^*_\alpha(\tau_1)\eta_\beta(\tau_2)$ compensates for the minus sign coming from the time ordering $T_\tau$. Moreover, we may include a fermionic source field $\xi_\alpha(\tau)^*$ and $\xi_\alpha(\tau)$ in the Lagrangian as $\xi^*\eta + \eta^*\xi$ with abbreviated notation. Then the second $\xi$-derivative of the partition function $Z(\xi^*, \xi)$ gives the Green function by analogy with Eq. (9.9). Namely, we obtain

$$\langle \eta^*_\alpha(\tau_1)\eta_\beta(\tau_2)\rangle = -\lim_{\xi^*,\xi\to 0} \frac{1}{Z}\frac{\delta^2}{\delta\xi_\alpha(\tau_1)\delta\xi^*_\beta(\tau_2)} Z(\xi^*,\xi) = G_{\beta\alpha}(\tau_2-\tau_1). \tag{9.33}$$

Thus we recognize the correspondence

$$\eta_\alpha(\tau) \to \delta/\delta\xi^*_\alpha(\tau), \quad \eta^*_\alpha(\tau) \to -\delta/\delta\xi_\alpha(\tau), \tag{9.34}$$

in taking the average of Grassmann numbers.

Let us take the special case of a free fermion Hamiltonian

$$H_0 = \sum_{\alpha\beta} h_{\alpha\beta} f^\dagger_\alpha f_\beta, \tag{9.35}$$

for which we can derive the Green functions exactly. In this example the path-integral formalism provides the simplest proof of the Wick's theorem, which is the key element in performing the perturbation theory for the general interacting case. The Green function is derived by the Gaussian integration, with close similarity to Eq. (9.8). The matrix $A$ in the Gaussian integration is interpreted as

$$A_{ij} \to \left(\delta_{\alpha\beta}\partial/\partial\tau + h_{\alpha\beta}\right)\delta\left(\tau-\tau'\right). \tag{9.36}$$

The matrix basis $(i, j)$ is now extended from the single-particle states $(\alpha, \beta)$ to include the continuous imaginary time $(\tau, \tau')$. Correspondingly, the Green function matrix $G = -A^{-1}$ also has double basis sets: $\alpha$ and $\tau$, and the matrix element is written as in Eq. (9.33). The Fourier transform to Matsubara frequencies results in the replacement $\partial/\partial\tau \to -\mathrm{i}\epsilon_n$, and the Green function matrix $G$ becomes diagonal in $\epsilon_n$. The explicit form as a matrix with the $\alpha$, $\beta$ basis is given by

$$G_0(\mathrm{i}\epsilon_n) = (\mathrm{i}\epsilon_n - h)^{-1}, \tag{9.37}$$

where the suffix 0 is put for emphasizing the free-fermion system.

Using Eq. (9.7) for $Z(\xi^*, \xi)$, it is now straightforward to derive the identity:

$$\begin{aligned}&\left\langle T_\tau f^\dagger(\tau_1) f(\tau_{n+1}) \ldots f^\dagger(\tau_n) f(\tau_{2n})\right\rangle_0 \\ &\quad = \sum_P \mathrm{sgn} P\, G_0\left(\tau_{n+1} - \tau_{P(1)}\right) \ldots G_0\left(\tau_{2n} - \tau_{P(n)}\right),\end{aligned} \tag{9.38}$$

where the average $\langle \cdots \rangle_0$ is taken for the free-fermion system, and the index $\alpha$ for single-particle states is omitted for notational simplicity. This result corresponds to Eq. (9.10) for Grassmann numbers, and is called the Wick's theorem. The original version of the theorem is for the field theory at zero temperature, and the extension to finite temperature has been achieved by Bloch and de Dominicis [2]. Hence the theorem is also called by their names.

We now derive the time-independent form of the Wick's theorem that is convenient to use in the Goldstone diagram method. Only in this paragraph we use the notations $c^\dagger(\tau)$ and $c(\tau)$ for creation and annihilation operators, and $F(\tau)$ to represent either of them. Presence of plural single-particle eigenstates $\alpha_j$ is taken into account simply as $F_j(\tau)$. We obtain the limiting case of Eq. (9.38) by relabeling the imaginary time as $\tau_1 > \tau_2 > \cdots > \tau_{2n} > 0$:

$$\begin{aligned}&\lim_{\tau_1 \to 0} \langle T_\tau F_1(\tau_1) F_2(\tau_2) \ldots F_{2n}(\tau_{2n})\rangle_0 = \langle F_1 F_2 \ldots F_{2n}\rangle_0 \\ &\quad = \sum_P{}' \mathrm{sgn} P\, \langle F_{P1} F_{P2}\rangle_0 \ldots \langle F_{P(2n-1)} F_{P(2n)}\rangle_0,\end{aligned} \tag{9.39}$$

where the summation over $P$ is restricted to such permutations that satisfy $P(2j-1) < P(2j)$ for $j = 1, \ldots, n$ and $P1 < P3 < \cdots < P(2n-1)$, which implies $P1 = 1$. The average $\langle F_{Pi} F_{P(i+1)}\rangle_0$ becomes either the particle occupation number $f(h_\alpha)$ for if $F_{Pi} = c_\alpha^\dagger$, $F_{P(i+1)} = c_\alpha$, or the hole occupation number $f(-h_\alpha) = 1 - f(h_\alpha)$ if $F_{Pi} = c_\alpha$, $F_{P(i+1)} = c_\alpha^\dagger$. Otherwise the corresponding permutation $P$ gives zero to the average. Here $f(h_\alpha)$ is the Fermi distribution function. In evaluating the Goldstone diagram, we encounter the $T = 0$ version of Eq. (9.39). A simple case with $n = 2$ has appeared in Fig. 5.4 of Chap. 5. Because of different spin states involved in the average, the partition in Eq. (9.39) is unique in this case. Then the Wick's theorem becomes trivial, and the product of number operators is evaluated as the product of occupation numbers of particles and holes.

In the presence of two-body interaction $V(f^\dagger, f)$, the partition function $Z$ cannot be derived explicitly in general. Nevertheless, we may formally represent $Z$ starting from the non-interacting counterpart $Z_0(\xi^*, \xi)$. With use of Eq. (9.34), we obtain [2]

$$Z = -\lim_{\xi^*,\xi\to 0} \exp\left[-\int_0^\beta \mathrm{d}\tau\, V\left(\frac{\delta}{\delta\xi}, -\frac{\delta}{\delta\xi^*}\right)\right] Z_0(\xi^*, \xi), \tag{9.40}$$

where $V(f^\dagger, f)$ has the N-product form. Expanding the RHS in terms of $V$, and applying the Wick's theorem, we recover the perturbation expansion of $Z$ to arbitrary order.

The perturbation expansion can be performed also for the Green functions. For the single-particle Green function, this is achieved by combination of Eqs. (9.33) and (9.40). The resultant quantity is expressed in terms of diagrams. Some examples will be shown in Fig. 9.1. Such pictorial representation of the many-body perturbation theory is first invented by Feynman [3], and is called the Feynman diagrams. The constituents of a Feynman diagram are (single-particle) Green functions and interactions. In contrast with Goldstone diagrams we have dealt with so far, a Feynman diagram combines all cases of the time ordering of operators into a single diagram. Thus the Green function of a particle and that of a hole are given by the same diagram. In contrast, the Goldstone diagram distinguishes the particle and hole by the occupation numbers $1 - f(\epsilon)$ and $f(\epsilon)$, as has been mentioned in p.98.

The Feynman diagrams are classified into linked (connected) and unlinked (unconnected) ones. A linked diagram cannot be separated into unlinked diagrams without breaking interaction line(s) and/or Green function lines. Unlinked diagrams need not be considered in the perturbation expansion of Green functions since the Wick's theorem guarantees cancellation of unlinked diagrams in the final average. Such simplifying property of the many-body perturbation theory is referred to as the linked cluster expansion. Since we have Wick's theorem both for fermion operators as Eq. (9.38) and for Grassmann numbers as Eq. (9.10), we may use either of them to make the perturbation expansion. Utilizing the absence of the time-ordering $T_\tau$ in the path-integral scheme, we may simply split the action as $S_0 + S_1$, where $S_0$ corresponds to the free action with $H = H_0$ in Eq. (9.29), while $S_1 = \int_0^\beta \mathrm{d}\tau\, V(\tau)$ with $V(\tau)$ expressed in terms of Grassmann fields. The partition function is written as

$$Z = Z_0 \left\langle \exp\left(-S_1\right)\right\rangle_0, \tag{9.41}$$

where $Z_0$ derives from $S_0$, and $\langle\cdots\rangle_0$ indicates the average with respect to $S_0$. Taking the logarithm of Eq. (9.41), we can derive the perturbative part $\Omega_1$ for the thermodynamic potential as

$$\Omega_1 = -T\left[\left\langle \exp\left(-S_1\right)\right\rangle_c - 1\right], \tag{9.42}$$

where $\langle\cdots\rangle_c$ indicates the cumulant average with respect to $S_0$. By definition, the cumulant average is contributed only by the linked Feynman diagrams.

## 9.4 Sum Rule for Green Functions and Variational Principle

Most quantum many-body systems are impossible to be solved exactly. One of the established methods to proceed is the perturbation theory. Even for systems with large two-body interaction, one may find a hidden small parameter to construct a reliable approximation scheme. We have encountered such an example in Sect. 6.4. On the other hand, it is sometimes possible to derive formal information that is free from approximation. In this section we discuss such exact relation in Fermi liquids using the Green function formalism. The variational principle plays an important role in the derivation.

As a preliminary, we recall the variational principle, or simply the stationary property, in thermodynamics. For example, the Helmholtz free energy $F$ has an infinitesimal change $\mathrm{d}F = -S\mathrm{d}T - p\mathrm{d}V + \mu\mathrm{d}N$ according to variations of its natural thermodynamic variables. Hence if we vary the other variables such as $p$ or $\mu$ with fixed $T, V, N$, we obtain the stationary property $dF = 0$. Similarly $\Omega$ is stationary against variation of $N$ or $p$ as long as its natural variables $T, V, \mu$ are fixed. Namely, we obtain

$$\left.\frac{\partial F}{\partial \mu}\right)_{T,V,N} = \left.\frac{\partial F}{\partial p}\right)_{T,V,N} = 0, \quad \left.\frac{\partial \Omega}{\partial N}\right)_{T,V,\mu} = \left.\frac{\partial \Omega}{\partial p}\right)_{T,V,\mu} = 0. \tag{9.43}$$

We shall develop analogous argument in the Green function formalism. For concise treatment, it is convenient to add

$$S_{\text{ext}} = \int_0^\beta \mathrm{d}\tau \int_0^\beta \mathrm{d}\tau' \eta_\alpha^*(\tau)\phi_{\alpha\beta}\left(\tau-\tau'\right)\eta_\beta\left(\tau'\right) \tag{9.44}$$

to the system action $S$ as given by Eq. (9.29). Here $\eta_\alpha^*$ and $\eta_\beta$ are Grassmann numbers corresponding to $f_\alpha^\dagger$ and $f_\beta$, respectively, and $\phi_{\alpha\beta}(\tau-\tau')$ is a fictitious external field. We use the notation Tr to represent the path integral $\int \mathcal{D}\eta^*\mathcal{D}\eta$, which is equivalent to trace operation over all degrees of freedom for fermions including Matsubara frequencies or imaginary time.

Let $\delta\phi$ an infinitesimal change of the external field. The corresponding change $\delta\Omega$ of the thermodynamic potential is given by

$$\beta\delta\Omega(\phi) = \mathrm{Tr}(G\delta\phi), \tag{9.45}$$

where $G$ is the Green function matrix with space-time indices. In order to utilize the variational principle, we now represent $\delta\Omega$ in terms of $\delta G$ instead of $\delta\phi$. The

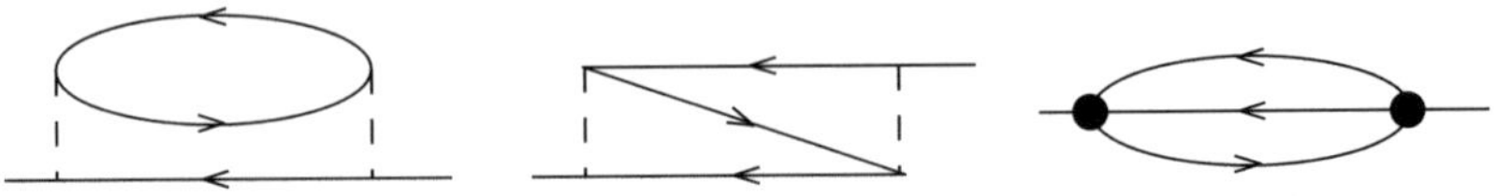


**Fig. 9.1** Examples of Feynman diagrams for the self-energy. The arrow in the Green function indicates the propagation of fermions, and the dashed line represents the two-body interaction. The second order self-energy include contributions from (**a**) and (**b**). The black circle in (**c**) is called the vertex part that includes not only the elementary interaction processes shown in (**a**) and (**b**), but also all higher order Feynman diagrams

analogy with thermodynamics is made by the correspondence $\delta G \to dN$ and $\delta\phi \to -d\mu$.

Let $g$ be the "bare" Green function, which refers to the one in the non-interacting system and without $\phi$. The "dressed", or exact Green function $G$ is related to $g$ by

$$G^{-1} = g^{-1} - \phi - \Sigma, \tag{9.46}$$

where $\Sigma$ is called the self-energy. Figure 9.1 shows some examples of $\Sigma$. Equation (9.46), which is called the Dyson equation, leads to another form of $\delta\Omega$ via the relation

$$G\delta\phi = -G\delta\left(G^{-1} + \Sigma\right) = -\delta\left(\ln G^{-1} + G\Sigma\right) + (\delta G)\Sigma. \tag{9.47}$$

The trace of the last term $(\delta G)\Sigma$ is written as

$$\delta\Phi = \mathrm{Tr}(\Sigma\delta G), \tag{9.48}$$

which defines the new quantity $\Phi$. In terms of Feynman diagrams, $\Phi$ consists of all "skeleton" diagrams that contribute to $\Omega$. The skeleton refers to such diagrams where the self-energy part is included only implicitly through constituent Green functions $G$ [2, 4]. Using the quantity $\Phi$, we can integrate Eq. (9.45) to obtain

$$\beta(\Omega\{G\} - \Omega_0) = \Phi\{G\} - \mathrm{Tr}(\Sigma G) - \mathrm{Tr}\ln\left(G^{-1}g\right), \tag{9.49}$$

where $\Omega_0$ is the thermodynamic potential for free fermions. The natural variables in Eq. (9.49) correspond to $\phi$, $T$, and $V$. Hence making $\delta G \neq 0$ with $\delta\phi = 0$ is analogous to varying $N$ with fixed $\mu$ in the thermodynamic potential. Note that Eq. (9.49) does not have explicit dependence on $\phi$. Therefore $\Omega$ with $\phi = 0$ should be stationary against variation of $G$ [5, 6]. Problem 9.3 deals with the explicit procedure of variation.

Using the variational property, we prove the following statement valid at zero temperature: *The volume inside the Fermi surface is independent of the mutual interaction as long as the symmetry of the system is unbroken.* This property is often called the Luttinger theorem, or sometimes Luttinger–Friedel sum rule [4]. Let us consider the zero-temperature limit of Eq. (9.49) with imaginary frequencies. For any Feynman diagram that constitutes $\Phi$, the result is invariant against uniform shift $i\epsilon$ in all the constituent Green functions. Namely, we have

$$\delta\Phi = \delta\epsilon \mathrm{Tr}\left(\Sigma\frac{\partial G}{\partial\epsilon}\right) = 0. \tag{9.50}$$

By partial integration over $\epsilon$ and noting that $\Sigma G$ vanishes at both ends of integration at $\epsilon = \pm\infty$, we obtain from Eq. (9.50)

$$\mathrm{Tr}\left(G\frac{\partial\Sigma}{\partial\epsilon}\right) = 0. \tag{9.51}$$

Next using the algebraic property

$$\delta\ln G = -\delta\ln\left(g^{-1} - \Sigma\right) = -G(\mathrm{i}\delta\epsilon - \delta\Sigma),$$

we obtain

$$\frac{\partial\ln G}{\mathrm{i}\partial\epsilon} = -G + G\frac{\partial\Sigma}{\mathrm{i}\partial\epsilon}, \tag{9.52}$$

where the second term in the RHS vanishes after taking the trace. The total number $N$ of fermions is then obtained as

$$N = \mathrm{Tr}\exp\left(\mathrm{i}\epsilon 0_+\right)G = -\mathrm{Tr}\exp\left(\mathrm{i}\epsilon 0_+\right)\frac{\partial\ln G}{\mathrm{i}\partial\epsilon}, \tag{9.53}$$

where we put the convergence factor $\exp(\mathrm{i}\epsilon 0_+)$. The details are discussed around Eq. (9.69).

For a spinless system with translational invariance, the result is most simply represented in the momentum space where the Green function matrix becomes diagonal. Namely, we use $i\epsilon = \omega$ and perform the trace by

$$\mathrm{Tr} \to \sum_{\boldsymbol{k}}\int_{-\infty}^{\infty}\frac{\mathrm{d}\epsilon}{2\pi} = \sum_{\boldsymbol{k}}\int_C\frac{\mathrm{d}\omega}{2\pi\mathrm{i}}, \tag{9.54}$$

where the integration contour $C$ is indicated in Fig. 9.2. Then Eq. (9.53) is rewritten as

$$N = \sum_{\boldsymbol{k}}\int_{-\infty}^{0}\frac{\mathrm{d}\omega}{\pi}\mathrm{Im}\frac{\partial\ln G\left(\boldsymbol{k},\omega+\mathrm{i}0_+\right)}{\partial\omega} = \frac{1}{\pi}\mathrm{Im}\sum_{\boldsymbol{k}}\ln\left[-G\left(\boldsymbol{k},\mathrm{i}0_+\right)\right], \tag{9.55}$$

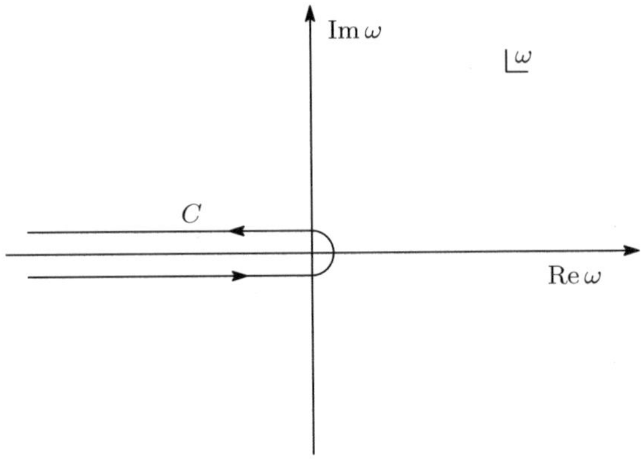


**Fig. 9.2** Integration contour $C$ in Eq. (9.54), which has been deformed from the original one along the imaginary axis

where the minus sign in front of $G(\boldsymbol{k}, \mathrm{i}0_+)$ is due to the contribution of the integrand at $\omega \to -\infty$.

From now on, we make explicit use of a Fermi liquid property. The lifetime of a quasi-particle tends to infinity at the Fermi surface. More generally we can prove that $\Sigma(\boldsymbol{k}, \mathrm{i}0_+)$ is a real number for general momentum $\boldsymbol{k}$. The proof of this property is the subject of Problem 9.4. With the real self-energy, $G(\boldsymbol{k}, \mathrm{i}0_+)$ is also real except for the case $\epsilon_{\boldsymbol{k}} + \Sigma(\boldsymbol{k}, \mathrm{i}0_+) = 0$ where the Green function is divergent. If $|\boldsymbol{k}|$ is small and $G(\boldsymbol{k}, \mathrm{i}0_+) > 0$, the phase given by $\mathrm{Im} \ln[-G(\boldsymbol{k}, \mathrm{i}0_+)]$ becomes $\pi$. On the other hand, with $|\boldsymbol{k}| \to \infty$, $G(\boldsymbol{k}, \mathrm{i}0_+)$ tends to zero from negative side. The set of momentum $\boldsymbol{k}$ where $G(\boldsymbol{k}, \mathrm{i}0_+)$ changes sign is nothing but the Fermi surface. Namely, Eq. (9.55) states that the total number $N$ of particles is given by the volume inside the Fermi surface. Hence with fixed $N$, the mutual interaction, how strong it may be, keeps intact the Fermi volume of the free Fermi gas. Thus the Luttinger theorem is proven. In the presence of periodic potential as in metals, the Fermi surface is not a sphere but takes a complicated shape in general. Accordingly, the Green function depends on the direction of $\boldsymbol{k}$, and so is the Fermi surface. Even in this case, the volume inside the Fermi surface is related to the total number of itinerant electrons in the same way as in the absence of mutual interactions. Hence the Luttinger theorem is practical in dealing with strongly correlated electrons [5].

We now turn to the Fermi liquid without the translational invariance nor lattice periodicity. The most important is a resonant impurity in metallic matrix. Let us consider the Anderson model as given by Eqs. (6.1) and (6.2). If the ground state is connected smoothly from the non-interacting limit $U = 0$, we obtain the result corresponding to Eq. (9.55) [7]:

$$n_f = \frac{2}{\pi} \mathrm{Im} \ln \left[ -G_f\,(\mathrm{i}0_+) \right], \tag{9.56}$$

where $n_f$ is the occupation number of the local state which may be a non-integer, and $G_f(\mathrm{i}0_+)$ is the Green function of local electrons at the Fermi level. The factor 2 in the RHS comes from the spin degrees of freedom. Equation (9.56) is called the

Friedel sum rule. Since both Eqs. (9.55) and (9.56) originate from the same Fermi liquid property of infinite lifetime at the Fermi level, it is reasonable to call them together as the Luttinger–Friedel sum rule.

The phase $\alpha$ of the Green function, as defined by $G_f(\mathrm{i}0_+) = |G_f(\mathrm{i}0_+)| \exp(\mathrm{i}\alpha)$, enjoys a simple relation $n_f = 2\alpha/\pi$. If $G_f(\mathrm{i}0_+)$ is pure imaginary, we obtain $\alpha = \pi/2$ and $n_f = 1$. The Kondo model represents this limiting case explicitly.

## 9.5 Effective One-Body Problem Under Fluctuating Fields

It is possible to convert the interaction among electrons into an effective external field, which is fluctuating both spatially and temporally. After the conversion, the path integral over fermionic degrees of freedom can be performed exactly since it is the Gaussian integral. Usually, of course, the remaining average over effective fields cannot be performed exactly. Thus the conversion may seem merely a replacement of the problem. Nevertheless, in some cases, this conversion of the problem turns out useful. For example, the fluctuating external fields may be treated more conveniently in numerical calculations. In approximate analytic treatment, on the other hand, various mean fields can be introduced easily in the framework of effective external fields. In this section, we describe this framework briefly by taking the Hubbard model.

In the partition function given by Eq. (9.21), the factor

$$\exp\left(-\delta\tau U n_\uparrow n_\downarrow\right), \tag{9.57}$$

describes the two-body interaction. Here we focus on a certain (arbitrary) site in the system and omit the site index for brevity. The operator $n_\uparrow n_\downarrow$ can be equivalently given in terms of occupation number $n$ and spin $S_z$ operators as

$$n_\uparrow n_\downarrow = \frac{1}{4}n^2 - (S_z)^2 \tag{9.58}$$

which is more convenient for Gaussian integration. We first assume $n = n_\uparrow + n_\downarrow$ as constant (=1) and concentrate on the spin part $(S_z)^2$. We use the identity including a positive constant $A$:

$$\exp\left(Ay^2\right) = \int_{-\infty}^{\infty} \frac{\mathrm{d}x}{\sqrt{2\pi A}} \exp\left(-\frac{x^2}{A} - 2xy\right), \tag{9.59}$$

which is analogous to Eq. (9.11). In the case of $A < 0$, we obtain similar identity by replacing the integration range of $x$ by $(-\mathrm{i}\infty, \mathrm{i}\infty)$. The spin part of Eq. (9.57) is thus converted to the one-body form in Eq. (9.59) with substitutions:

$$A = U, \quad y = \sqrt{\delta\tau}S_z, \quad x = \sqrt{\delta\tau}\phi. \tag{9.60}$$

The partition function $Z$ of the Hubbard model with the constraint $n = 1$ is then represented as

$$Z = \int \mathcal{D}\eta^* \mathcal{D}\eta \mathcal{D}\phi \exp\left\{-\int_0^\beta \mathrm{d}\tau \left[\mathcal{L}_0(\tau) + \mathcal{L}_2(\tau) + \mathcal{L}_\phi(\tau)\right]\right\}, \tag{9.61}$$

$$\mathcal{L}_2(\tau) = \frac{1}{U}\sum_i \phi_i(\tau)^2, \quad \mathcal{L}_\phi(\tau) = 2\sum_i \phi_i(\tau) S_i^z(\tau), \tag{9.62}$$

where $\mathcal{L}_0$ is the Lagrangian of the non-interacting system with $U = 0$. This procedure to rewrite $Z$ with the auxiliary field $\phi$ is called the Hubbard–Stratonovich transformation.

It is possible to carry out the Gaussian path integral over $\eta^*$ and $\eta$ exactly. The result is given by

$$Z = \int \mathcal{D}\phi \exp\left[-\int_0^\beta \mathrm{d}\tau \mathcal{L}_2(\tau) + \mathrm{Tr}\ln\left(g^{-1} - \phi\sigma_z\right)\right], \tag{9.63}$$

where Tr indicates trace over spin, spatial, and temporal degrees of freedom, and $g$ is the bare Green function matrix. We have encountered a simpler form equivalent to $\mathrm{Tr}\ln g^{-1}$ in Eq. (9.25). In this way, the system with the two-body interaction $U$ is replaced by the equivalent one-body system under the fluctuating effective field $\phi$.

We consider now the correlation function of spins. Using Eq. (9.62) we obtain

$$\left\langle T_\tau S_i^z(\tau) S_j^z(\tau')\right\rangle = Z^{-1}\int \mathcal{D}\eta^* \mathcal{D}\eta \mathcal{D}\phi \exp\left[-\int_0^\beta \mathrm{d}\tau \left\{\mathcal{L}_0(\tau) + \mathcal{L}_2(\tau)\right\}\right]$$
$$\times \frac{\partial^2}{4\partial\phi_i(\tau)\partial\phi_j(\tau')} \exp\left[-\int_0^\beta \mathrm{d}\tau \mathcal{L}_\phi(\tau)\right], \tag{9.64}$$

where $\partial/\partial\phi_i$ behaves as the spin operator $S_i^z(\tau)$. This is naturally understood by regarding $\phi_i$ as a fluctuating magnetic field conjugate to $S_i^z$. By partial integration to operate $\partial/\partial\phi_i$ on $\mathcal{L}_2(\tau)$ in Eq. (9.64), and by space-time Fourier transform, we obtain the spin susceptibility as

$$\chi(\boldsymbol{q}, \mathrm{i}\nu) = \frac{1}{U}\left[\frac{1}{U}\left\langle|\phi(\boldsymbol{q}, \mathrm{i}\nu)|^2\right\rangle - 1\right], \tag{9.65}$$

where $\phi(\boldsymbol{q}, \mathrm{i}\nu)$ is the Fourier transform of $\phi_i(\tau)$, and $\langle\ldots\rangle$ is the average over fluctuating effective field $\phi$. This result suggests that $\phi_i(\tau)$ represents the fluctuating magnetic moment. However, the fluctuation $\langle|\phi(\boldsymbol{q}, \mathrm{i}\nu)|^2\rangle$ tends to $U$ in the high-frequency limit, instead of going to zero. This limiting value is cancelled by the term $-1$ in the RHS of Eq. (9.65). Since the fluctuation of $\phi$ is complicated, it is impossible to evaluate Eq. (9.65) exactly. The simplest approximation is to use the Gaussian fluctuation for $\phi$, which amounts to the lowest-order expansion of

$\phi$ in the exponent of Eq. (9.63). Problem 9.5 deals with deriving the approximate susceptibility.

The spin–charge decomposition made in Eq. (9.58) is not unique. For example, the rotational invariance in the spin space allows the replacement $(S_z)^2 \to \boldsymbol{S}^2/3$. If one uses the latter representation, the effective field becomes a vector, and the lowest-order result for $\chi(\boldsymbol{q}, i\nu)$, which will be given in Eq. (9.75), involves $U/3$ in place of $U$. The discrepancy is traced to the approximation in the expansion of $\phi$. Since the factor $1/U$ in Eq. (9.65) lowers the degree of $U$ by one, it is necessary to expand the logarithmic function up to $O(\phi^4)$ in order to derive $\chi(\boldsymbol{q}, i\nu)$ correctly up to $O(U)$. Then $\langle|\phi(\boldsymbol{q}, i\nu)|^2\rangle$ is evaluated up to $O(U^2)$. It is a fortunate accident in using $S_z$ that the lowest-order theory can derive the susceptibility equivalent to the mean field theory. Finally we comment on including the fluctuation of $n$ by introducing the corresponding effective field. By a procedure similar to obtain Eq. (9.65), the charge susceptibility is represented as the second moment of the effective field.

## Problems

**9.1** Define $(\eta_1\eta_2)^*$ in a proper manner for general Grassmann numbers $\eta_1, \eta_2$.

**9.2*** Derive the physically meaningful result from Eq. (9.25) where the RHS is divergent as it stands.

**9.3** Demonstrate the variational property $\delta\Omega/\delta G = 0$.

**9.4*** How do the Luttinger theorem and Eq. (9.55) depend on perturbation theory in the two-body interaction $v$?

**9.5** Expand the logarithm in Eq. (9.63) up to $O(\phi^2)$, and derive the dynamical susceptibility from Eq. (9.65).

## Solutions to Problems

### Problem 9.1

The complex conjugate is defined so as to satisfy the equality $(\eta^*\eta)^* = \eta^*\eta$, which is analogous to the Hermite conjugate of the matrix product. Then we have to follow the definition

$$(\eta_1\eta_2)^* = \eta_2^*\eta_1^* = -\eta_1^*\eta_2^*, \quad (\eta_2^*\eta_1^*)^* = \eta_1\eta_2. \tag{9.66}$$

The minus sign in $-\eta_1^*\eta_2^*$ may appear strange if both $\eta_1$ and $\eta_2$ are "real" Grassmann numbers with $\eta^* = \eta$. However, the product of two real Grassmann numbers need not be real.

**Problem 9.2***
By integrating Eq. (9.26) we obtain

$$\frac{\partial \Omega}{\partial E} = f(E) + C_1, \tag{9.67}$$

where $f(E) = 1/[\exp(\beta E) + 1]$ and $C_1$ is a constant of integration. Further integration leads to

$$\Omega = -T \ln\left[1 + \exp(-\beta E)\right] + C_1 E + C_2, \tag{9.68}$$

where $C_2$ is another integration constant. The occupation number given by $\partial\Omega/\partial E$ should go to zero in the limit $E \to \infty$. This condition makes $C_1 = 0$. On the other hand, $C_2$ only sets the origin of $\Omega$, or the Hamiltonian $H$, and remains arbitrary. We put $C_2 = 0$ for simplicity.

We now try to avoid the divergence by modifying Eq. (9.25). Writing $0_+$ as positive infinitesimal, we make use of $\pi\delta(x) = -\mathrm{Im}\,(x + \mathrm{i}0_+)^{-1}$. Then we rewrite the thermodynamic potential $\Omega = \Omega(E)$ as

$$\begin{aligned}
\Omega &= \int_{-\infty}^{\infty} \frac{\mathrm{d}\epsilon}{2\pi \mathrm{i}} \Omega(\epsilon) \left( \frac{1}{-\epsilon + E + \mathrm{i}0_+} - \frac{1}{-\epsilon + E - \mathrm{i}0_+} \right) \\
&= \int_{-\infty}^{\infty} \frac{\mathrm{d}\epsilon}{2\pi \mathrm{i}} f(\epsilon) \left[\ln(-\epsilon + E + \mathrm{i}0_+) - \ln(-\epsilon + E - \mathrm{i}0_+)\right] \\
&= \int_C \frac{\mathrm{d}z}{2\pi \mathrm{i}} \exp(z0_+)\, f(z) \ln(-z + E) \\
&= -T \sum_n \exp(\mathrm{i}\epsilon_n 0_+) \ln(-\mathrm{i}\epsilon_n + E),
\end{aligned} \tag{9.69}$$

by partial integration. The integration contour $C$ is the same as that shown in Fig. 3.3. However, this time the contour runs originally parallel to the real axis along both directions, and is continuously deformed so as to pick up all poles of $f(z)$ at $z = i\epsilon_n = (2n+1)i\pi T$ lying on the imaginary axis of $z$. The convergence factor is inserted so that $\exp(z0_+) f(z)$ goes to zero with $z \to \pm\infty$. The last line of Eq. (9.69) is derived by taking the residues at all poles of $f(z)$. In other words, Eq. (9.25) is made finite by insertion of the factor $\exp(\mathrm{i}\epsilon_n 0_+)$.

**Problem 9.3**
Variation of each term composing $\Omega$ in Eq. (9.49) reads

$$\delta\Phi = \mathrm{Tr}\Sigma\delta G, \quad -\delta(\Sigma G) = -(\delta\Sigma)G - \Sigma\delta G, \quad -\delta\ln\left(G^{-1}g\right) = (\delta\Sigma)G. \tag{9.70}$$

These terms clearly add up to zero. Hence we obtain $\delta\Omega = 0$. The stationary property does not depend on whether the natural variable is $\delta G$ or $\delta\Sigma$, as long as the relation $G^{-1} = g^{-1} - \Sigma$ holds.

**Problem 9.4***
In the original derivation [5] the perturbation series in $v$ is expressed by Feynman diagrams, and $\Phi$ is derived by analysis of the diagrammatic structure. Therefore the proof needs convergent perturbation series. On the other hand, the present derivation of Eqs. (9.49) and (9.55) has assumed only the continuity against the change of external field $\phi$. The perturbation theory in $v$ has not been used [6]. Therefore, Eq. (9.55) should be valid even for one-dimensional systems where the perturbation theory in $v$ is not valid [8]. On the other hand, the Luttinger theorem requires the real self-energy at the Fermi level, which relies on perturbation theory in $v$.

Let us explain the logic for the real self-energy by taking the Feynman diagram in Fig. 9.1c. Similar argument applies to general diagrams with more internal lines. After analytic continuation to real frequencies, the right-going internal line has the energy $\epsilon_1$, and two left-going lines have $\epsilon_2, \epsilon_3$. Correspondingly, each line has the momentum $\boldsymbol{k}_i$ with $i = 1, 2, 3$ and the spectral function $\rho(\boldsymbol{k}_i, \epsilon_i)$ as defined by Eq. (3.100). The imaginary part is then given by

$$-\mathrm{Im}\Sigma\,(\boldsymbol{k}, \epsilon + \mathrm{i}0_+) \sim \int_{1,2,3} |\Gamma_{123}|^2 \rho(\boldsymbol{k}_1, \epsilon_1)\rho(\boldsymbol{k}_2, \epsilon_2)\rho(\boldsymbol{k}_3, \epsilon_3) f(\epsilon_1) f(-\epsilon_2) f(-\epsilon_3)$$
$$\times\, \delta(\epsilon + \epsilon_1 - \epsilon_2 - \epsilon_3)\delta(\boldsymbol{k} + \boldsymbol{k}_1 - \boldsymbol{k}_2 - \boldsymbol{k}_3), \tag{9.71}$$

where $\Gamma_{123}$ is the vertex part, and the integration is written symbolically. The Fermi function $f(\epsilon)$ at zero temperature is reduced to the step function $\theta(-\epsilon)$. For a nonzero product of Fermi functions, it is necessary to have $\epsilon_1 < 0$ and $\epsilon_2, \epsilon_3 > 0$. With this combination, however, it is impossible to have nonzero delta function for $\epsilon = 0$. Thus we obtain $\mathrm{Im}\Sigma(\boldsymbol{k}, \mathrm{i}0_+) = 0$, which means the real self-energy at the Fermi surface. Note that the momentum $\boldsymbol{k}$ is arbitrary. Moreover, since the momentum conservation does not play any role, the argument applies equally to impurity systems, resulting in the Friedel sum rule.

If the self-energy has five or more internal lines, similar argument leads to the zero product of Fermi functions and the energy-conserving delta function. Hence we can safely take $G(\boldsymbol{k}, \mathrm{i}0_+)$ real in Eq. (9.55).

**Problem 9.5**
We make the expansion by writing $\phi_z \equiv \phi\sigma_z$

$$\ln\left(g^{-1} - \phi_z\right) \sim \ln g^{-1} - g\phi_z + \frac{1}{2}g\phi_z g\phi_z + O\left(\phi^3\right), \tag{9.72}$$

where $g$ is the bare Green function matrix. Taking the trace and making the Fourier transform, the $O(\phi^2)$ term in Eq. (9.72) gives $\chi_0(\boldsymbol{q}, \mathrm{i}\nu)$ as the coefficient of $\phi^2$. This process is similar to the one in Eq. (3.140). Hence the coefficient $\phi^2$ in the action is modified as

$$\frac{1}{U} \to \frac{1}{U} - \chi_0(\boldsymbol{q}, \mathrm{i}\nu), \tag{9.73}$$

which determines the Gaussian fluctuation of $\phi$. Namely, we obtain

$$\left\langle |\phi(\boldsymbol{q}, \mathrm{i}\nu)|^2 \right\rangle = \left(\frac{1}{U} - \chi_0(\boldsymbol{q}, \mathrm{i}\nu)\right)^{-1} \tag{9.74}$$

and consequently

$$\chi(\boldsymbol{q}, \mathrm{i}\nu) = \frac{1}{U}\left(\frac{1}{1 - U\chi_0(\boldsymbol{q}, \mathrm{i}\nu)} - 1\right) = \frac{\chi_0(\boldsymbol{q}, \mathrm{i}\nu)}{1 - U\chi_0(\boldsymbol{q}, \mathrm{i}\nu)}. \tag{9.75}$$

This result is the same as the one obtained in the random phase approximation (RPA), which we have encountered in Eq. (6.80).

## References

1. Feynman, R.P., Hibbs, A.R., Styer, D.F.: Quantum Mechanics and Path Integrals. Dover, New York (2010)
2. Zinn-Justin, J.: Quantum Field Theory and Critical Phenomena. Oxford University Press, Oxford (1993)
3. Feynman, R.P.: Rev. Mod. Phys. **20**, 367 (1948)
4. Abrikosov, A.A., Gorkov, L.P., Dzyaloshinskii, I.E.: Methods of Quantum Field Theory in Statistical Physics. Dover, New York (1975)
5. Luttinger, J.: J. Ward Phys. Rev. **118**, 1417 (1960)
6. Baym, G.: Phys. Rev. **127**, 1391 (1962)
7. Langreth, D.C.: Phys. Rev. **150**, 516 (1966)
8. Oshikawa, M.: Phys. Rev. Lett. **84**, 3370 (2000)

# Chapter 10
# Dynamical Mean Field Theory

**Abstract** In the ordinary mean field theory, the effective field is static. If one allows temporal variation of the effective field, fluctuation effects of the corresponding field can be included. The correlation problem in a large system is then replaced by a single-site problem surrounded by a dynamical effective medium. The effective impurity system is regarded as zero-dimensional. It can be shown that the replacement by an effective impurity is exact in the limit of large number of neighboring sites, which is the case in infinite dimensions. Hence the zero and infinite dimensions are connected continuously. In this way the dynamical mean field theory (DMFT) approaches actual three-dimensional systems from the infinite-dimensional limit. The DMFT has achieved remarkable success in understanding those many-body effects which come from strong local correlations. This chapter explains the DMFT starting with the background of its development.

## 10.1 Mean Field and Fluctuations

The concept of mean field, also called molecular field, has a long history in physics. One of the most successful applications in condensed matter is the BCS theory of superconductivity where the two-body interaction is replaced by a form given by Eq. (5.18). The BCS theory is highly accurate if the transition temperature $T_c$ is much smaller than the Fermi energy. Another remarkable success is the Fermi liquid theory described in Chap. 4 where the interaction effect is taken into account in terms of the Landau parameters. In both cases, mean fields comprise a large number of degrees of freedom. Hence individual fluctuation affects the magnitude only slightly.

The mean field is most simply defined in the momentum space in these examples for itinerant fermions. In strongly correlated electron systems, on the other hand, it is sometimes more convenient to start from the localized picture, even though the ground state becomes a Fermi liquid. In this section we study the mean field theory from the localized picture. Let us first take the Ising model as the simplest model with localized degrees of freedom. The Ising model has the variable $\sigma_i$ taking $\pm 1$



Y. Kuramoto, *Quantum Many-Body Physics*, Lecture Notes in Physics 934,
https://doi.org/10.1007/978-4-431-55393-9_10

at each site $i$. Under a magnetic field $h$ in the $z$ direction, the model can be written as

$$H = -J \sum_{\langle ij \rangle} \sigma_i \sigma_j - h \sum_i \sigma_i, \tag{10.1}$$

where $\langle ij \rangle$ are the nearest-neighbor sites, and we consider the ferromagnetic case $J > 0$. The molecular (Weiss) field $h_{\mathrm{W}}$ at arbitrary site $i$ is defined by

$$h_{\mathrm{W}} = J \sum_{j \in n(i)} \langle \sigma_j \rangle = J Z_n m, \tag{10.2}$$

where $n(i)$ denote the set of nearest-neighbor sites whose number is $Z_n$, and $\langle \sigma_j \rangle = m$ is the average of $\sigma_j$ to be determined self-consistently. One has $Z_n = 2d$ in the $d$-dimensional hypercubic lattice. The mean field Hamiltonian $H_{\mathrm{MF}}$ is given by

$$H_{\mathrm{MF}} = -(h_{\mathrm{W}} + h) \sum_i \sigma_i, \tag{10.3}$$

which gives the magnetization as

$$m = \tanh[\beta(h_{\mathrm{W}} + h)], \tag{10.4}$$

with $\beta = 1/T$. Together with Eq. (10.2), $h_{\mathrm{W}}$ is determined self-consistently. Even without $h$, nontrivial solution $m \neq 0$ emerges if $T$ is smaller than the Curie temperature $T_c = J Z_n$.

Let us now ask how reliable is the mean field approximation (MFA). If $h_{\mathrm{W}}$ consists of contribution of many ($Z_n \gg 1$) spins, fluctuation effect from each spin should be slight. To be more quantitative, we consider the deviation from the mean field as defined by

$$\Delta h_{\mathrm{W}} \equiv J \sum_{j \in n(i)} (\sigma_j - m), \tag{10.5}$$

which becomes zero on average. The variance is evaluated as

$$\left\langle (\Delta h_{\mathrm{W}})^2 \right\rangle = J^2 \sum_{j \in n(i)} \left\langle (\sigma_j - m)^2 \right\rangle = Z_n^{-1} T_c^2 \left(1 - m^2\right), \tag{10.6}$$

with use of $\langle (\sigma_j - m)^2 \rangle = 1 - m^2$. Thus we obtain the ratio

$$\frac{\langle (\Delta h_{\mathrm{W}})^2 \rangle}{h_{\mathrm{W}}^2} = \frac{1}{Z_n} \left( \frac{1}{m^2} - 1 \right), \tag{10.7}$$

which is of $O(1/Z_n)$ as long as $m$ is not too close to 1, nor to 0. This is the case if either the nonzero external field of $h \sim O(T_c)$ is present in the paramagnetic phase, or $1 - T/T_c \sim O(1)$ in the ordered phase. In the case of $m^2 < 1/Z_n$, however, the fluctuation becomes significant.

By imposing the self-consistency of thermal fluctuations, the improved MFA becomes applicable even to the paramagnetic phase with $m = 0$. As a result, for instance, the specific heat for $T > T_c$, which is zero in the simplest MFA, becomes finite with inclusion of fluctuations. Let us consider the susceptibility which reflects spin fluctuations in the disordered phase. Assuming that the external field $h_j$ may depend on site $j$, the magnetization at site $i$ is given in the MFA as

$$m_i = \chi_0 \left( h_i + J \sum_{j \in n(i)} \langle \sigma_j \rangle \right). \tag{10.8}$$

Here $\chi_0 = \beta$ is the susceptibility of an isolated spin. If the infinitesimal external field has the wave number $\boldsymbol{q}$, the corresponding susceptibility $\chi(\boldsymbol{q}) = \partial m_{\boldsymbol{q}}/\partial h_{\boldsymbol{q}}$ is given in the MFA by

$$\chi_{\text{MFA}}(\boldsymbol{q}) = \frac{\chi_0}{1 - J_{\boldsymbol{q}}\chi_0}, \tag{10.9}$$

where $J_{\boldsymbol{q}}$ in the hypercubic lattice is given by

$$J_{\boldsymbol{q}} = 2J \sum_{i=1}^{d} \cos q_i. \tag{10.10}$$

The homogeneous susceptibility $\chi_{\text{MFA}}(\boldsymbol{q} = 0)$ is divergent at the Curie temperature.

We note that the simplest MFA does not satisfy the basic consistency relation between response and fluctuation. Namely, the local susceptibility, which is given by the $\boldsymbol{q}$-average of $\chi(\boldsymbol{q})$, should be equal to $\chi_0$ for any value of $J$. This is the special property of the Ising model that comes from conservation of each spin $\sigma_i$. In the MFA, however, the LHS in Eq. (10.9) does not give $\chi_0$ on average. The reason for this inconsistency is traced to the overcounting in $h_W$ as the effective field acting on the given spin. Namely, $h_W$ includes a part contributed by the given spin itself. This part, which is called the reaction field, should be removed in considering the action on the given spin. The reaction field must be proportional to the local magnetization, and is written as $-\lambda m$. The corrected field $h_W - \lambda m$ is called the cavity field. The inhomogeneous magnetization $m_{\boldsymbol{q}}$ under the external field $h_{\boldsymbol{q}}$ is now given by

$$m_{\boldsymbol{q}} = \chi_{\boldsymbol{q}} h_{\boldsymbol{q}} = \chi_0 \left[ h_{\boldsymbol{q}} + \left( J_{\boldsymbol{q}} - \lambda \right) m_{\boldsymbol{q}} \right], \tag{10.11}$$

which gives

$$\chi_{\boldsymbol{q}} = \frac{\chi_0}{1 + \left(\lambda - J_{\boldsymbol{q}}\right)\chi_0}, \tag{10.12}$$

or

$$\chi_{\boldsymbol{q}}^{-1} = \chi_0^{-1} + \lambda - J_{\boldsymbol{q}} \equiv \bar{\chi}_0^{-1} - J_{\boldsymbol{q}}. \tag{10.13}$$

Here $\bar{\chi}_0$ is called the irreducible susceptibility which includes the local field correction in terms of $\lambda$. From the self-consistency condition, the $\boldsymbol{q}$-average of $\chi_{\boldsymbol{q}}$ must be equal to the local susceptibility $\chi_0$, which leads to the relation:

$$1 = \frac{1}{N}\sum_{\boldsymbol{q}} \frac{1}{1 + \left(\lambda - J_{\boldsymbol{q}}\right)\chi_0}, \tag{10.14}$$

where $N$ is the total number of lattice sites, being equal to the number of $\boldsymbol{q}$ inside the Brillouin zone. Equation (10.14) determines $\lambda$ self-consistently.

Because of its nature of correcting the molecular field, the reaction field has a tendency to suppress the transition temperature below the mean field value. In the extreme example in which the exchange interaction is independent of $\boldsymbol{q}$ as given by $J_{\boldsymbol{q}} = J_0$, Eq. (10.14) demands that $\lambda = J_0$. Then we obtain finite $\chi_{\boldsymbol{q}}$ $(= \chi_0)$ for any finite $T$. This is in strong contrast with the mean field results giving $T_c = J_0$.

Another contrast with the ordinary MFA is that $T_c$ becomes zero in one- and two-dimensional systems. This is because Eq. (10.14) is never satisfied with finite $T_c$. To see this, we assume certain finite $T_c$, which leads to $T_c + \lambda - J_{\boldsymbol{q}} \propto \boldsymbol{q}^2$. Then integration over $\boldsymbol{q}$ gives divergence in the form $q_c^{-1}$ $(d = 1)$ or $\ln q_c$ $(d = 2)$ where $q_c = L^{-1} \to 0$ is the lower cut-off determined by the size $L$ of the system. The result for $d = 1$ is qualitatively correct, but that in $d = 2$ is different from the exact solution [1]. Thus the correction to the MFA in $d = 2$ is too much. As $d$ increases beyond 2, it improves the original MFA in a reasonable way. It is possible to interpret the reaction field as incorporating the mode-coupling effect of fluctuations with different wave numbers.

## 10.2 Dynamical Effective Fields

The effective Hamiltonian for the Ising model takes the single-site form in the MFA. Furthermore the local feature remains after with inclusion of the reaction field. We shall now turn to electrons with strong local correlations. At a given site, the effect of other sites is represented by a local effective field which is now dynamical. Such extension keeping the local nature is called the dynamical mean field theory (DMFT). An important feature distinct from the Hartree–Fock theory is that the self-consistency is imposed on each energy of the Green function. Among the vast literature of relevant papers, we recommend the extensive review [2] including the historical background and basic ideas. The DMFT turns out very useful for studying dynamical as well as thermodynamic properties of quantum particles with strong local correlation. For simplicity, we confine the discussion in this chapter to the paramagnetic phase.

To be specific, we take the Hubbard model

$$H = \sum_{k\sigma} \varepsilon_{k} c_{k\sigma}^{\dagger} c_{k\sigma} + U \sum_{i} c_{i\uparrow}^{\dagger} c_{i\uparrow} c_{i\downarrow}^{\dagger} c_{i\downarrow}, \tag{10.15}$$

where the energy spectrum $\varepsilon_{k}$ is the Fourier transform of $t_{ij}$ as given by

$$\varepsilon_{k} = \sum_{i} t_{ij} \exp\left(-\mathrm{i}k \cdot R_{ij}\right), \tag{10.16}$$

with vectors for the lattice sites $R_{ij} \equiv R_i - R_j$. The on-site interaction ($U$) term has been written in the site representation.

As the crucial feature with large number $Z_n$ of neighbors, we show the dominance of the site-diagonal part of the Green function. We first take the free fermions, and derive the upper bound of the off-diagonal part of the Green function. In the momentum space, the Green function $g_{k}[\tau] = -\langle \mathrm{T}_{\tau} c_{k\sigma}(\tau) c_{k\sigma}^{\dagger}(0)\rangle_0$ is obtained explicitly as

$$g_{k}[\tau] = \begin{cases} -\left[1 - f\left(\epsilon_{k}\right)\right] \exp\left(-\epsilon_{k}\tau\right), & (\tau > 0) \\ f\left(\epsilon_{k}\right) \exp\left(-\epsilon_{k}\tau\right), & (\tau < 0), \end{cases} \tag{10.17}$$

which is real and bounded as

$$-1 \le g_{k}[\tau] \le 1, \tag{10.18}$$

for any value of $\tau$. We have used the notation $g_{k}[\tau]$ with angular bracket to distinguish the object from the Green function $g_{k}(z)$ in the energy ($z$) domain. Similar notation has been used in Eq. (3.45). By Fourier transform to the real space, we obtain the site representation of the Green function

$$g_{ij}[\tau] = -\left\langle \mathrm{T}_{\tau} c_{i\sigma}(\tau) c_{j\sigma}^{\dagger}(0)\right\rangle_0 = \frac{1}{N} \sum_{k} g_{k}[\tau] \exp\left(\mathrm{i}k \cdot R_{ij}\right). \tag{10.19}$$

By the inverse Fourier transform we obtain

$$g_{k}[\tau]^2 = \sum_{jl} g_{ij}[\tau] g_{li}[\tau] \exp\left(\mathrm{i}k \cdot R_{jl}\right), \tag{10.20}$$

where only terms with $j = l$ remain after the summation over $\boldsymbol{k}$. Using the property $g_{ji} = g_{ij}^*$, which follows from $g_{\boldsymbol{k}}$ being real, we obtain the inequality:

$$\frac{1}{N}\sum_{\boldsymbol{k}} g_{\boldsymbol{k}}[\tau]^2 = \sum_j |g_{ij}[\tau]|^2 \leq 1. \tag{10.21}$$

If there are $Z_n$ equivalent sites $j$ as seen from $i$, we obtain $|g_{ij}[\tau]| \sim O(1/\sqrt{Z_n})$. On the other hand, the diagonal element $g_{ii}[\tau]$ has the order of unity. Thus the off-diagonal elements are small for large $Z_n$. In this way the thermodynamics is derived in a controlled manner with large $Z_n$. However, in dynamics, the $1/Z_n$ classification requires more care. Namely, as seen from the divergence at $z = \epsilon_{\boldsymbol{k}}$ of the Green function $g(\boldsymbol{k}, z) = (z - \epsilon_{\boldsymbol{k}})^{-1}$, the $\boldsymbol{k}$-dependence cannot be neglected in dynamics. Hence selection of Feynman diagrams according to the $1/Z_n$ classification requires partial inclusion of higher order processes. This means possible relevance of off-diagonal elements $g_{ij}(z)$ in dynamics.

We now extend the inequality given by Eq. (10.21) to the Green function of interacting systems. We use the representation:

$$G[\boldsymbol{k}, \tau] = \int_{-\infty}^{\infty} \mathrm{d}\epsilon \rho(\boldsymbol{k}, \epsilon) g_\epsilon[\tau], \tag{10.22}$$

where $\rho(\boldsymbol{k}, \epsilon)$ is the spectral function and

$$g_\epsilon[\tau] = e^{-\tau\epsilon}\left\{-\theta(\tau)[1 - f(\epsilon)] + \theta(-\tau) f(\epsilon)\right\}. \tag{10.23}$$

With use of the sum rule Eq. (3.103) for $\rho(\boldsymbol{k}, \epsilon)$, we obtain

$$|G[\boldsymbol{k}, \tau]| \leq \int_{-\infty}^{\infty} \mathrm{d}\epsilon \rho(\boldsymbol{k}, \epsilon)\, |g_\epsilon[\tau]| \leq 1 \tag{10.24}$$

because $|g_\epsilon[\tau]| \leq 1$. Hence we have generalized Eq. (10.18), and further generalization of Eq. (10.21) for $G_{ij}$ is straightforward. Namely, the magnitude of off-diagonal elements $G_{ij}[\tau]$ is smaller by $O(1/\sqrt{Z_n})$ as compared with the diagonal element.

We proceed to discuss the self-energy for large $Z_n$ [2, 3]. Figure 10.1 shows the simplest Feynman diagrams which contribute to the off-diagonal elements in

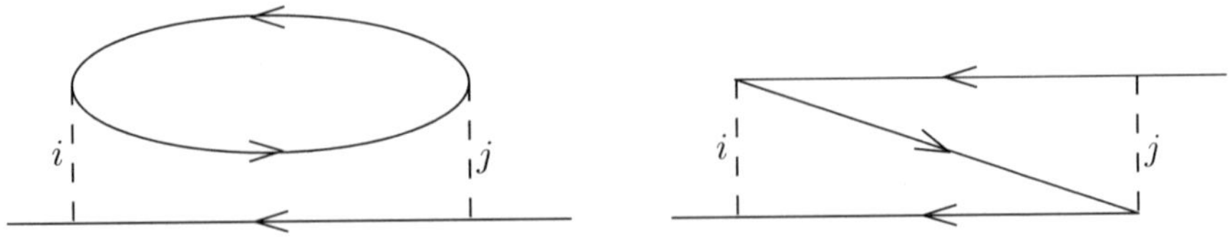


**Fig. 10.1** Lowest-order processes for the off-diagonal elements $\Sigma_{ij}$ of the self-energy

the lowest order. Since both diagrams have three off-diagonal elements $G_{ij}$, the upper bound of $\Sigma_{ij}[\tau]$ is $O(Z_n^{-3/2})$. Account of higher order self-energies does not change the order estimate. Thus any off-diagonal element $\Sigma_{ij}[\tau]$ of the self-energy is smaller by $O(Z_n^{-3/2})$ as compared with the diagonal element. Note that this argument does not depend on temperature and energy scales. If a theory is free from approximations other than neglecting the $\boldsymbol{k}$-dependence of the self-energy, the theory becomes exact in the limit of $Z_n \to \infty$, or in infinite dimensions. The DMFT has precisely this feature.

The Green function in the DMFT is thus given by

$$G(\boldsymbol{k}, z) = [z - \varepsilon_{\boldsymbol{k}} - \Sigma(z)]^{-1}, \tag{10.25}$$

with neglect of $\boldsymbol{k}$-dependence of the self-energy [2, 3]. The site-diagonal part $\bar{G}(z)$ of the Green function is related to $G(\boldsymbol{k}, z)$ as

$$\text{(i)} \quad \bar{G}(z) = \frac{1}{N}\sum_{\boldsymbol{k}} \frac{1}{z - \varepsilon_{\boldsymbol{k}} - \Sigma(z)} = \int_{-\infty}^{\infty} d\epsilon \frac{\rho(\epsilon)}{z - \Sigma(z) - \epsilon} = g(z - \Sigma(z)), \tag{10.26}$$

where $N$ is the total number of lattice sites, and $g(z)$ is the local Green function in the non-interacting system. For simple form of the density of state $\rho(\epsilon)$, integration over $\epsilon$ can be performed analytically. Some examples will be discussed later. Equation (10.26) is the first of the DMFT equations.

In order to derive $\bar{G}(z)$ as the solution of the effective impurity problem, we introduce the cavity Green function $\mathcal{G}(z) \equiv [z - \lambda(z)]^{-1}$ which serves as the zeroth order Green function for the effective impurity. Namely, $\mathcal{G}(z)$ includes the effect of effective medium, while it has not included the effect of $U$ at the impurity site. The latter should be included as the self-energy. Thus we obtain the second equation of the DMFT:

$$\text{(ii)} \quad \bar{G}(z) = \left[\mathcal{G}(z)^{-1} - \Sigma(z)\right]^{-1} = [z - \lambda(z) - \Sigma(z)]^{-1}. \tag{10.27}$$

Because of the translational invariance of the lattice system, the self-energy $\Sigma(z)$ should be common to the impurity and the medium. Figure 10.2 illustrates the steps to replace the lattice system to the effective impurity.

The two relations set by Eqs. (10.26) and (10.27) are not yet sufficient to determine the Green function, since we have three unknowns $\lambda(z)$, $\Sigma(z)$, $\bar{G}(z)$ for given $z$. By solving the effective impurity problem explicitly, we obtain another relation between $\lambda(z)$ and $\Sigma(z)$. Namely, the DMFT is completed in the following third step:

(iii) By solving the effective impurity problem, $\bar{G}(z)$ is derived in terms of $\lambda(z)$ and $U$.

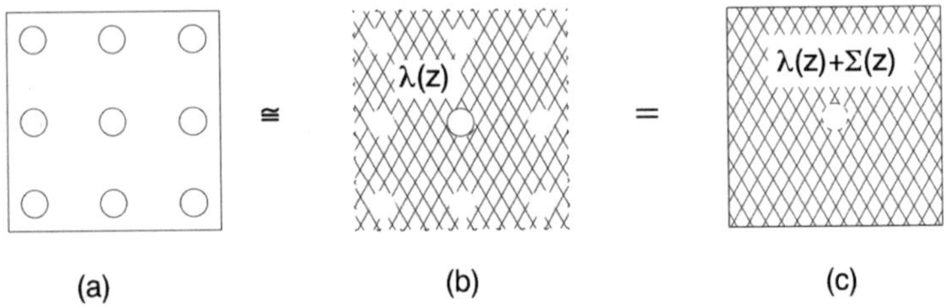


**Fig. 10.2** Construction of the effective impurity problem. The original lattice (**a**) is simulated by an effective medium characterized by $\lambda(z)$ as shown in (**b**). The local interaction in the impurity system is accounted for by the self-energy $\Sigma(z)$ as shown in (**c**)

Any method can be used in the step (iii), which is called the impurity solver. In the Hubbard model, and in more complicated models, one employs mostly numerical methods as the impurity solver.

As seen in Eq. (10.26), once we know $g(z)$ for a given lattice, the renormalized Green function is obtained by the replacement: $z \to z - \Sigma(z)$. The density of states $\rho(\epsilon)$ determines $g(z)$ by the spectral integral, which is equivalent to the Hilbert transform. We now discuss representative results for $g(z)$ and $\rho(\epsilon)$, assuming the nearest-neighbor hopping. Let us first consider the generalized cubic lattice in large spatial dimensions $d$ with the hopping parameter $-t$ ($<0$). The spectrum is given by

$$\varepsilon_{\boldsymbol{k}} = -2t \sum_{i=1}^{d} \cos k_i \equiv \sum_{i=1}^{d} \varepsilon_i, \tag{10.28}$$

with the lattice constant unity. Writing the momentum average as $\langle\langle \cdots \rangle\rangle$, we obtain

$$\langle\langle \varepsilon_{\boldsymbol{k}} \rangle\rangle = \langle\langle \varepsilon_i \rangle\rangle = 0, \tag{10.29}$$

$$\left\langle\left\langle \varepsilon_{\boldsymbol{k}}^2 \right\rangle\right\rangle = \sum_{i=1}^{d} \left\langle\left\langle \varepsilon_i^2 \right\rangle\right\rangle = t^2 d \equiv D_{\mathrm{G}}^2. \tag{10.30}$$

If one fixes the magnitude of $t$, and makes $d$ larger, both the bandwidth $4td$ and the variance $D_{\mathrm{G}}^2$ diverge. On the other hand, if one lets $d$ larger while keeping the $D_{\mathrm{G}}^2 = t^2 d$ fixed, the resultant density of states tends to the Gaussian form [4]

$$\rho_{\mathrm{G}}(\epsilon) = \frac{1}{\sqrt{2\pi}\, D_{\mathrm{G}}} \exp\left(-\frac{\epsilon^2}{2D_{\mathrm{G}}^2}\right). \tag{10.31}$$

with finite width $D_{\mathrm{G}}$. Although the band tails extend to $\epsilon = \pm\infty$, the main weight of the density of states is concentrated in the finite energy range of $O(D_{\mathrm{G}})$. Thus physical quantities such as free energy remain finite with the Gaussian density of states.

The emergence of the Gaussian distribution corresponds to the central limit theorem in probability theory, which can be proved most concisely with use of the generating function:

$$M(x) \equiv \int_{-\infty}^{\infty} \mathrm{d}\epsilon \rho(\epsilon) e^{-\mathrm{i}\epsilon x} = \langle\langle \exp(-\mathrm{i}\varepsilon_{\boldsymbol{k}} x) \rangle\rangle. \tag{10.32}$$

In terms of the moments $\mu_n \equiv \langle\langle \varepsilon_i^n \rangle\rangle$, which is the same for all $i$, we obtain

$$M(x) = \left(1 - \frac{1}{2}x^2\mu_2 + \frac{1}{4!}x^4\mu_4 + \cdots\right)^d \underset{d\to\infty}{\to} \exp\left(-\frac{1}{2}x^2 D_{\mathrm{G}}^2\right), \tag{10.33}$$

where odd moments all vanish because $\epsilon_i$ is an even function of $k_i$. Thus inverse Fourier transform gives Eq. (10.31). More details of the derivation is the subject of Problem 10.1. The real part of the Green function is obtained by the Hilbert transform of the Gaussian as

$$D_{\mathrm{G}} \,\mathrm{Re}\, \bar{g}\,(\epsilon + \mathrm{i}0_+) = \sqrt{2} e^{-x^2/2} \int_0^{x/\sqrt{2}} \mathrm{d}y\, e^{y^2} = \sqrt{2} D\left(x/\sqrt{2}\right), \tag{10.34}$$

with $x = \epsilon/D_{\mathrm{G}}$. Here $D(y)$ is known as the Dawson function:

$$D(y) \equiv e^{-y^2} \int_0^y \mathrm{d}u\, e^{u^2} = \frac{2}{\sqrt{\pi}} e^{-y^2}\, \mathrm{erfi}(y), \tag{10.35}$$

with $\mathrm{erfi}(y) \equiv -\mathrm{i}\,\mathrm{erf}(\mathrm{i}y)$ being the imaginary error function. Figure 10.3a shows the real and imaginary parts of $\bar{g}(\epsilon + \mathrm{i}0_+)$. Derivation of Eq. (10.34) is the subject of Problem 10.2.

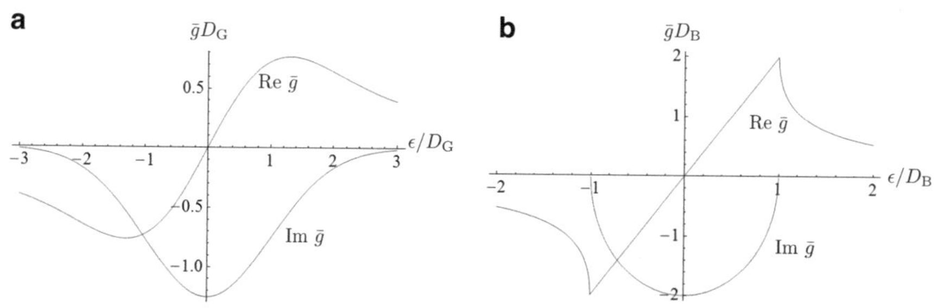


**Fig. 10.3** Real and imaginary parts of the Green function $\bar{g}(\epsilon + \mathrm{i}0_+)$ for (**a**) Gaussian and (**b**) semi-elliptic density of states. The unit of energy is taken to be the respective bandwidth parameter $D_{\mathrm{G}}$ or $D_{\mathrm{B}}$

Another useful density of states is given by

$$\rho_{\mathrm{B}}(\epsilon) = \frac{2}{\pi D_{\mathrm{B}}}\sqrt{1-\left(\frac{\epsilon}{D_{\mathrm{B}}}\right)^2}, \tag{10.36}$$

for $|\epsilon| \le D_{\mathrm{B}}$, and zero otherwise. This semi-elliptic density of states emerges if one considers a hypothetical lattice which has no connected path to return to the original site without retracing, and takes the limit $Z_n \gg 1$ in the lattice. The lattice without such loop is called the Bethe lattice or the Cayley tree. Let us consider the homogeneous Bethe lattice with the nearest-neighbor hopping $-t$. An electron situated at the origin can hop to one of the $Z_n$ neighbors, then the electron may either return to the origin or hop to one of the $Z_n - 1$ neighbors other than the origin. Thus the site-diagonal part $\bar{g}(z)$ of the Green function of non-interacting electrons satisfies the relation

$$\bar{g}(z) = \frac{1}{z - t^2 Z_n \tilde{g}(z)}, \tag{10.37}$$

where $\tilde{g}(z)$ is the modified site-diagonal Green function which excludes hopping to the origin. The term with $\tilde{g}(z)$ acts as a self-energy to the bare Green function $1/z$. If one takes the limit $Z_n \gg 1$ keeping $t^2 Z_n \equiv 4D_{\mathrm{B}}$ fixed, the difference between $\bar{g}(z)$ and $\tilde{g}(z)$ becomes negligible. Hence from Eq. (10.37) we obtain the quadratic equation for $\bar{g}(z)$ with the solution

$$\bar{g}(z) = \frac{2}{D_{\mathrm{B}}^2}\left(z - \mathrm{i}\sqrt{D_{\mathrm{B}}^2 - z^2}\right), \tag{10.38}$$

where the branch of the square root should be chosen so that $\bar{g}(z)$ tends to $1/z$ with $|z|/D_{\mathrm{B}} \gg 1$. The semi-elliptic density of states given by Eq. (10.36) corresponds to $-\pi^{-1}\mathrm{Im}\bar{g}(\epsilon + \mathrm{i}0_+)$. Figure 10.3b shows the real and imaginary parts of $\bar{g}(\epsilon + \mathrm{i}0_+)$ for the Bethe lattice.

Let us now turn to a simpler model with a local interaction where the DMFT step (iii) in p.242 can be carried out easily [5]. The model is given by

$$H_{\mathrm{FK}} = \sum_{ij} t_{ij} c_i^\dagger c_j + \sum_i \epsilon_f f_i^\dagger f_i + U \sum_i f_i^\dagger f_i c_i^\dagger c_i, \tag{10.39}$$

which is called the Falicov–Kimball (FK) model. The FK model is obtained from the Hubbard model by setting $t_{ij} = 0$ only for down-spin electrons, and shift their energy to $\epsilon_f$. The spin-up electrons are rewritten as spinless $c$-electrons, while spin-down electrons as spinless $f$ electrons. Because of the absence of hopping, the $f$-electron number $n_f$ is conserved at each site. In terms of the cavity Green function

$\mathcal{G}(z) = [z - \lambda(z)]^{-1}$ for $c$-electrons, $\bar{G}(z)$ is written as

$$\bar{G}(z) = \frac{1 - n_f}{z - \lambda(z)} + \frac{n_f}{z - \lambda(z) - U}, \tag{10.40}$$

which takes exact account of $U$. Thus the step (iii) has been performed immediately.

We proceed to derive $\lambda(z)$ and $\Sigma(z)$ by taking the Lorentzian density of states:

$$\rho(\epsilon) = \frac{1}{\pi} \frac{D_{\mathrm{L}}}{\epsilon^2 + D_{\mathrm{L}}^2} = -\frac{1}{\pi} \mathrm{Im}\bar{g}\,(\epsilon + \mathrm{i}0_+)\,, \tag{10.41}$$

with $\bar{g}(z) = (z + \mathrm{i}D_{\mathrm{L}})^{-1}$. Although the Lorentzian form is hardly realized in actual lattices, this form is useful for obtaining nontrivial results analytically. With a little manipulation through steps (i) and (ii), we obtain $\lambda(z) = -\mathrm{i}D_{\mathrm{L}}$ with Im $z > 0$. This independence of $\lambda(z)$ on $z$ results in the simplifying relation: $\mathcal{G}(z) = \bar{g}(z)$. The resultant $\bar{G}(z)$ gives the self-energy in the closed form:

$$\Sigma(z) = U n_f + \frac{U^2 n_f \left(1 - n_f\right)}{z + \mathrm{i}D_{\mathrm{L}} - U\left(1 - n_f\right)}. \tag{10.42}$$

The first term $Un_f$ in the RHS corresponds to the Hartree–Fock (or the simplest MFA) result, which becomes dominant for large $|z|$. In the opposite case of $U \gg |z|, D_{\mathrm{L}}$, the second term tends to cancel the first term, making $\Sigma(z) \sim [n_f/(1 - n_f)](z + \mathrm{i}D_{\mathrm{L}})$. The second term thus corrects the MFA like the reaction field in the Ising model. Details of deriving Eq. (10.42) is the subject of Problem 10.3.

## 10.3 Variational Principle and Optimum Effective Field

The fundamental DMFT equations (i) and (ii), given in Eqs. (10.26) and (10.27), are not only intuitively appealing, but are also based on the variational principle [6, 7]. We shall apply the argument in Sect. 9.4 and derive (i) and (ii). The starting point is Eq. (9.49) which represents $\Omega$ as a functional of the Green function. We shall show near the end of this section that the functional $\Phi\{G\}$ in Sect. 9.4 is actually determined by the site-diagonal element $\bar{G}$ in the limit of many neighboring sites. The self-energy is independent of the momentum in the same limit. The stationary condition $\delta\Omega = 0$ is equally valid no matter whether we vary $G$ or $\Sigma$, as long as we keep the relation $G^{-1} = g^{-1} - \Sigma$. In Eq. (9.49), it is convenient to regard

$$\beta\Phi\{\bar{G}\} - \mathrm{Tr}(\Sigma\bar{G}) \equiv \beta\Psi(\Sigma) \tag{10.43}$$

as a Legendre transformation to change the natural variable to $\Sigma$.

We regard the cavity Green function $\mathcal{G}(z)$ or, equivalently, $\lambda(z)$ as the reference for the effective impurity. Correspondingly, the renormalized Green function with account of $U$ is written as $D_\lambda(z) \equiv [z - \lambda(z) - \Sigma_\lambda(z)]^{-1}$ where the $\lambda$-dependence of the self-energy is emphasized. The thermodynamic potential $\Omega_\lambda$ of the impurity is then given, in accordance with Eq. (9.49), by

$$\beta[\Omega_\lambda - \Omega_0(\mathcal{G})] = \Psi(\Sigma_\lambda) - \mathrm{Tr}\ln D_\lambda^{-1}\mathcal{G}, \tag{10.44}$$

where $\Omega_0(\mathcal{G})$ is for the reference state. The crucial observation is that $\Psi(\Sigma_\lambda)$ has the same functional form as the original $\Psi(\Sigma)$ in $\Omega$, as long as $\Sigma$ is independent of the momentum. Hence we can eliminate $\Psi$ to write [7]

$$\beta\Omega(\Sigma_\lambda) = \beta\Omega_\lambda(\Sigma_\lambda) + \mathrm{Tr}\ln\left(\mathcal{G}^{-1} - \Sigma_\lambda\right) - \mathrm{Tr}\ln\left(g^{-1} - \Sigma_\lambda\right). \tag{10.45}$$

We consider varying $\Sigma_\lambda$ with $\lambda(z)$ fixed at some trial value. Although $\Omega_\lambda$ is stationary against variation by construction, $\Omega(\Sigma_\lambda)$ is not so in general because the optimum self-energy for $\Omega$ is different from $\Sigma_\lambda$ for this choice of $\lambda(z)$. With correct choice of $\lambda(z)$, on the other hand, we obtain $\Sigma_\lambda = \Sigma$ which realizes not only $\delta\Omega_\lambda/\delta\Sigma_\lambda = 0$ but also $\delta\Omega/\delta\Sigma_\lambda = 0$. In this case the stationary condition gives $\mathrm{Tr}(D_\lambda - G)\delta\Sigma_\lambda = 0$ or, equivalently,

$$D_\lambda(z) = \frac{1}{z - \lambda(z) - \Sigma(z)} = \frac{1}{N}\sum_{\boldsymbol{k}} \frac{1}{z - \epsilon_k - \Sigma(z)}, \tag{10.46}$$

which combines the first and second DMFT equations given by (10.26) and (10.27).

## 10.4 Anderson Model as an Effective Impurity

As we have discussed in p.242, the DMFT has the step (iii) to derive $\Sigma(z)$ for given $\lambda(z)$ and $U$. The effective medium characterized by $\lambda(z)$ can be simulated by the Anderson model which has been discussed in Chap. 6 [8]. Using the analyticity of $\lambda(z)$ in the upper half plane of $z$, we employ the spectral representation

$$\lambda(z) = \epsilon_f + \int_{-\infty}^{\infty} \mathrm{d}\epsilon \frac{\eta(\epsilon)}{z - \epsilon}, \tag{10.47}$$

where $\epsilon_f$ corresponds to the local electron level. The spectral function is simulated in terms of the hybridization parameter $V$ and the energy $\epsilon_c(k)$ of hypothetical conduction band as

$$\eta(\epsilon) = \frac{V^2}{N}\sum_k \delta\left[\epsilon - \epsilon_c(k)\right]. \tag{10.48}$$

Thus a hypothetical Anderson Hamiltonian $H_A$ is specified in terms of $\epsilon_f$, $V$, and $\epsilon_c(k)$. We take the one-dimensional model for $\epsilon_c(k)$.

In the Hubbard model with realistic density of states, an insulating state with antiferromagnetism can be realized as the ground state with unit occupation per site. The antiferromagnetism disappears with increasing temperature, but an insulating paramagnetic state may persist. This is a typical situation of Mott insulators. In low-dimensional systems, on the other hand, the insulating state may remain down to zero temperature. We have discussed such cases in Sect. 7.8 for one-dimensional systems. In the Mott paramagnetic state toward zero temperature, the spin entropy is actually removed by intersite spin correlations. If there are competing interactions between the spins, any magnetic order may be prevented even in higher dimensional case, but the entropy is removed by correlations. In such situations, the DMFT has a serious difficulty in describing the low temperature state. Namely, within the DMFT framework, a finite amount of entropy remains in the insulating ground state. Thus although the DMFT is powerful in dealing with strong local correlations, it loses the power if the dominant intersite correlations are beyond the mean field theory. In order to remedy the defect by including intersite correlations more accurately, various approaches have been tried either from the real space [9], or from the momentum space [10].

On the other hand, the DMFT has no difficulty of entropy in the metallic ground state, which is realized in the presence of a substantial band tail. In the following we confine the discussion to the metallic ground state without a long-range order such as magnetism and superconductivity. In the special case where the density of states $\rho(\epsilon)$ of the Hubbard model has a Lorentzian shape, the step (iii) is much simplified. Namely, using Eq. (10.26) we obtain

$$\bar{G}(z) = \frac{1}{z - \epsilon_0 + \mathrm{i}D_\mathrm{L} - \Sigma(z)}, \tag{10.49}$$

where the Lorentzian is characterized by the center energy $\epsilon_0$, and the width $D_\mathrm{L}$. This means $\lambda(z) = \epsilon_0 - \mathrm{i}D_\mathrm{L}$. The corresponding Anderson model has the conduction band with the constant density of states $\rho_c$. The relation to the Lorentzian is given by

$$V^2 \rho_c = D_\mathrm{L}. \tag{10.50}$$

The Anderson model $H_A$ with the constant $\rho_c$ can be solved exactly by use of the Bethe ansatz [11]. The absence of the cut-off in the hypothetical conduction band is obviously unrealistic. However, this does not cause a further mathematical trouble since the natural cut-off for the many-body effect is provided by $U$. Hence the bandwidth much larger than $U$ does not influence the solution. Although we do not go into details, exact thermodynamics has been derived for arbitrary set of $(D_\mathrm{L}, U)$ in the Anderson model [11].

On the basis of knowledge about the impurity Anderson model, we now discuss the dynamical property of the Hubbard model in infinite dimensions. The ground state with the Lorentzian density of states is a Fermi liquid, which derives from the local Fermi liquid property in the effective impurity model. We expand $\Sigma(z)$ in Eq. (10.49) as

$$\Sigma(z) = \Sigma(0) - \alpha z + O\left(z^2\right), \tag{10.51}$$

where both $\Sigma(0)$ and $\alpha$ are real in the Fermi liquid. Then the local Green function for small $|z|$ is arranged in the form

$$\bar{G}(z) \sim \frac{1}{(1+\alpha)z - \epsilon_0 + \mathrm{i}D_\mathrm{L} - \Sigma(0)} \equiv \frac{a}{z - a\left[\epsilon_0 - \mathrm{i}D_\mathrm{L} - \Sigma(0)\right]}, \tag{10.52}$$

where $a = (1+\alpha)^{-1}$ is the renormalization factor, which is smaller than unity. Similar calculation has been done around Eq. (6.74). Using Eq. (10.52) we obtain the renormalized density of states $\rho_R(\epsilon)$ near the Fermi level in the Hubbard model as

$$\rho_R(\epsilon) \equiv -\frac{1}{\pi}\mathrm{Im}\bar{G}(\epsilon + \mathrm{i}0_+) \sim \frac{a}{\pi}\frac{aD_\mathrm{L}}{(\epsilon - a\tilde{\epsilon}_0)^2 + (aD_\mathrm{L})^2} \tag{10.53}$$

with $\tilde{\epsilon}_0 = \epsilon_0 + \Sigma(0)$. The energy shift is estimated as $\Sigma(0) \sim Un/2$ in the Hartree–Fock approximation, with $n$ being the occupation number per site. We should have $\tilde{\epsilon}_0 = 0$ for $n = 1$ because of the particle–hole symmetry. Namely, the fully occupied state with $n = 2$ and the vacant state $n = 0$ turn into each other by the particle–hole transformation, which corresponds to the sign change of the energy level. The state with $n = 1$ turns into itself by the transformation. Hence the energy level must be zero. Note that occupation number per spin is 1/2.

The width of $\rho_R(\epsilon)$ has been reduced by the factor $a$ as compared with the non-interacting value $D_\mathrm{L}$. Accordingly, the renormalized center energy $a\tilde{\epsilon}_0$ comes closer to the Fermi level for $n \neq 1$. On the contrary, in the density of states at $\epsilon = 0$, the renormalization factor cancels each other in the numerator and the denominator, resulting in $\pi\rho_R(0) = D_\mathrm{L}/[\tilde{\epsilon}_0^2 + D_\mathrm{L}^2]$. In the special case $n = 1$ with $\tilde{\epsilon}_0 = 0$, one obtains $\pi\rho_R(0) = 1/D_\mathrm{L}$ independent of $U$. In other words, the density of states at the Fermi level is the same as the non-interacting one. Remarkably, $\rho_R(0)$ can be evaluated exactly for any occupation $n$ with the help of the Friedel sum rule. As discussed in Problem 10.4, the Fermi liquid constraint leads to

$$\rho_R(0) = \frac{1}{\pi D_\mathrm{L}}\sin^2\left(\frac{\pi}{2}n\right). \tag{10.54}$$

In the energy range far from the Fermi level, we can no longer use the approximate form given by Eq. (10.53). A complementary route is to start from the Green function

$$\bar{G}(z) \sim \frac{1 - n/2}{z - \epsilon_0} + \frac{n/2}{z - \epsilon_0 - U}, \tag{10.55}$$

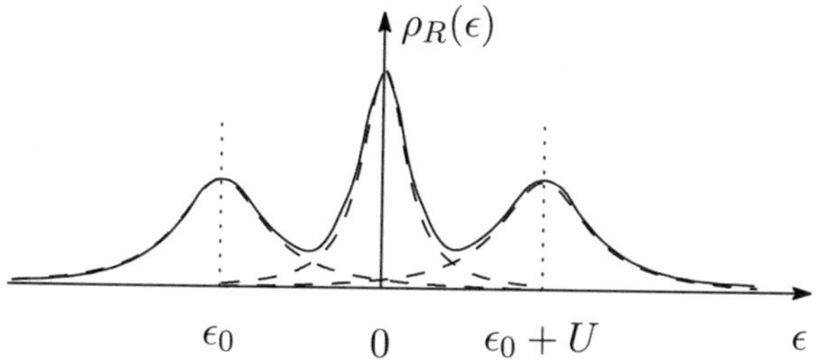


**Fig. 10.4** Illustration of the renormalized density of states in the infinite-dimensional Hubbard model with $n = 1$. The central peak derives from quasi-particles in the Fermi liquid, while the two side peaks derive from the correlation in the atomic limit

which is relevant in the strong-coupling case. Because of the finite hopping in the Hubbard model, $\bar{G}(\epsilon + i0_+)$ with $\epsilon \sim \epsilon_0$ and $\epsilon_0 + U$ should acquire the imaginary part of the order of $1/D_L$. Thus, together with the one given by Eq. (10.53), there appear three peaks in the renormalized density of states $\rho_R(\epsilon)$. Figure 10.4 illustrates the situation. Since each numerator in Eq. (10.55) becomes about 1/2 for $n = 1$, the magnitude of $\rho_R(\epsilon)$ at each peak is about half of that given by Eq. (10.54). In the extreme case of $U \gg D_L$, the width of the central peak is of the order of Kondo temperature $T_K$ with

$$T_K \sim U \exp\left(-\frac{U}{8D_L}\right), \tag{10.56}$$

by reference to Eq. (6.18) with the replacement $D \rightarrow U$. In this case the renormalization factor is very small: $a = T_K/D_L \ll 1$. In actual transition metal systems for which the Hubbard model is commonly used, it is rare to have such combination for magnitudes of the Coulomb repulsion and the bandwidth. In rare-earth systems, on the other hand, the situation corresponding to the extreme case is often realized as discussed in the next section.

## 10.5 Heavy Electrons

According to the Fermi liquid theory, the specific heat at low temperature is determined by the density of states $\rho^*(\mu)$ at the Fermi level $\mu$ of quasi-particles, as shown in Eq. (4.16). With a given density of electrons, $\rho^*(\mu)$ is proportional to the effective mass $m^*$ as given by Eq. (4.5). It has been found in a number of $f$-electron systems such as metallic rare-earth and actinide compounds that the specific heat coefficient $\gamma$ is more than 100 times of that in simple metals such as Al and Cu. It shows the presence of electrons with very large effective mass, which are called heavy electrons, or heavy fermions.

The Hubbard model is not sufficient to understand $f$-electron systems. In each rare-earth ion, $4f$ electrons are located inside the dominant amplitudes of $5d$ and $6s$ wave functions so that $4f$ wave functions keep the strong local character as in isolated atoms. On the other hand, broad energy bands originating from $5d$, $6s$ and surrounding ligand orbitals can partially hybridize with $4f$ states. Similar situation arises for $5f$ electrons in actinide systems. The local character of $5f$ electrons is weaker than the $4f$ case because of the larger spatial extent. If the relevant energy level $\epsilon_f$ is close to the Fermi level, the hybridization causes fluctuating occupation number of $f$ electrons. In such a case, $f$ electrons acquire a partially itinerant character. Correspondingly the average occupation deviates from an integer. This phenomenon is called valence fluctuation since the $f$ occupation number is related to valence of the ion. Note that valence fluctuation accompanies spin fluctuations in addition to charge fluctuations. As the $f$-electron level goes away from the Fermi level, charge fluctuations become less important. However, the spin fluctuations can remain significant, and causes the Kondo effect at each site.

To simulate the situation described above, we consider the Anderson lattice which consists of the periodic arrangement of impurity Anderson models. The Hamiltonian $H_{\mathrm{AL}}$ is given by

$$H_{\mathrm{AL}} = \sum_{\boldsymbol{k}\sigma} \epsilon_{\boldsymbol{k}} c_{\boldsymbol{k}\sigma}^{\dagger} c_{\boldsymbol{k}\sigma} + V \sum_{\boldsymbol{k}\sigma} \left( c_{\boldsymbol{k}\sigma}^{\dagger} f_{\boldsymbol{k}\sigma} + V f_{\boldsymbol{k}\sigma}^{\dagger} c_{\boldsymbol{k}\sigma} \right) + \sum_{i\sigma} \left( \epsilon_f + \frac{1}{2} U n_{i\bar{\sigma}}^f \right) n_{i\sigma}^f, \qquad (10.57)$$

where $n_{i\sigma}^f = f_{i\sigma}^{\dagger} f_{i\sigma}$ is the number operator of $f$ electrons at site $i$ and spin $\sigma$, with $\bar{\sigma} = -\sigma$. The operator $f_{\boldsymbol{k}\sigma}$ represents the Fourier transform of $f_{i\sigma}$.

In the impurity Anderson model, we have learned in Sect. 10.4 that the renormalization factor $a$, given by $a \sim T_{\mathrm{K}}/V^2\rho_c$, can be very small if $U \gg V^2\rho_c$. In $f$ electron systems, this situation is often realized in actual systems. Hence if $f$ electrons become itinerant by hybridization, we obtain the estimate

$$m^*/m \sim V^2\rho_c/T_{\mathrm{K}} \gg 1. \qquad (10.58)$$

In this way, emergence of heavy electrons in $f$ electrons is understood in terms of the Kondo effect. On the other hand, the origin of heavy mass in $d$-electron systems such as $LiV_2O_4$ seems different, and frustrated intersite interactions should play an important role [12, 13]. The heavy electrons show up not only as paramagnetic metals but also as superconducting state, which was first found in $CeCu_2Si_2$ [14]. Here we restrict ourselves to the paramagnetic state and sketch the DMFT for the Anderson lattice.

Let us begin with the $U = 0$ case which can be easily solved. The Hamiltonian is simply diagonalized by taking the Bloch states for both $c$ and $f$ electrons. The eigenvalues $E_\pm(\boldsymbol{k})$ are the solution of

$$\det \begin{pmatrix} E - \epsilon_{\boldsymbol{k}} & -V \\ -V & E - \epsilon_f \end{pmatrix} \equiv \det \hat{g}(\boldsymbol{k}, E)^{-1} = 0, \tag{10.59}$$

with the result

$$E_\pm(\boldsymbol{k}) = \frac{1}{2}\left(\epsilon_{\boldsymbol{k}} + \epsilon_f\right) \pm \frac{1}{2}\sqrt{\left(\epsilon_{\boldsymbol{k}} - \epsilon_f\right)^2 + 4V^2}. \tag{10.60}$$

The two bands result from hybridization of the conduction band with energy $\epsilon_{\boldsymbol{k}}$ and the flat band deriving from $f$-states. Each band accommodates up to two electrons per site with the spin degeneracy. Hence if the total electron number $n$ is equal to 2, the ground state can be an insulator with the lower band completely filled. The details of this insulating state is discussed in the next section. In this section we assume a metallic ground state. Namely, the electron number per site should be a non-integer. In actual systems, presence of other conduction bands, is a source of non-integer number in the Anderson lattice. However, these bands are often not included explicitly in the model.

The Green function with $U = 0$ is easily obtained as the matrix $\hat{g}(\boldsymbol{k}, z)$ which has appeared in Eq. (10.59). The $f$-electron component is given by

$$g_f(\boldsymbol{k}, z) = \left(z - \epsilon_f - \frac{V^2}{z - \epsilon_{\boldsymbol{k}}}\right)^{-1}. \tag{10.61}$$

The DMFT concentrates on the site-diagonal part $\bar{G}_f(z)$ of the renormalized $f$-electron Green function, which is given by

$$\begin{aligned} \bar{G}_f(z) &= \frac{1}{N}\sum_{\boldsymbol{k}} \frac{1}{z - \varepsilon_f - \Sigma_f(z) - V^2/(z - \varepsilon_{\boldsymbol{k}})} \\ &= \left[z - \epsilon_f - \lambda_f(z) - \Sigma_f(z)\right]^{-1}, \end{aligned} \tag{10.62}$$

where the self-energy $\Sigma_f(z)$ accounts for the interaction effect, and $\lambda_f(z)$ characterizes the cavity Green function. Following Eq. (10.47) we employ the spectral resolution as

$$\lambda_f(z) = \Delta\epsilon_f + \int_{-\infty}^{\infty} \mathrm{d}\epsilon \frac{\eta_f(\epsilon)}{z - \epsilon}. \tag{10.63}$$

Thus we define another hypothetical impurity Anderson model by

$$\eta_f(\epsilon) = \frac{\tilde{V}^2}{N}\sum_k \delta\left[\epsilon - \tilde{\epsilon}_c(k)\right]. \tag{10.64}$$

The ingredients for the hypothetical impurity is the spectrum $\tilde{\epsilon}_c(k)$ of the conduction band, fictitious hybridization $\tilde{V}$, and the local level $\epsilon_f^* = \epsilon_f + \Delta\epsilon_f$. The structure of $\eta_f(\epsilon)$ is more complicated than that in the Hubbard model. Since a set of parameters that allows analytic solution is difficult to find, one has to resort to numerical methods in most cases.

The standard process of solution with a proper impurity solver goes as follows. Suppose we can derive the self-energy $\Sigma_f(z)$ for system parameters associated with a trial $\lambda_f(z)$. Substitution of the resultant $\Sigma_f(z)$ into the first line of Eq. (10.62) defines a new $\lambda_f(z)$ in the second line, which is in general different from the initial one. If they agree, the self-consistent solution is obtained. Otherwise, one iterates derivation of $\Sigma_f(z)$ using a new $\lambda_f(z)$ until the convergence is achieved.

## 10.6 Kondo Insulators

In the Anderson lattice, charge fluctuations are suppressed with large $U$, and both $\epsilon_f$ and $\epsilon_f + U$ far from the Fermi level. Then $f$ electrons are left with only the spin degrees of freedom at each site. This is the same situation with the impurity case when we have introduced the Kondo model in Sect. 6.1. The periodic lattice of such Kondo centers is called the Kondo lattice. The Hamiltonian is given by

$$H_{\rm KL} = \sum_{k\sigma} \epsilon_k c_{k\sigma}^\dagger c_{k\sigma} + J\sum_i \boldsymbol{S}_i \cdot \boldsymbol{s}_i^{\rm c}, \tag{10.65}$$

where each site $i$ accommodates a localized spin $\boldsymbol{S}_i$ and a conduction-electron spin $\boldsymbol{s}_i^{\rm c}$. The latter consists of conduction electrons as

$$\boldsymbol{s}_i^{\rm c} = \frac{1}{2}\sum_{\sigma\sigma'} c_{i\sigma}^\dagger \boldsymbol{\sigma}_{\sigma\sigma'} c_{i\sigma'},$$

where $c_{i\sigma}$ is the Fourier transform of $c_{k\sigma}$. With $i$ at the origin, the same quantity is given by Eq. (6.11) in terms of $c_{k\sigma}^\dagger$ and $c_{k'\sigma'}$.

Since the $f$-electrons in the Kondo lattice do not have the charge degrees of freedom, the $f$-electron Green function cannot be defined properly. However, an approach like the DMFT should be still applicable if we use the $T$-matrix of conduction electrons instead of the $f$-electron Green function. Namely, we can assume that the $T$-matrix is dominated by the site-diagonal element. As the main topic in this section, we shall show that the ground state is insulating if the number of conduction electron is one per site. At first sight this result looks contradicting to the energy band theory, which demands that the partially filled conduction band leads to the metallic ground state. As a continuation of the Anderson lattice, on the other hand, the insulating ground state is reasonable since the sum of $f$- and conduction electrons adds up to two per site. In contrast to ordinary band insulators, the band gap in the Kondo lattice depends strongly on temperature $T$, and vanishes for $T \gg T_{\mathrm{K}}$. Such insulating state is called the Kondo insulator. In contrast to the Mott insulating state, the DMFT has no difficulty in describing the Kondo insulator where the spin entropy vanishes by the local correlation due to Kondo effect.

We focus on the $T$-matrix $t(z)$ of conduction electrons at each site, which is reduced to $V^2\bar{G}_f(z)$ in the Anderson lattice as given by Eq. (10.62). In the DMFT, $\bar{G}_f(z)$ is simulated by the effective Anderson impurity, where the ground state is the local Fermi liquid. According to the argument in Sect. 6.6, the impurity Green function $G_f(z)$ at $T = 0$ near the Fermi level is parameterized as

$$G_f(z) = \frac{a_f}{z - \tilde{\epsilon}_f + \mathrm{i}\tilde{\Delta}}, \tag{10.66}$$

where the renormalization factor $a_f$ relates the renormalized width $\tilde{\Delta}$ to the bare one $\Delta = \pi V^2 \rho_c$ by $\tilde{\Delta} = a_f \Delta \sim T_{\mathrm{K}}$. Then the $T$-matrix takes the following form

$$t(z) = \frac{1}{\pi\rho_c} \cdot \frac{\tilde{\Delta}}{z - \tilde{\epsilon}_f + \mathrm{i}\tilde{\Delta}}. \tag{10.67}$$

Note that the scattering strength $\tilde{\Delta}$ is the same as the width of the resonance. This is consistent with the relation

$$\mathrm{Im}\, t(0) = -\pi\rho_c |t(0)|^2, \tag{10.68}$$

and is equivalent to the unitarity of the $S$-matrix as discussed around Eq. (6.24). The $T$-matrix given by Eq. (10.67) with the condition $n_f = 1$, namely $\tilde{\epsilon}_f = 0$, is applicable to the Kondo lattice. Note that Kondo effect makes up the $T$-matrix of the resonance type, even without charge degrees of freedom in localized spins. We simply assume Eq. (10.67) as a reasonable final form without following the self-consistent loop of the DMFT.

The self-energy $\Sigma_c(z)$ of conduction electrons in the Kondo lattice is obtained from the $T$-matrix by using the relation

$$t(z) = \Sigma_c(z) + \Sigma_c(z) g_c(z) t(z), \tag{10.69}$$

with $g_c(z)$ the site-diagonal part of the free Green function. If the conduction band is symmetric about the Fermi level, we may approximate $g_c(z) \sim -\mathrm{i}\pi\rho_c$ for the energy range much smaller than the bandwidth. The real part vanishes as a result of momentum average for the symmetric density of states. Then we obtain

$$\Sigma_c(z)^{-1} = t(z)^{-1} - \mathrm{i}\pi\rho_c = z/\tilde{V}^2, \tag{10.70}$$

with $\tilde{V}$ defined by the relation $\tilde{\Delta} = \pi\tilde{V}^2\rho_c$. Note that $\Sigma_c(z)$ does not have an imaginary part for real $z$, as a consequence of the unitarity constraint Eq. (10.68). The renormalized Green function $G_c(\boldsymbol{k}, z)$ is related to the bare one $g_c(\boldsymbol{k}, z)$ through the Dyson equation

$$G_c(\boldsymbol{k}, z)^{-1} = g_c(\boldsymbol{k}, z)^{-1} - \Sigma_c(z) = z - \epsilon_{\boldsymbol{k}} - \tilde{V}^2/z. \tag{10.71}$$

This Green function behaves as if each site had a localized $f$-electron level at the Fermi level, even though there is no charge degrees of freedom for $f$-electrons. The renormalized spectrum in this case is derived as

$$E_\pm(\boldsymbol{k}) = \frac{1}{2}\left(\epsilon_{\boldsymbol{k}} \pm \sqrt{\epsilon_{\boldsymbol{k}}^2 + 4\tilde{V}^2}\right), \tag{10.72}$$

in the same way as Eq. (10.60). Figure 10.5 illustrates the resultant spectrum. Since the Fermi level is zero, a completely filled band emerges corresponding to $E_-(\boldsymbol{k})$.

According to the energy band theory, there must be two electrons per site to have a completely filled band. However, we have only one electron per site. How can we resolve this paradox? The resolution comes from the quasi-particle weight of the spectrum. Namely, we rewrite the Green function as

$$G_c(\boldsymbol{k}, z) = \frac{a_-(\boldsymbol{k})}{z - E_-(\boldsymbol{k})} + \frac{a_+(\boldsymbol{k})}{z - E_+(\boldsymbol{k})}, \tag{10.73}$$

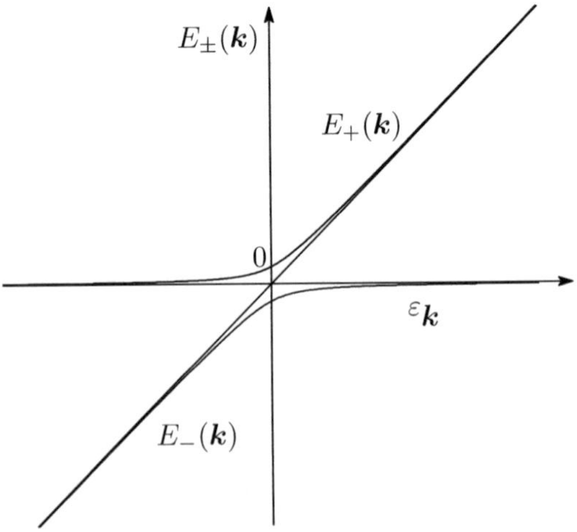


**Fig. 10.5** Schematic energy spectrum of conduction electrons in the Kondo lattice. The spectrum depends on $\boldsymbol{k}$ only through the bare energy $\epsilon_{\boldsymbol{k}}$ which is taken in the horizontal axis. The Fermi level is located at the zero energy

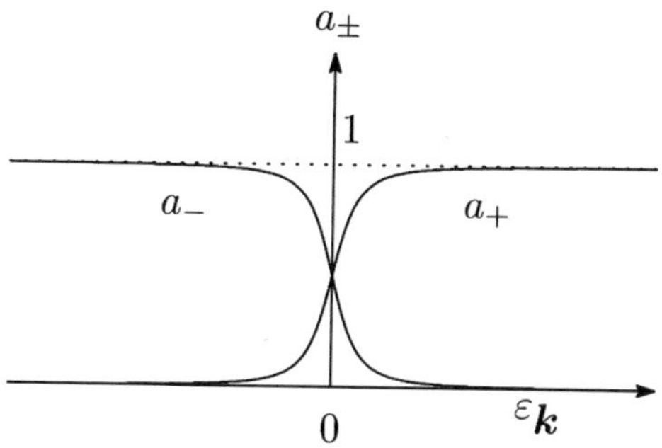


**Fig. 10.6** Weight of quasi-particles in hybridized conduction bands

where the weights $a_\pm(\boldsymbol{k})$ are given by

$$a_\pm(\boldsymbol{k}) = \frac{1}{2}\left(1 \pm \frac{\epsilon_{\boldsymbol{k}}}{\sqrt{\epsilon_{\boldsymbol{k}}^2 + 4\tilde{V}^2}}\right). \tag{10.74}$$

Figure 10.6 shows the weights for both branches. It is clear from Eq. (10.74) that the sum rule $a_+(\boldsymbol{k}) + a_-(\boldsymbol{k}) = 1$ holds for each momentum. Namely, the total weight of the spectrum is the same as that of a single band, although there are two branches. With the symmetric weight above and below the Fermi level, the occupation number per site of conduction electrons is given by

$$\frac{2}{N}\sum_{\boldsymbol{k}} a_-(\boldsymbol{k}) = 1, \tag{10.75}$$

with the spin multiplicity 2. In the quasi-particle picture, on the other hand, the number of filled quasi-particle states is 2 per site, since the weights $a_\pm$ should be replaced by unity for the Green function of quasi-particles. In this way the reduced weight in the spectrum resolves the conflict with the energy band theory. With increasing temperature, the resonance structure given by Eq. (10.67) becomes obscure, and vanishes for $T \gg T_{\mathrm{K}}$. Then the system becomes a metal. This is in strong contrast with ordinary insulators where the energy gap is almost independent of $T$.

## Problems

**10.1** Derive the central limit theorem as expressed by Eq. (10.33).

**10.2** Derive the real part of the Green function given by Eq. (10.34) for the Gaussian density of states.

**10.3** Derive the self-energy of the Falicov–Kimball model in the DFMT as given by Eq. (10.42).

**10.4** Derive the Fermi liquid constraint as given by Eq. (10.54).

## Solutions to Problems

**Problem 10.1**

The lower-order moments are given by $\langle\langle \varepsilon_i^2 \rangle\rangle = D_G^2/d$ and $\langle\langle \varepsilon_i^4 \rangle\rangle = O(t^4) = O(D_G^4/d^2)$. Other moments are of higher order in $1/d$ with $D_G$ fixed. Hence we obtain

$$\ln M(x) = d \ln \left[ 1 - \frac{1}{d} x^2 D_G^2 + O\left( \frac{1}{d^2} x^4 D_G^4 \right) \right] \underset{d \to \infty}{\to} -\frac{1}{2} x^2 D_G^2, \tag{10.76}$$

which leads to Eq. (10.33).

**Problem 10.2**

We employ the representation

$$\frac{1}{z - \epsilon} = -\mathrm{i} \int_0^\infty \mathrm{d}t \exp\left[\mathrm{i}(z - \epsilon)t\right], \tag{10.77}$$

for $\mathrm{Im}\, z > 0$. With $\rho_G(\epsilon)$ given by Eq. (10.31), we perform the integral over $\epsilon$ first to obtain

$$\bar{g}(z) = -\mathrm{i} \int_0^\infty \mathrm{d}t \exp\left( -\frac{1}{2} t^2 + \mathrm{i} z t \right). \tag{10.78}$$

Next we introduce the dimensionless variable $\zeta = z/D_G$, and change the integration variable from $t$ to $y = (\zeta + \mathrm{i} t D_G)/\sqrt{2}$. Then the resulting form is expressed as

$$D_G \bar{g}(z) = -\sqrt{2} \exp\left( -\zeta^2/2 \right) \int_{\zeta/\sqrt{2}}^{\mathrm{i}\infty} \mathrm{d}y \exp\left( y^2 \right). \tag{10.79}$$

The integral range is decomposed into

$$\int_{\zeta/\sqrt{2}}^{\mathrm{i}\infty} \mathrm{d}y = \int_0^{\mathrm{i}\infty} \mathrm{d}y - \int_0^{\zeta/\sqrt{2}} \mathrm{d}y. \tag{10.80}$$

In the case of $z \to \epsilon + \mathrm{i}0+$ with real $\epsilon$, the first integral contributes a pure imaginary number to the LHS. Hence we obtain with $x = \epsilon / D_{\mathrm{G}}$

$$D_{\mathrm{G}} \mathrm{Re} \bar{g}(\epsilon + \mathrm{i}0+) = \sqrt{2} \exp\left(-x^2/2\right) \int_0^{x/\sqrt{2}} \mathrm{d}y \exp\left(y^2\right), \tag{10.81}$$

which gives Eq. (10.34).

**Problem 10.3**
We first derive $\lambda(z)$. In the case of the Lorentzian density of states for $c$-electrons, integration in Eq. (10.26) leads to

$$\bar{G}(z) = \frac{1}{z + \mathrm{i}\Delta - \Sigma(z)}, \tag{10.82}$$

with Im $z > 0$. Hence comparison with Eq. (10.27) gives immediately the result $\lambda(z) = -\mathrm{i}\Delta$. On the other hand, taking the inverse of Eq. (10.27) and comparing with Eq. (10.40), we obtain

$$\Sigma(z) = z - \lambda - \bar{G}(z)^{-1} = z - \lambda - \frac{(z-\lambda)(z-\lambda-U)}{z - \lambda - \left(1 - n_f\right) U}. \tag{10.83}$$

In the rightmost side, we separate the constant term $Un_f$ from the other contributions that tend to zero as $z$ goes to infinity. Then putting $\lambda = -\mathrm{i}\Delta$, we obtain Eq. (10.42).

**Problem 10.4**
In order to use the Friedel sum rule, we take the Green function $G_f(z)$ of the effective Anderson model, which is the same as the site-diagonal component $\bar{G}(z)$ in the Hubbard model. Since the self-energy $\Sigma(0)$ is real at the Fermi level $z = 0$, Eq. (10.49) leads to

$$\mathrm{Im}\, G_f\left(\mathrm{i}0_+\right)^{-1} = D_L = -\left|G_f\left(\mathrm{i}0_+\right)\right|^{-2} \mathrm{Im}\, G_f\left(\mathrm{i}0_+\right) \tag{10.84}$$

with $0_+$ being positive infinitesimal. From the Friedel sum rule given by Eq. (9.56), the complex number $-G_f(\mathrm{i}0_+)$ should have the argument $\pi n_f/2$. Then we obtain

$$-\mathrm{Im}\, G_f\left(\mathrm{i}0_+\right) = \left|G_f\left(\mathrm{i}0_+\right)\right| \sin\left(\pi n_f/2\right). \tag{10.85}$$

Comparison of Eqs. (10.84) and (10.85) gives $\left|G_f(\mathrm{i}0_+)\right| = \sin(\pi n_f/2)/D_L$. By identifying $G_f = \bar{G}$ and $n_f = n$ in the Hubbard model, we obtain Eq. (10.54).

## References


1. Onsager, L.: Phys. Rev. **65**, 117 (1944)
2. Georges, A., Kotliar, G., Krauth, W., Rozenberg, M.J.: Rev. Mod. Phys. **68**, 13 (1996)
3. Müller-Hartmann, E.: Z. Phys. B **74** 507 (1989)
4. Metzner, W., Vollhardt, D.: Phys. Rev. Lett. **62**, 324 (1989)
5. Brandt, U., Mielsch, C.: Z. Phys. B **75**, 365 (1989)
6. Kuramoto, Y., Watanabe, T.: Physica **148B**, 80 (1987)
7. Potthoff, M.: Eur. Phys. J. B **36**, 335 (2003)
8. Georges, A., Kotliar, G.: Phys. Rev. B **45**, 6479 (1992)
9. Maier, T., Jarrell, M., Pruschke, T., Hettler, M.H.: Rev. Mod. Phys. **77**, 1027 (2005)
10. Rohringer, G., Hafermann, H., Toschi, A., Katanin, A.A., Antipov, A.E., Katsnelson, M.I., Lichtenstein, A.I., Rubtsov, A.N., Held, K.: Rev. Mod. Phys. **90**, 025003 (2018)
11. Tsvelick, A.M., Wiegmann, P.B.: Adv. Phys. **32**, 453 (1983)
12. Urano, C., Nohara, M., Kondo, S., Sakai, F., Takagi, H., Shiraki, T., Okubo, T.: Phys. Rev. Lett. **85**, 1052 (2000)
13. Kaps, H., Brando, M., Trinkl, W., Buttgen, N., Loidl, A., Scheidt, E.-W., Klemm, M., Horn, S.: J. Phys. Condens. Matter **13**, 8497 (2001)
14. Steglich, F., Aarts, J., Bredl, C.D., Lieke, W., Meschede, D., Franz, W., Schafer, H.: Phys. Rev. Lett. **43**, 1892 (1979)

# Index




Y. Kuramoto, *Quantum Many-Body Physics*, Lecture Notes in Physics 934,
https://doi.org/10.1007/978-4-431-55393-9

Printed in the United States
By Bookmasters